Chemistry for Sustainable Development in Africa

Ameenah Gurib-Fakim
Jacobus Nicolaas Eloff
Editors

Chemistry for Sustainable Development in Africa

Co-ordinated by:
Daniel Nyanganyura, ICSU Regional Office for Africa, Pretoria, South Africa
Edith Madela-Mntla, ICSU Regional Office for Africa, Pretoria, South Africa

Springer

Editors
Ameenah Gurib-Fakim
University of Mauritius
Reduit
Mauritius

Jacobus Nicolaas Eloff
Phytomedicine Programme
Faculty of Veterinary Science
University of Pretoria
Pretoria
South Africa

ISBN 978-3-642-44206-3 ISBN 978-3-642-29642-0 (eBook)
DOI 10.1007/978-3-642-29642-0
Springer Heidelberg New York Dordrecht London

Printed on acid-free paper

Springer is part of Springer Science+Business Media (www.springer.com)

Foreword

The International Year of Chemistry was a year-long initiative, organized by the International Union of Pure and Applied Chemistry (IUPAC) and the United Nations Educational, Scientific and Cultural Organization (UNESCO), was designed to "celebrate the achievements of chemistry and its contributions to the well-being of humankind".

Another major goal was to examine ways to promote international collaboration for the purposes of enhancing training and research in countries that currently lack the capacity to engage as fully-fledged partners in the field—either at an individual or institutional level.

The growing global interest in turning to chemistry as a significant tool for sustainable development, especially in developing countries, is just one more reason why this book, *Chemistry for Sustainable Development in Africa*, is such a welcome addition to the academic literature focusing on the relationship between scientific capacity and sustainable development in the developing world.

The book has not been written in isolation. Instead it serves as an important addition to the growing emphasis that has been placed on putting science to work for sustainable development in poor countries. This series of articles, written by some of Africa's most prominent chemists, rightfully places the field of chemistry at the center of such efforts.

Advances in chemistry hold great promise to address a broad range of critical issues facing Africa as it seeks to build secure and sustainable pathways for enhancing the well-being of its people. These issues, in many cases closely tied to the millennium development goals (MDGs), include greater access to safe drinking water and adequate sanitation, higher crop yields, improved nutrition and public health, larger and more dependable sources of energy (particularly renewable energy), and the development of new materials for the creation of products and services of enormous value for the domestic economy and export.

Spurred on by the recent experiences of Brazil, China, India, Turkey, and other emerging economies that have successfully pursued strategies for science-based development, science has become a cornerstone of sustainable development efforts across the developing world.

Yet, as CNR Rao, Linus Pauling Research Professor and Honorary President of the Jawaharlal Nehru Center for Advanced Scientific Research in India and immediate past President of TWAS, recently noted in a commentary in *Nature Chemistry*: "Chemistry creates agony and hope in less developed countries".

"Hope" is generated by the growing interest that developing countries have displayed for incorporating chemistry into their sustainable development agendas.

It is encouraging to note that Ethiopia led the global efforts to create the International Year of Chemistry, and that 19 of the 25 countries which officially sponsored the IYC initiative were developing countries.

It is also encouraging to note that Chinese scientists now rank first in the world in the number of articles published on nanotechnology in international peer-reviewed journals, and that a growing number of developing countries, including Brazil, India, and South Africa, are investing significant sums of money in nanoscience and nanotechnology.

And it is encouraging as well to observe that over the past decade, a growing number of regional and national associations and networks designed to promote training and research in chemistry have emerged across the developing world. In Africa, these organizations include the Federation of African Societies of Chemistry, the Pan African Chemistry Network, the Southern and Eastern Network of Analytical Chemists, and national chemical societies, for example, in Botswana and Malawi.

But the "agony" that developing countries, particularly the poorest developing countries, face when it comes to enhancing the role that chemistry can play in sustainable development involves this stark reality: broad knowledge and applications of chemistry to address critical challenges in sustainability remain far short of their potential. Moreover, the capacity to take advantage of this potential remains woefully inadequate due to poor training and antiquated laboratory facilities.

Chemistry in Africa, for instance, suffers from a lack of access to reagents and instruments, which inhibits the ability of researchers and students to conduct experiments. And, while the internet has improved access to the most recent literature in the field (an initiative launched by the Royal Society in 2006, 'Archive for Africa', has provided African chemists with free electronic access to hundreds of thousands of articles), much more needs to be done to ensure that the continent's chemists can keep abreast of the most recent findings in the field.

A shortage of well-trained professors and laboratory technicians, poorly equipped laboratories, and lingering obstacles to timely access to literature, despite the expanded use of the internet, pose serious challenges for advocates of chemistry in Africa as they seek to gain support, and funding for building sufficient capacity in the field.

Efforts to address such fundamental issues, moreover, are compounded by profound shifts that are unfolding within the discipline itself.

As Atta-ur-Rahman, TWAS Vice President and Coordinator General of the Organization of Islamic Cooperation's (OIC) Ministerial Standing Committee on Scientific and Technological Cooperation (COMSTECH), recently noted: "The kind of research currently taking place in many developing countries largely focuses on traditional fields of chemistry—for example, the study of simple chemical structures and compounds".

"Cutting-edge chemistry", he went on to say, "encompasses a much wider range of subject areas". Indeed some of the most exciting areas of science today lie at the interface of chemistry and biology. In addition to nanotechnology and molecular medicine, these fields, include neuroscience, bioinformatics, and structural biology. As the lines between the various fields of science continue to blur, chemistry plays a critical role in broadening the knowledge base by providing a "platform" for understanding and investigating the fundamental properties of atoms and molecules.

Current trends within the field will mean that Africa cannot simply mimic what developed countries have successfully done in the past to build strong capacity in chemistry. As Atta-ur-Rahman points out, Africa cannot focus solely on traditional subfields in chemistry that were once at the center of the discipline but are no longer.

Consequently, the research and training agenda for chemistry in Africa must be innovative in its methodologies and relevant and up-to-date in its subject matter if the continent hopes to build its capacity to international levels of excellence. Efforts must concentrate on training the next generation of African chemists and on pursuing research agendas designed to integrate laboratory findings into broader sustainable development initiatives. Support for chemistry in Africa (and elsewhere) should therefore be viewed as a process, not a goal, driven by funding strategies that evolve as circumstances change in this rapidly developing discipline.

"Chemistry: Our Life, Our Future" served as the driving refrain of the International Year of Chemistry. It is a refrain that is increasingly resonating among advocates for chemistry in Africa as well. In the articles that follow, the authors describe how chemistry can—and indeed must—become a primary tool for poverty reduction and sustainable development across the continent.

As the Executive Director of TWAS and the former Minister of Science and Technology in Rwanda, I extend my congratulations to those who have contributed to this collection. I also urge policy makers and representatives of nongovernmental groups, private industry and chemistry associations, and unions to examine and embrace the significant opportunities and challenges that are outlined in this volume to help advance the ways in which chemistry can benefit both science and society in Africa.

We are living at a historic moment in the history of science. The prospects for positive change have never been brighter. Policy makers in developing countries have rarely expressed greater support for the role that science can play in promoting sustainable development. The number of concrete examples of how science can improve societal well-being continues to grow, not only in terms of individual programmes and projects, but also in terms of national policies that are lifting tens of thousands out of poverty each year.

All of these trends make "chemistry for sustainable development in Africa" not just a goal to which we should aspire, but also a realistic pathway for improving the lives of millions of Africans in the years ahead.

The articles that follow provide an analytical platform for bringing science and society closer together in Africa in mutually reinforcing ways. It is only fitting that chemistry, which is increasingly viewed as a "platform" discipline, serves as the focal point of this discussion.

TWAS
The Academy of Sciences for the Developing World
Trieste, Italy

Romain Murenzi

Preface

The African continent entered the twenty-first century as the world's poorest continent. The economies of most of the countries of the African Union were either growing slowly or declining. This is despite the abundance of natural resources in the continent. Several factors could have been responsible for the poverty and low growth. There have been many studies e.g. by the World Bank on aspects influencing poverty in Africa and how changes in policies and governance can lead to a turn around.

Some encouraging changes have taken place over the past decade. Since 2000, six of the fastest growing countries, were from Africa with Angola being the fastest growing in the world. This change may be ascribed to many aspects. War and political strife was a major factor in causing poverty. In the new century there have been many changes to a more democratic situation. Since then there has also been better economic policies and there was a boom in commodity prices. The per capita income was equally low and falling. Since 2004, there has been dramatic change and the economies of many countries grew on average of 4.6%—the highest rate in the decade. It has been reported that improved macroeconomic management has been the major driver of the recovery. However, looking at GDP alone as a marker for prosperity is misleading as the number of people living in absolute poverty remains higher compared to past decades.

The application of science, technology, and innovation (STI) has led to enormous growth in countries with limited resources. One of the limitations of many countries in Africa is that resources are exported without any beneficiation to create more work and to increase the general quality of life of the people. Yet, it is the most neglected sectors in the development drive of countries even though STI, has an important role to play in the attainment of the continent's sustainable development objectives.

Africa's continued low investment in science and technology is also manifested in the declining quality of science and engineering education at all levels of educational systems. Throughout the 1980s and 1990s, science and technology investments were not prioritized despite considerable empirical evidence from

South–East Asia and other regions showing that investment in science and technology yields direct and indirect benefits to national economies. Of all the world regions, Africa as a whole has the lowest human development index and highest poverty indicators. Food security, nutrition, healthcare, and environmental sustainability are among Africa's biggest challenge.

In the last part of the twentieth century, southern Africa, for example, was reported to have the highest prevalence of HIV and AIDS. The devastating impact of HIV and AIDS is not only exacerbated by the increase in levels of poverty; it is also a manifestation of the breakdown in the African healthcare system. Preventable diseases such as malaria are in fact one of the biggest blights afflicting the people of Africa. Yet low cost solutions are available, such as Vitamin A supplements, insecticide-treated nets, oral-hydration therapy could significantly reduce these deaths but are largely unavailable. Burden of disease and economic growth are, of course, closely related.

Apart from mineral riches Africa also has a large and valuable biodiversity that is not adequately used. It is surprising that although Africa contains 25% of the world's plant species diversity only 8% of the herbal medicines commercialized come from Africa. In a remarkable international collaboration of scientists, growers, exporters, and importers of medicinal plants from 14 different countries the publication of the African Herbal Pharmacopoeia is an example of how collaboration can lead to useful products.

Fortunately, more African leaders now view science, technology, and innovation as critical to human development. A series of developments at the international and regional levels from 2000 to date provide new sources of optimism and action. Time and time again policy-makers have underlined the importance of science-based decision-making, by inter alia calling for: integrating scientists' advice into decision-making bodies; partnerships between scientific, public and private institutions; improved collaboration between natural and social scientists, and establishing regular channels for requesting and receiving advice between scientists and policy makers; making greater use of integrated scientific assessments, risk assessments and interdisciplinary and inter-sectoral approaches, and increasing the beneficial use of local and indigenous knowledge. Strengthening and creating centers for sustainable development in developing countries are encouraged, as well as networking with and between centers of scientific excellence and between science and education for sustainable development.

Chemistry, as a central science, deals with all these areas of human activity. It touches everyone. It pervades our lives and in 1987, Jean Marie Lehn, Nobel Prize winner stated that 'A world without chemistry would be a world without synthetic material as chemistry is behind most of the innovations that have improved our lives.' The past two decades have witnessed university researchers and industrial chemists competing to use science especially chemistry, to find ingenious responses to climate change and environmental degradation. Sustainable development may have been conceptualized in different ways, but the most widely used definition, as articulated by the World Commission on Environment and Development, is "development that meets the needs of the present without compromising the ability of future generations to meet their own needs". As such, chemistry remains the cornerstones for sustainable development, not only in Africa but also worldwide.

Yet the true impact of chemistry for sustainable development and for impacting livelihoods will be visible when different fields related to chemistry are brought together sometimes in ways that were previously not envisaged. Today the marriage of chemistry with biology to computing is key to the development of new crops, drugs, vaccines, diagnostic kits for diseases, contraceptives, and much more. Nutrition and healthcare are not the only winners from this alliance, industrial competitiveness is also a winner.

The alliance of computing to the biochemical sciences has opened up whole new areas of research and development, such as combinatorial chemistry, genomics, bioinformatics, and structural biology. Raw computing power is being harnessed to test the potential of new drugs and vaccines (combinatorial chemistry), to unfold the map of the human, animal and plant genomes (bioinformatics), and to do this in record time. Add nanotechnology to this and one begins to see the future of drug discovery and production through products, such as biosensors, biochips, smart drug delivery systems, bioelectronics, and biomaterials.

For Africa to be able to make a difference in these areas, there is a need to develop and retain a critical mass of trained and experienced researchers in all areas of science especially as scientific research is going multidisciplinary with chemistry and all its sub-disciplines as major components. This book showcases the attempts being made by some African researchers who are trying to address the development priorities of the continent. Publications deal with varied topics like nanotechnology, climate change, natural product chemistry, and biotechnology amongst others.

Expectation is high as Africa has potential and has a great future. It is expected that by 2020, Africa will have a collective GDP of 2.6 trillion dollars and with 1.1 billions Africans under the age of 20–50% are expected to be living in the cities by 2030. Africa's economic pulse has quickened and is infusing the continent with a new commercial vibrancy and with a GDP rising to around 5% from 2000 to 2009. One factor that could explain this is Africa's increased trade both internationally and regionally. Increasingly member states of the continent are spending on infrastructure and further increasing collaboration and cooperation in science and innovation.

Apart from political issues, the sustainable development of the African continent rests squarely on priority areas within the scientific domains. Critical capabilities need to be developed and will include human capacity building, reinventing African universities to retain highly qualified scientists, if not within the country of origin at least within Africa, enhancing collaboration of universities within Africa.

Other aspects are developing continent-wide regulatory measures that are effective, transparent and efficient, and aimed at promoting innovation, engaging the African diaspora, designing effective collaborations with regional, and international partners are also key considerations.

Ameenah Gurib-Fakim
J. N. Eloff

Contents

Part I
Health; Biodiversity Utilisation

1. Amuka: An overview in support of continued research into phytomedicine: Past, Present and future (17 MS pages)
2. Masimirembwa: The Metabolism of Antiparasitic Drugs and Pharmacogenetics in African populations – from molecular mechanisms to clinical applications (17)
3. Mapope: Role of flavonoids and isoflavonoids molecules in symbiotic functioning and host plant defence in the Leguminosae (25 MS pages)
4. Mutanda: Sustainable biodiesel production using waste water streams and microalgae in South Africa (34 MS pages)
5. Onyengecha: Antifungal properties of plant extract and density on some fungal diseases and yield of cowpea (17 pages)

An Overview in Support of Continued Research into Phytomedicine: Past, Present, and Future

Omari Amuka

Abstract The role played by plants in the livelihood of humankind is unimaginable. There would be no existence of higher animals on the planet earth without plants. Most of the substances used for therapeutic purposes have been and continue to be directly derived from plants. A good percentage of the current pharmaceuticals are of partially or wholly plant origin.

1 Introduction

There is a general belief that traditional forms of treatments involving the use of plants and plant extracts are archaic and ineffective. The notion is an ill-conceived one. This may not be necessarily true. To remove this myth from the minds of scholars, a small, fast disappearing community through assimilation into other stronger ones has been chosen for study. There is need for new drugs to manage emerging and re-emerging diseases. Plants have in the past been the source of remedy for many diseases. The older generation possessing traditional knowledge is fast disappearing. Thus, there is fear that such knowledge could soon be lost unless proper documentation is done. This review is a critical analysis of plants as source of medication in the ancient past, present, and distant future.

Plants have been utilized for medicinal purposes for many years. Some of such records are found in the Indus civilization dating back to 900 BC and the second millennium BC [2]. These facts are contained in hymns found in the Rigveda and the Atharvaveda which contain records of useful plants [30]. In Indian classical

O. Amuka (✉)
Department of Botany and Horticulture,
Maseno University, Private BagMaseno, Kenya
e-mail: amukaomari@yahoo.co.uk

A. Gurib-Fakim and J. N. Eloff (eds.), *Chemistry for Sustainable Development in Africa*, DOI: 10.1007/978-3-642-29642-0_1,

medicine, that is the *Ayurveda* (strictly the science of life), several concrete proofs or examples may be traced in these texts and one can then say that plants form an important and integral part of Ayurvedic pharmacopoeia [26, 33, 37].

A total of 341 different plant species are listed in the *Charaka Samhita,* (900 BC), as useful in the management of human health [2]. In the *Susruta Samhita* there are a total of 395 plant species listed for the same purpose [23]. It is also evident from other treatise authors, from this field, where there were over 70 species with the list being expanded to 600 of plants that are used in Ayurvedic [24]. Such a culture depending on Mother Nature has been practiced for over 2000 years [24]. Similarly, as the Indus civilization took root, the Chinese were also evolving the culture of phytomedicine, the *Kanpo*, which was systematized in the *Shang Han Tsu Ping Lun,* a16-volume compendium. It is believed that the compendium which was compiled by Chang Chung Ching must have been done in the seconnd century (456–536 AD). The compilation *Shiri-Nung Pen Tso Ching* of Tao Hung-Ching comprises about 365 crude drugs, all of which are of plant origin [35].

It is only in the last three decades that scientific evaluation of their efficacy has generated interest [14]. With the advent of scientific methods of analysis, many of these reported medicinal plants came under scrutiny leading to the elucidation of their active principles. In the Amazonia, the early South and Central American culture dating back to 1000 BC, there were systematic studies of the indigenous flora and documented knowledge of the advantage of the local inhabitants confirming that a pharmacopoeia existed for the Indian population [16, 42].

The ancient Egyptian culture in Africa around 1600 BC contains enormous literature pertaining to the use of plants as food and for curative purposes. An Egyptian medical treatise (papyrus), drawn up in Thebes during the aforementioned period has an inventory of 700 plants used in medicine [27]. There are also Egyptian motifs depicting appreciation and celebrations of bountiful harvests after a successful agricultural cropping year [12]. In West Africa in more recent times, such as amongst most communities, for example the Yoruba prior to the European colonization it was mandatory that a young boy before being initiated into adulthood had to learn the names of all the useful plants in relation to future uses by the young boy in life [31]. The Greeks and the Romans, subsequent cultures that emerged after the Egyptian, contain all that it inherited from the latter. This is evidenced by the works of Hippocrates (370–287 BC) and Diokorides [39] that had extensive knowledge of medicinal plants [20]. Diokorides, a Roman soldier physician, made the first taxonomic compendium of useful medicinal plants in the Roman empire [34].

In several parts of the world there have been continued use of plants in the folklore medicine and a good number of the allopathic medicine originated from medicinal plants [44]. This was made possible with the advent of scientific methods of screening to establish their chemical constituents [11]. The chemical scrutiny came into effect in the nineteenth century, and preference was given to plants of known medicinal values. Their active principles were extracted and characterized. An example is morphine which was isolated in 1805 from opium [39]. There are examples of several important plants which gave pharmacologically active compounds, which were isolated and elucidated during this period. Thereafter, compounds became an integral part of the pharmacopoeias of several countries. As this phase of modern medicine developed, chemists and pharmacologists were embroiled in the evaluation of new molecules. In the process of chemical evaluation, new compounds were also synthesized based on the active compounds from the plants. It was imperative that more constructive and comprehensive work be done on natural products. This was achieved when Paul Erhlich at the Institute for Experimental Therapy, Frankfurt,was one of the pioneer scientists to propound his theories on drug action [39]. Since then the use of drugs in the management of ailments is ever increasing and plants continue to be an important source of drugs.

Over 80 % of the world's population relies on traditional medicine, most of which are plants or plant extract-based drugs [7]. Thus, plants dominate the scenario to about 80 % [43]. An analysis of prescriptions from community pharmacies in the USA carried out in 1973 revealed that over 38 % of the prescriptions contained one or more products of plant origin as the therapeutic agent [5, 14]. Approximately 25 % were therapeutic agents derived from higher plants. The major diseases managed by preparations from higher plants include chronic diseases like diabetes, cancers, hypertension, asthma, HIV-related problems, epilepsy, and such other conditions in which allopathic medicines are less successful [13].

One reason for choosing plants is that, they are readily available either for free or at minimal cost which the majority of the rural poor communities in the developing and the developed world can afford [32]. Of the entire world flora, 250,000 species have been identified and used for curative purposes [10, 26]. This number represents only 15 % of what has been effectively investigated and found useful [25]. Consequently, there is a staggering over 85 % of higher plants to be investigated. Through ethno-botany, the useful plants can be deciphered from a large list of higher plants numbering a total of 850,000 plant species. Most of these plants occur in the tropical and subtropical floral diversity [13], and a large number of this floral diversity has so far not been prospected [6].

Reliance on ethno-botany to carry out or do bio-prospecting has enabled humans to identify, recognize, and incorporate certain compounds into various pharmacopoeias of the world. A few such plants and their identifiable compounds are codeine, ephedrine, digoxin, atropine, quinidine, theophylline, and caffeine [38]. Some of the aforementioned compounds are now used in modern allopathic medicine without any modification. However, some drugs are plant-based sources and can now be synthesized in laboratories due to low cost [11].

Based on the ethno-botanical information there are some plants that have been found scientifically and economically important and are currently used in modern medicine. Examples include *Prunus africana* used in the management of hyperplasia and *Artemisia annua*, an important source of artemisinin currently used in malaria management. Some of the compounds included are not medicines *per se*, but are important raw materials that provide skeletons that are used in the manufacture of several pharmaceutical compounds. Diosgenin, a sapogenin, is a steroidal compound which is used in the synthesis of steroid and hormonal drugs.

Some plants which yield alkaloidal-based drugs, are widely distributed amongst higher plants, especially in the dicotyledons, that number in excess of 10,000 genera of which, 9 % contain such compounds [3]. The families: Amaryllidaceae, Apocynaceae, Liliacaceae, Rubiaceae, Rutaceae, and Solanaceae are known to possess alkaloids and 4000 alkaloids have been isolated from them [36].

The Madagascar periwinkle *Catharanthus roseus* G. Dn, an Apocynaceae, has potential and its usefulness came into prominence in 1959–1960. While studying the plant's ability to treat diabetes, some scientists accidentally found it effective in the treatment of human maladies like leukemia and caposis sarcoma [29]. Folklore stories from Jamaica indicate that its leaf infusion can be used in diabetes mellitus management. The hypoglycemic principles could not however be substantiated. Some alkaloidal fractions contained in the extracts caused bone-marrow depression in studies with rats. Scientists who were studying the extracts of the plant were able to isolate, from the alkaloidal fractions, vinblastine which has anti-leukemia activity. Scientists from Eli Lilly, an American pharmaceutical multinational company, succeeded in isolating vinblastine and other potent anticancer alkaloids. Since these alkaloids are present in the plant at very low concentrations and in a mixture of 90 other alkaloids, their acquisition has only been possible with judicious systematic separation using appropriate pharmacological assays leading to elucidation of structures of ajmalicine (**1**), vincristine (**2**), and vinblastine (**3**) [38].

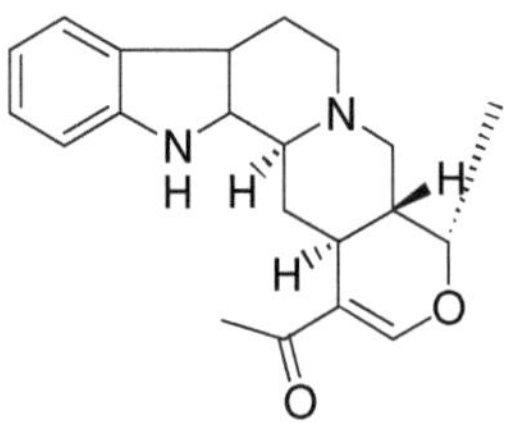

1 Ajmalicine

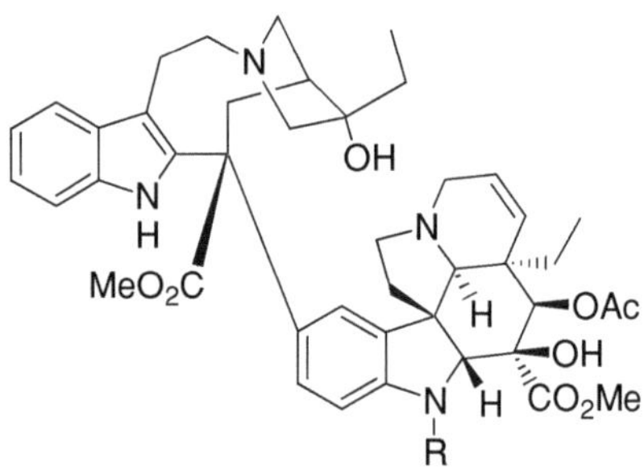

R = CHO 2 Vincristine
R = Me 3 Vinblastine

Vincristine and vinblastine are therapeutically amongst the most useful antineoplastic agents. The sulfate of the latter is used to treat Hodgkin's disease while the sulfate of the former is used in pediatric leukemia and lymphatic leukemia. They are administered intravenously. More often the drugs are used in cocktail with other therapeutic agents [17].The drugs are produced from two plant species which are erect shrubs with opposite, oblong leaves, growing up to 1 m high, branching at the base with a spread of up to 70 cm in diameter. The plant has two varieties based on the flower colors: *C. roseus* produces pink flowers and *C. alba* white flowers. The two varieties grown as ornamental plant flowers for commercial cultivation are found in India, Israel, and USA [17].

In Central America there are several plant species whose extracts have been incorporated into various pharmacopoeias of several countries. They include *Cephaelis* spp; *Cinchona* spp; *Papaver somniferum (L)*; *Rhamnus purshiana* DC; *Digitalis* spp, and *Dioscorea* spp (Schultes and Farnworth, 1980). *Celphaelis* spp (Rubiaceae), which is a straggling evergreen shrub, produces rhizomes that have been used to cause vomiting, and has been used for treating dysentery for centuries by the South American Indians and tribes. By the seventeenth century the plant preparation was in use in Europe against amoebic dysentery [39]. To date emetine hydrochloride, from *Celphaelis* spp, is still considered important in the treatment of amoebiasis through both subcutaneous and intramuscular injections and is effective against hepatic and bowel infections. Emetine–bismuth iodide is however, administered orally. The low doses of the drug preparations are used in cough and whooping cough [39].

There are certain plants that have been used for curative purposes by various civilizations. Henbane (*Hyoscyamus niger* L) seeds were used by the Babylonians to relieve problems of toothache. Belladona (*Atropa belladona* L) has been used in Europe for centuries to relieve pain and was recorded in the London pharmacopoeia in 1809. Belladona roots and leaves are reliable sources of atropine (**4**), which is important in the treatment of eye diseases (mydriasis).

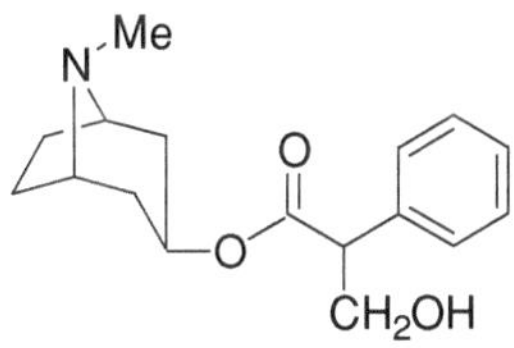

4 Atropine

Other plants yielding these important alkaloids are *Hyocymus muticus* (L), *Datura inoxia* Miller, *Datura metel* (L), *Datura stromanium* (L.), and *Duboisia lechahardtii* F. v. Muell. Other uses of the drugs from this group of plants are in the treatment of asthma, whooping cough, and as an antidote to poisoning by cholinesterase inhibitors. Scopolamine and Hyoscine are used in the treatment of duodenal ulcer [9].

The world requirement of the drug emetine is met by synthetic sources. However, *Belladona* is still commercially produced in Brazil and India with USA being the major importer. The demand for preparations based on the whole crude drug, such as ipecacuanha, is expected to remain stable [17]. Below is the structure of the alkaloid (**5**), emetine (**5**) and cephaeline (**6**).

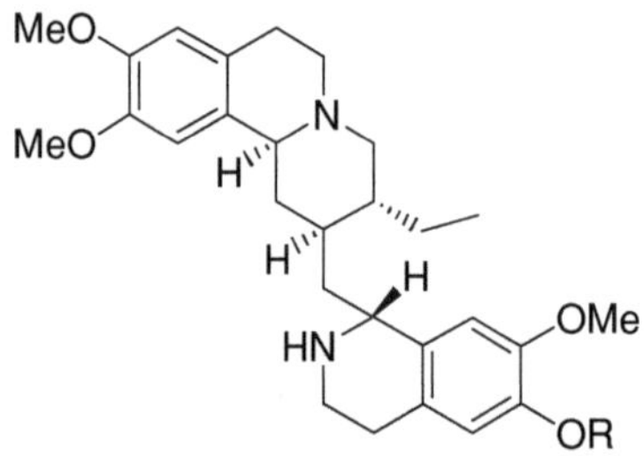

R = Me 5 Emetine
R = H 6 Cephaeline

In malaria infested countries, quinine is a household name. These are alkaloids that occur naturally in *Cinchona* spp, (Rubiaceae), and are indigenous to the slopes of the Eastern Andes. The bark of the plant was used by Peruvian Indians to cure fever [11]. The drug reached European medicine and appeared in the British pharmacopoeia in 1677. Currently, there are commercial plantations in Africa, Asia, and South America.

During the last half of the twentieth century quinine was replaced with chloroquine as a drug of choice but has also been withdrawn as the first line of treatment; consequently, quinine (**7**) was reintroduced in the 1990s as a reliable source of treatment against malaria [19].

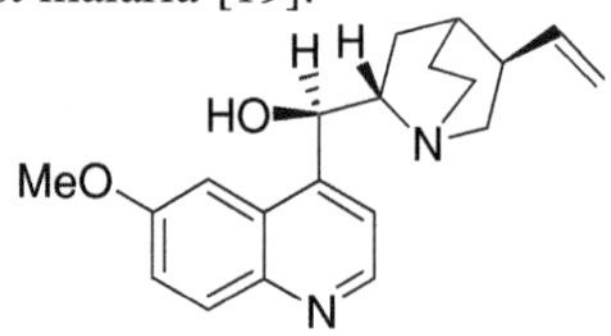

7 Quinine

Over 40 species of cinchona are known but the most important are *C. succirubra* Pavon ex klotzsch; *C. calisaya* Weddel; *C. afficinalis* (L); *C. ledgeriana* Moens ex Tremen, and some hybrids that do exist [4]. Quinidine (**8**) isolated from the cinchona is a natural antiarithiimic drug.

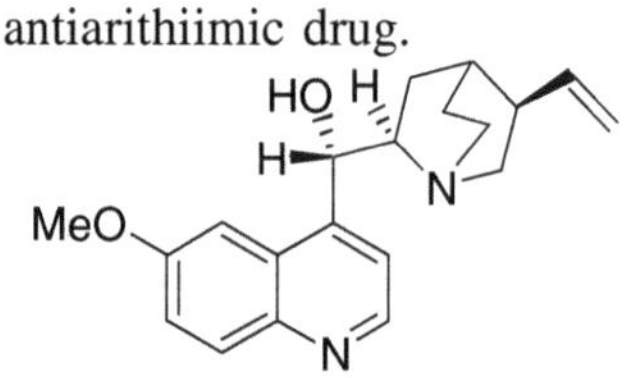

8 Quinidine

Cinchona spp are trees growing up to 20 m high and prefer a cool climate approaching montane, soil (pH 4.2–5.6), and precipitation of 190–500 cm annually [17]. The plant contains over 30 alkaloids from the bark of various species of *Cinchona*. The most recognized is quinine, which has antimalarial activities and antipyretic properties. A combination of quinine and 8-aminoquinolone is recommended for malaria and relief from nocturnal leg cramps [19]. Quinine sulphate is used in food and drink preparations while quinidine sulfate is used in cardiac arrhythmias [19]. Currently, there is a worldwide increase in the demand for *Cinchona* products [17].

Anthraquinone glycosides are found in several higher plant species and plant choice is variable. Some important plants currently used in various countries include *Aloe* spp, *Cassia* spp, *Rhamnus purshiana* DC (cascara), and *Rheum* spp (rhubarb) [11]. Basically, the major constituents of such drug plants are hydroxyan thraquinone derivatives and their glycosides, which have a huge market as laxatives, reaching an annual sale of $300 million as imports into USA [17, 28]. *Aloe barbadensis* Liliaceae, Mill. (Syn *A. vera* (L), and *A. ferox* Mill. have become important and all over the world farmers are turning to establish their plants. Mature plants are squizzed, and sap exported for use in pharmaceutical and cosmetic industries. Rhubarb that comprises rhizomes of *Rheum palmatum* (L) is an ancient drug in China since 2700 BC [8]. Major chemical constituents include emodin, aloe-emodin, rhein, chysophanol, and their glycosides. Cascara was at one time a major drug for constipation and was found from the back of *Rhamnus purshiana* DC (Rhamnaceae). There is a reasonable world output of the raw material from which cascarosides A.B.C and D are extracted.

The sennoside (**9**) extracted from *Senna angustifolia* Delile (Fabaceae) was used as a laxative in lidoginom of Alexandria (Codd, 1972). Currently, seeds of the plant are exported to Europe. The main constituents are glycosides sennoside A and B used as “tea”. Calcium sennosides are manufactured in Switzerland, USA, and India and used as carminative [17]).

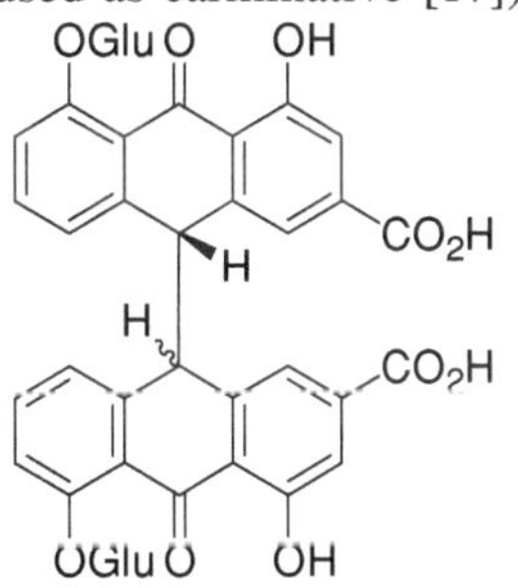

9 Sennoside A (H = α)
Sennoside B (H = β)

Steroids are some of the natural products extensively used in pharmaceuticals. Unfortunately, it was believed that animal source was the only available avenue for their acquisition. In 1936, Marker, discovered a sapogenin from the roots *Dioscorera* spp (Dioscoreraceae). In the same year it was converted to progesterone. Further, researchers found the Mexican *Dioscorea* as an abundant source of diosgenin. Diosgenin (**10**) soon became a competitive source for steroid synthesis [19].The side-chain degradation of cholesterol and sitosterol is now possible , which has reduced the demand for such raw materials [19]

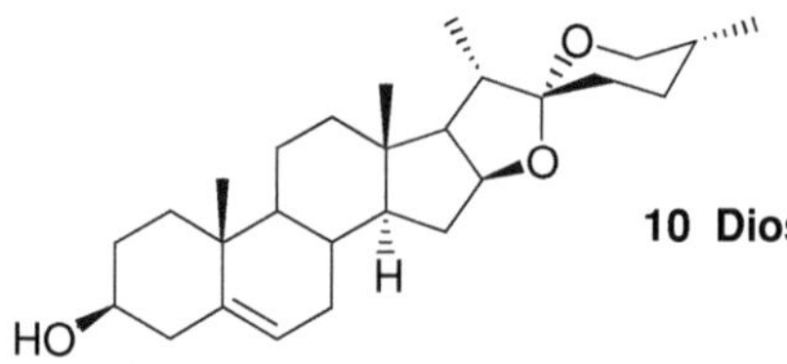

10 Diosgenin

Yams still remain an important source of raw materials for steroid drugs synthesis and the major species used are *D composita* Hemst; *D zingiberensis* C. H. Wright; *D. deltoidea* Wall, *D. panthacia* Prain and Bark. More recently, plantations of *D. floribunda* (Mart and Gal) have been established in India [17]).

Steroidal are natural compounds capable of acting directly on the heart. Such glycosides are referred to as cardiac glycosides as they have specific properties that increase the heart's excitability and contractibility [15, 19]. These plant products are invaluable in the treatment of heart disorders. Examples are cardenolides (**11**) and bufadienolide (**12**), which are considered pivotal in the management of heart problems. The occurrences of these two categories of compounds are restricted to angiosperms and especially *Digitalis purpurea* (Scrophulariaceae), *Nerium, strophanthus* and *Acokanthera* (Apocynaceae), *Asclepias* (Asclepadiaceae), and *Erysimin* (Brassicaceae). Most of these species are found in the tropical regions of South America and Africa and have been used by the indigenous people as arrow poisons or drugs [11].

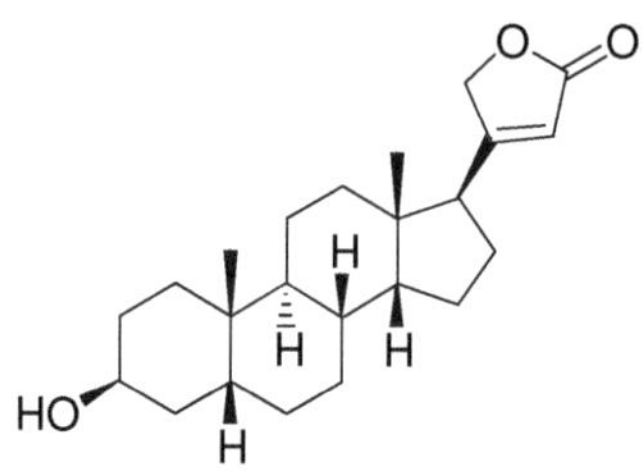

11 Cardenolide

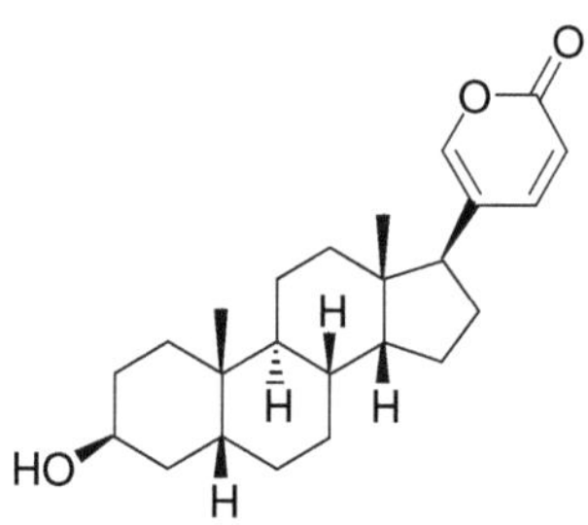

12 Bufadienolide

Digitalis spp., which is a native of Europe, is worth mentioning. The plant has been used for therapeutic purposes since medieval times for the purpose of poison preparations. However, it has also been used for dropsy [39] and was introduced for heart treatment in the mid-twentieth century management+ [39]. The most important species, which are used in production of current pharmaceutical drugs, are *D. purpurea* Ehrh which produce cardenolides digoxin A and B. Gitoxin, gitaloxin glucoverodoxin, and odoroside are cardenolides; while *D. ianata* (Ehrh) is rich in cardio active glycosides whose digogenin and diginatigenin are important drugs of choice for heart ailments [19].

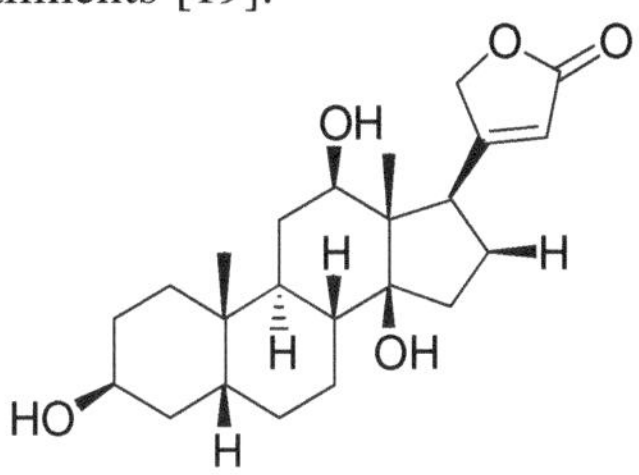

13 Digoxigenin (Series C)

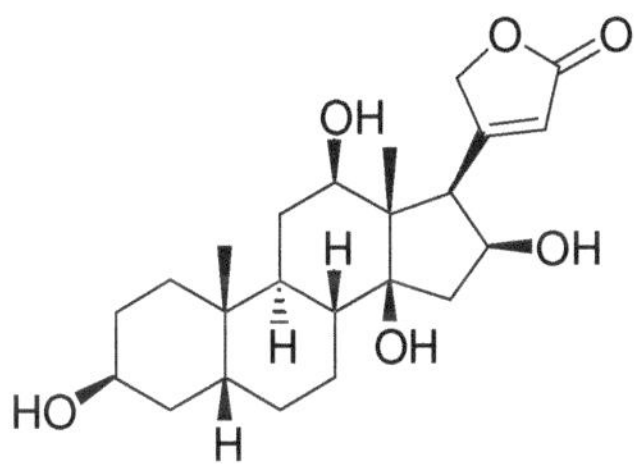

14 Digoxigenin (Series D)

Certain cases where plants have been used for curative purposes across several cultures are (Opium poppy), *Papaver somniferum* L. (Papaveraceae) which is a native of the Western Mediterranean region but has been grown in Egypt, Turkey, Greece, Asia Minor, Balkans, and Italy since ancient treatment times. Pliny recommended its products for the treatment of arthritis, headaches, and for curing wounds [22, 39]. Its importance is contained in history books with specific reference to the British–Chinese opium wars [39]. The milky extract from the fruit of the plant contains a complex mixture of compounds of triterpenoids and alkaloids and other compounds. The most important 40 species have yielded alkaloids. Morphine, whose structure was established in 1952, though isolated in 1803 by Derosne, remains the most important and the strongest narcotic [15]. Narcotine (**18**) occurs as an admixture and a mild antitussive, and is used in the preparation of cough lintus. Morphine (**15**) is converted into codeine (**16**) as an antitussive, which is widely used in medicine as an analgesic, a central nervous system stimulant, and antitussive. There are also other important alkaloids which are also used for curative purposes. They include papaverine (**17**), which is a smooth muscle relaxant, a cerebral vasodilator, and treats asthma. Thebaine is a convulsant poison and is only used as a raw material in the manufacture of codeine or other semi-synthetic analgesics and narcotic antagonists like nalorphine and etorphine.

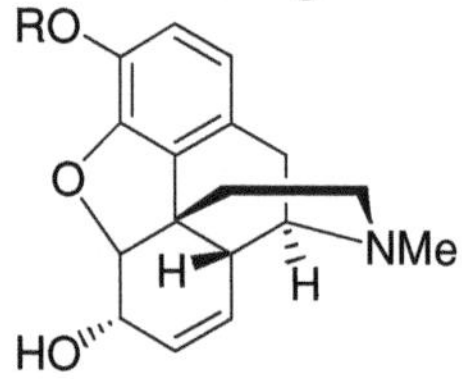

R = H 15 Morphine
R = Me 16 Codeine

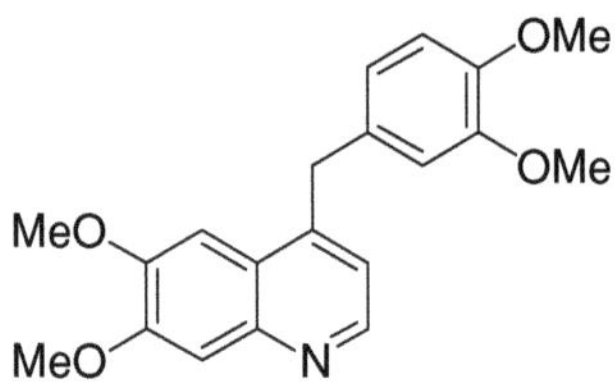

17 Papaverine

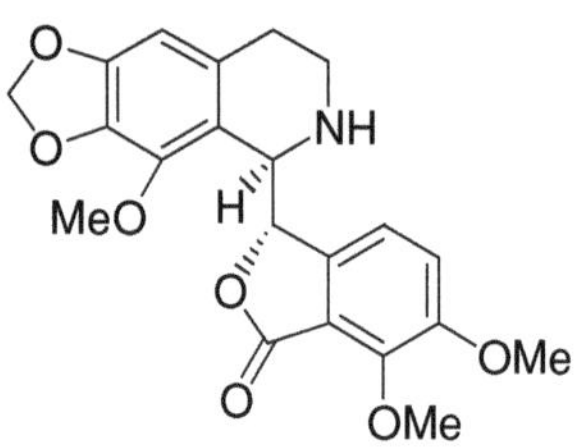

18 Narcotine

There is a trend of emerging new diseases like AIDS, as a result of HIV infection and candidiasis. Re-emergence of diseases like Tuberculosis and Legionella, which were believed to be under control, are becoming almost epidemic and require urgent attention to save the human race from such scourges [25]. The main reason for resistance to antibiotics is due to the fast spread of resistant organisms due to the lack of firm control of dispensation of antibiotics and lack of compliance to the treatment regimen [41]. The other reason for the emergence of resistances to synthetic antibiotics is due to the emergence of spontaneous mutants in vitro that confer resistance to virtually any antibiotic [45]. Solutions to such problems are embedded in the diverse plant resources and other innovations like biotechnology.

An increasing number of patients with HIV infection and/or AIDS cannot use the currently approved anti-HIV drugs. Due to poverty, HIV is ever-increasing and there are adverse side effects and the emergence of resistant strains of pathogens [3]. This has been experienced more adversely in TB cases [25]. The only avenue of treatment considered safe and sustainable for a reasonable period of time is the use of plant products or their derivatives (Li et al., 2004). Compounds extracted from many plants have been found to be effective in chronic and terminal cases like cancer [21]. However, the goals of finding curative agents from plant sources can only be achieved through consistent screening of plants supported from the background by basic ethnobotanical studies of the indigenous people.

It is estimated that only 1 % of medicinal plants are known by scientists and accepted for commercial purposes [1]. This would mean that the market for herbal medicine is still a goldmine and stands at $60 billion [40]. There would be a further impetus to this research if only the world could provide intellectual property protection to the natural product discovery, particularly traditional herbal medicine and herbal medicine products [18]. This overstatement may mean that very little has been done pertaining to medicinal plants and a lot remains outside there undiscovered and untapped. After all, a large proportion of drugs have been discovered with the aid of ethnobotanical knowledge of the traditional uses of plants [18]. A lot of useful compounds still remain to be prospected outside there. After all hardly 2% of the Earth's flora has been exploited and harnessed for pharmaceutical uses.

References

1. Aguilar G (2001) Access to Genetic Resources and Protection of Traditional Knowledge in the Territories of Indigenous Peoples. Environ Sci Policy 4:241–256
2. Ali M (2008) Pharmacognocy (pharmacognocy and phytochemistry). CBS publishers and Distributors, New Delhi
3. Asres K, Seyoum A, Veeresham C, Bucar F, Gibbon S (2005) Naturally derived anti-HIV agents. Phytother Res 19:557–581
4. Atal CK, Kapur BM (eds) (1982) Cultivation and Utilisation of Medicinal Plants, Regional Research Laboratory C. S. I. R. New Delhi p 177
5. Barnes J (2003) Quality, efficacy and safety of complementary medicines: fashions, facts and future. Part II: Efficacy and Safety. Br J Clin Pharmacol 55:331–340
6. Buchman DD (1980) Herbal medicine. Gramercy Publishing Company, New York, pp 31–36
7. Busia K (2005) Medical provision in africa. Past and Present Phytother Res 19:919–923
8. Codd LW (ed) (1972) Materials and Technology, Vol 5. Longman, London, pp 707–757
9. Cordell GA (1981) Introduction to alkanoids. Wiley Interscience, New York, p 1055
10. Dev S (1983) Natural Products in medicine-Present Status and Future Prospects. Current Science 52:949–956
11. Dev S (1989) Higher Plants as a source of drugs. In: Plants and Society. Macmillan Publishers Ltd, London, pp 267–292
12. Diop CA (1989) Africa's contribution to world civilisation: exact sciences in Nile valley civilizations. J Afr Civiliz 6(2)
13. Ernst E (2005) The efficacy of herbal medicine–an overview. J Fundam Clin Pharmacol 19:405–409

14. Farnworth NR, Bingel AS (1977) Problems and prospects of discovering new drugs from higher plants by pharcological biological or therapeutical activity. In: Wagner H, Wolf, P (eds) Springer Verlag, Berlin, pp 1–22
15. Foye WO (ed) (1981) Principles of medicinal chemistry. Lea and Febiger, Philadelphia, p 931
16. Gottlieb O (1979) Chemical studies on medicinal Myristicaceae from Amazonia. J Ethnopharm 1:309–343
17. (ITC) International Trade Centre UNCTAD/GATT (1982) Markets for selected medicinal plants and their derivations TTC UNCTAD/GATT Geneva pg. 200–216
18. Kartal M (2007) Intellectual property protection in the natural product drug discovery, traditional herbal medicine and herbal medicinal products. Phytother Res 21:113–119
19. Kirk O (1978) Encyclopaedia of chemical technology, vol 1. Wiley–Inter Science, New York pp 645–729
20. Kochhar SL (1989) Plants as Stimuli for Exploration and Exploitation. In: Swaminathan MS, Kochhar SL (eds) Plants and Society. Macmillan Publishers, London, pp 44–85
21. Laus G (2004) Advances in chemistry and bioactivity of the genus. Uncaria Phytother Res 18:259–274 (Review)
22. Lewis WH, Elvin-Lewis MPF (1977) Medicinal Botany. Wiley—Interscience, NewYork, p 513
23. Majumdar RC (1971) Medicine in a concise history of science in India. In: Bose DM (ed) Indian national science academy, New Delhi pp 217–273
24. Namjoshi A (1979) Ayurvedic pharmacopoeia and drug standardisation. In: Sharma S (ed) Realms of ayurveda. Arnold–Heinemann, New Delhi, pp 217–273
25. Okeke IN, Klugman KP, Bhutta ZA, Duse AG, Jenkins P, O'Brien TF, Mendez AP, Laxminarayan R (2005) Antimicrobial resistance in developing countries. Part II: Strategies for containment. Lancet Infect Dis 5:568–580
26. Patwardhan B, Warude D, Pushapangandan P, Bhatt N (2005) Ayurveda and traditional chinese medicine: a comparative overview. ECAM 2005 2(4): 465–473E (Mail: bhushan@unipune.ernet.in)
27. Pelt JM (1979) Medicines green revolution. The UNESCO courier, July, 8
28. Phytochemical Dictionary of leguminosae (1994) 1st edn. vol. 1. Plants and their constituents. Pub. 1994, Chapman and Hall, London, pp 165–166. Phytother Res 20:378–391
29. Protzen KD (1993) Produktion und Marktbedentung aetherischer Oele. In: Carle R (ed) Aetherische Oele Anspruch und Wicklichkeit Stuttgart: Wisserischaftliche Verlagsgelleschaft Sc. 93
30. Rajan S, Sethuraman M, Mukherjee PK (2002) Ethnobiology of the nilgiri hills. India Phyther Res 16:96–116
31. Rodney W (1971) How Europe under developed Africa. Tanzania Publishing House, Dar es Salaam
32. Samie A, Obi CL, Bessong PO, Namrita L (2005) Activity profiles of fourteen selected medicinal plants from Rural Venda communities in South Africa against fifteen clinical bacterial species. Afr J Biotechnol 4(12):1443–1451
33. Sharma S (ed) (1979) Realms of ayurveda. Arnold Heinemann, New Delhi, p 336
34. Sharma S (1982) Realms of Ayuverda. Arnold—Heinemann, New Delhi, p 336
35. Shibata S (1981) Chinese drug constituents: isolation of the biologically active principles in advances. In: Natori S (ed) Natural products chemistry. Kodasha, Tokyo, pp 398–429
36. Such D (1983) Natural products medicines–present status and future prospects. C Sci 52:949–956
37. Sushruta S (1963) Sutra sthana kaviraj kunjalal bhishagratna. Chowkhamba a Sanskrit Series Office, Varanasi, pp 20–26
38. Swaminathan MS, Kochhar SL (eds) (1989). In plants and society. Macmillan Publishers, London pp 278, 293–310, 351–417 and 471
39. Taylor N (1965) Plant drugs that changed the world, 3rd edn. George Allan and Urwin, London, p 275
40. Timmermans K (2003) Intellectual Property Rights and Traditional Medicine: Policy Dilemas at the Interface. J Soc Sci Med 57:745–756

41. Vicente M, Hodgson J, Massida O, Tonjum T, Henriques-Normark B, Ron EZ (2006) The Fallacies of Hope: Will We Discover New Antibiotics to Combat Pathogenic Bacteria in Time? FEMS Microbiology Review xx 000–0021
42. Weiner MA (1972) Earth Medicine, Earth Foods Collier-Macmillan, London, p 214
43. WHO (1978) The promotion and development of traditional medicine. Technical report series 1978:622
44. Williamson EM (2002) Plant and animal kingdom as a source of drugs. In: Trease and evans pharmacognosy saunders. Elsevier Science, London 30–35:15–41
45. Woodford N, Ellington, MJ (2006) The Emergence of antibiotic resistance by mutation. doi:10.1111/j/1469-0691. 01492

The Metabolism of Antiparasitic Drugs and Pharmacogenetics in African Populations: From Molecular Mechanisms to Clinical Applications

Collen Masimirembwa

Abstract We characterised over 20 antiparasitic drugs with respect to the enzymes responsible for their metabolism. We showed that CYP2C8 is responsible for the metabolism of amodiaquine (ADQ) to desethylamodiaquine and identified a novel reactive metabolite catalysed by extrahepatic CYP1A1 and CYPIB1 which is giving us insights into possible ways of synthesising safer analogues of ADQ. Praziquantel (PZQ) was shown to be metabolised by CYP1A2 and 3A4, knowledge which is being used to explore the possibility of coadministering PZQ with known inhibitors of these enzymes in order to increase its bioavailability. From evaluating over 30 antiparasitic drugs for inhibition of major drug metabolising enzymes, 10 were shown to be potent inhibitors with a potential risk to cause metabolism based drug–drug interactions. The inhibitory effects of artemisinin and thiabendazole on CYP1A2 where further investigated in vivo and the effect of thiabedazole resulted in clinically relevant drug–drug interactions. We studied the genetic polymorphism of drug metabolising enzymes in African populations. We screened genes of 8 drug metabolising enzymes (CYP2B6, 2C9, 2C19, 2D6, FMO, NAT-2, GSTT and GSTM) for over 15 single nucleotide polymorphisms (SNPs) in 9 ethnic groups from across Africa (Ibo, Hausa and Yoruba of Nigeria, Luo, Kikuyu and Masai of Kenya, mixed Bantu volunteers from Tanzania, the Venda of South Africa, the Shona and San of Zimbabwe). Multivariate cluster analysis showed that Caucasian, Oriental and African populations show differential cluster groups, an indication that these major population groups are likely to metabolically handle medicines differently. Further studies led to the discovery of new genetic variants unique to populations of African origin such as CYP2D6*17. Clinical studies on the metabolism and elimination of efavirenz by the polymorphic CYP2B6 showed that

C. Masimirembwa (✉)
African Institute of Biomedical Science & Technology, P.O.Box 2294,
LAPF Center Corner Jason Moyo and Chinhoyi Street, Harare, Zimbabwe
e-mail: Collen.masimirembwa@aibst.com

A. Gurib-Fakim and J. N. Eloff (eds.), *Chemistry for Sustainable Development in Africa*, DOI: 10.1007/978-3-642-29642-0_2,

African populations had a reduced capacity to dispose efavirenz and that patients homozygous for the CYP2B6*6 variant would require as low as half the dose given to Europeans to achieve the same safe and efficacious concentrations.

1 Introduction

In the discovery, development and clinical use of medicines, pharmacokinetics determines how much and how often drugs should be administered to patients for safe and efficacious outcomes. Pharmacokinetics is essentially the time course of a drug and its metabolites in the body and is characterised by the processes of absorption, distribution, metabolism and excretion (ADME). Of these processes, metabolism is the determinant of the clearance of over 75 % of drugs on the market. This makes understanding the biotransformation of drugs and mechanisms of regulation of drug metabolising enzymes important in chemotherapy. Major regulatory mechanisms of enzyme expression and activity have been shown to be induction, inhibition and genetic variability of genes coding for some of the enzymes. Genetic variability that results in variable response to medicines, pharmacogenetics, is pushing the practise of medicine from one-treatment-fits-all to individualised treatment where drugs will be given to people they are predicted to work and at doses they will be safe and efficacious based on their genetic status. The knowledge base and clinical applications of drug metabolism and pharmacogenetics is very advanced in developed countries and significantly lags behind in Africa.

In the 1980s, the laboratory of Professor Julia Hasler at the University of Zimbabwe identified this as a niche for research which could make important contributions towards the safe use of medicines in African populations. With funding from the International Science Programme (ISP), Sweden, www.isp.uu.se, her drug metabolism group started characterising the metabolism of antiparasitic drugs [16–18, 21, 22, 24] and the genetic polymorphism of drug metabolising enzymes in Zimbabweans [17, 18, 20, 21]. The theme of metabolic sciences has been sustained by IPICS funded activities by graduates from the drug metabolism group. Professor Yogi Naik went on to specialise in ecotoxicology at the National University of Science & Technology (NUST) in Zimbabwe. Dr. Stanley Mukanganyama focused on cancer chemotherapy and remained at the University of Zimbabwe. After many years in the pharmaceutical industry, Professor Collen Masimirembwa went on to establish the African Institute of Biomedical Science & Technology, AiBST, www.aibst.com which has drug metabolism and pharmacokinetics (DMPK) as one of its focus areas of research. The seed funding provided by IPICS catalysed a chain reaction of developments in Zimbabwe that has resulted in the training of many postgraduates, establishment of cutting edge research platforms and the conduct of biomedical research, which is beginning to have a clinical impact on Zimbabweans in particular and Africans in general.

The sciences of drug metabolism and pharmacogenetics represent a unique interplay of chemistry, enzymology and molecular biology. The results from our work presented in this chapter, therefore cover aspects of enzyme kinetics, molecular biology, molecular modelling through to clinical evaluations of some of our findings.

2 Metabolism of Antiparasitic Drugs

Most drugs used for the treatment of parasitic diseases were discovered more than 50 years ago, way before pharmacokinetics (PK) was an integral part of the drug discovery and development process. Most of them, therefore carry a number of PK related inadequacies as evidenced by complex dosing regimens and a host of adverse drug reactions associated with their use. Since few new antiparasitic drugs are coming on the market due to poor funding, as a stopgap measure, we decided to invest the modest resources we have into PK characterisation of the available antiparasitic drugs with a view to improving their clinical use. Over 20 antiparasitic drugs (Fig. 1) were evaluated with respect to major enzymes involved in their metabolism and their ability to inhibit the activity of drug metabolising enzymes. The in silico, in vitro and in vivo studies were conducted according to current practices in major pharmaceutical industry [26].

2.1 Identification of Enzymes Responsible for the Metabolism of Antiparasitic Drugs

This was done by a combination of three methods: (i) incubating each compound against a panel of seven major drug metabolising recombinant cytochrome P450 (rCYPs) (1A2, 2A6, 2C8, 2C9, 2C19, 2D6, 3A4), (ii) incubating the compounds in human liver microsomes (HLM) and using selective and potent diagnostic inhibitors of each of the various major CYPs, and (iii) using a relative activity factor approach that combines the use of HLMs and rCYPs. These reaction phenotyping studies [14] indicated that 1–3 major CYPs can be involved in the elimination of each drug (Table 1). Knowing which enzyme(s) are involved in the elimination of a drug helps us understand how individuals vary in their ability to eliminate the drug based on our knowledge of the variability of expression and activity of the involved enzyme.

We have shown that amodiaquine is metabolised to desethylamodiaquine by the enzyme, CYP2C8, with high affinity, turnover, and selectivity [13, 14]. This work has resulted in FDA recommending amodiaquine N-deethylation as an acceptable marker reaction for in vitro studies of CYP2C8 in industry (http://www.fda.gov/cder/guidance/index.htm). Using a combination of in vitro and electrochemical oxidation

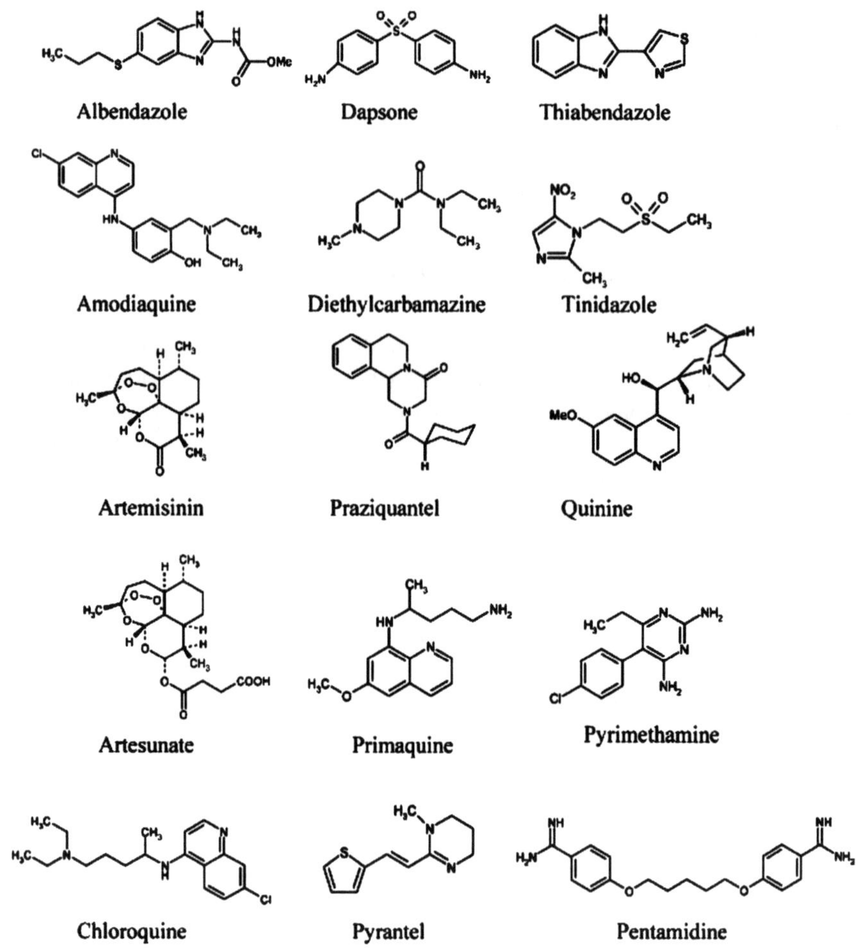

Fig. 1 Structures of some antiparasitic drugs whose metabolism was investigated in this study

methods followed by structural elucidation by MSMS and NMR, we also showed that amodiaquine is metabolised by extrahepatic CYP1A1 and 1B1 to a reactive aldehyde metabolite, Fig. 2, [9, 10, 13]. The reactivity of the metabolites was determined by trapping experiments with nucleophiles such as glutathione, N-acetyl cycsteine (NAC), and methoxyl amine (MOA). This led us to postulate that the two major adverse effects of amodiaquine, liver toxicity and agranulocytosis are caused by separate tissue specific metabolic bioactivation processes, liver toxicity by the bioactivation to the quinonimine metabolite and agranulocytosis by both the quinonimine and the aldehyde metabolite. We are, therefore, synthesising potentially safer ADQ analogues in which we aim to block these 2 bioactivation pathways.

The knowledge of enzymes involved in a drug's metabolism also helps us to understand and anticipate drug–drug interactions when a drug is given together

Table 1 Predicted hepatic clearance and relative contribution of major drug metabolising enzymes to the hepatic metabolic clearance of some antiparasitic drugs from in vitro experiments

Drug	Major disease treated	Predicted hepatic clearance CL_H = ml/min/kg	Percent (%) contribution of major drug metabolising cytochrome P450s in the metabolism of the antiparasitic drugs							
			1A2	2A6	2B6	2C8	2C9	2C19	2D6	3A4
Albendazole	Helminths	18.2	**53**	<10	<10	<10	<10	<10	<10	<10
Amodiaquine	Malaria	19.1	<10	<10	<10	**67-100**	<10	<10	<10	<10
Artemisinin	Malaria	9.8	<10	<10	**10**	<10	<10	<10	<10	**25**
Artesunate	Malaria	12.3	<10	**100**	<10	<10	<10	<10	<10	<10
Chloroquine	Malaria	1.4	<10	<10	<10	**54**	<10	<10	**53**	**13**
Dapsone	Malaria	5.0	<10	<10	<10	**10**	**48**	12	<10	**31**
Praziqunatel	Schistosomiasis	16	**39**	<10	<10	<10	<10	**14**	<10	**30**
Primaquine	Malaria	5.3	**60**	<10	<10	<10	<10	<10	**23**	<10
Pyrental	Helminths	3.4	<10	<10	<10	<10	<10	<10	**90**	<10
Quinine	Malaria	6.1	<10	<10	<10	<10	<10	<10	<10	**70**
Thiabendazole	Helminths	17.9	**100**	<10	<10	<10	<10	<10	<10	<10
Tinidazole	Protozoa	1.1	<10	<10	12	<10	<10	<10	<10	**77**

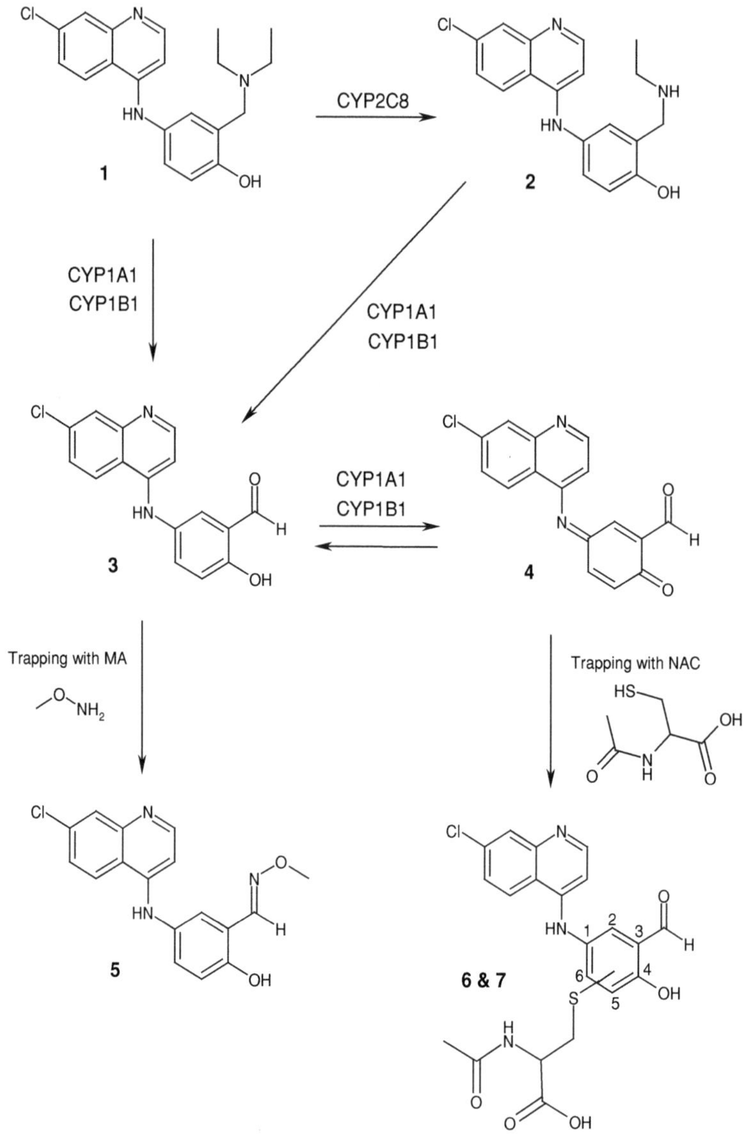

Fig. 2 Metabolic routes of amodiaquine and desethylamodiaquine observed in RLMs and HLMs and recombinant CYP2C8, CYP1A1 and CYP1B1. Trapping reactions were performed on metabolites formed in the rCYP incubations. All the metabolites and trapped adducts were also obtained in the electrochemical system

with another drug which inhibits or induces the major enzyme(s) that eliminate it. Our results on the role of CYP1A2 and CYP3A4 in the metabolism and elimination of praziquantel (PZQ), can explain why earlier combinations of PZQ with dexamethasone, a now-known CYP3A4 inducer, resulted in reduced PZQ levels in patients being treated for neurocystecosis [32]. Our findings are also providing the rationale for current efforts to increase the oral bioavailability of PZQ by giving it together with grapefruit juice [5], a known inhibitor of CYP3A4. The use of metabolic drug–drug interactions to improve oral bioavailability has precedence in the coadministration of the CYP3A4 substrate drug cyclosporine with ketoconazole, a CYP3A4 potent inhibitor [11]. The clinical benefits of which have been the administration of a lower dose of the expensive immunosuppressant but achieving clinically effective concentrations. Another example is the use of protease inhibitor boosted regimens in which two protease inhibitors (both substrates and inhibitors of CYP3A4) are given together, one as an inhibitor of CYP3A4, retinovir, and the other as the therapeutic agent, e.g. amprenavir, resulting in prolonged exposure of the latter at therapeutically effective concentrations [33]. This strategy facilitated the development of once a day dosing regimens of the otherwise very rapidly cleared protease inhibitors.

2.2 *Inhibition of Drug Metabolising Enzymes*

In Africa, most patients are on more than one drug at any one time. This polypharmacy is usually due to the need to treat coinfections, the need to use drug combinations for improved efficacy and to avoid the emergence of drug resistance in the treatment of diseases such as malaria, TB and HIV/AIDS. Simultaneous exposure to many drugs predisposes a patient to drug–drug interactions, most of which are based on the inhibition or induction of drug metabolising enzymes. In the case of inhibition, one drug (the perpetrator) will inhibit the enzyme responsible for the elimination of another (the victim) resulting in the latter increasing in plasma concentration which can lead to increased adverse drug reactions. In the case of induction, the perpetrator increases the expression and activity of the enzyme(s) responsible for the elimination of the victim drug which can lead to the latter failing to reach therapeutic plasma levels. Subtherapeutic levels not only affect efficacy, but promote the emergence of drug resistance in the treatment of HIV, TB and malaria.

Our studies [2] indicate that of the over 30 drugs screened for inhibitory effect against 5 major CYPs, 1A2, 2C9, 2C19, 2D6, and 3A4, 10 showed in vitro potential for inhibition of either CYP1A2 or CYP2D6 (Table 2). Thiabendazole's inhibitory effects on CYP1A2 were studied in detail and were shown to cause mixed mode inhibition (competitive, non-competitive and time dependent (TDI) mechanism based inhibition (MBI)) [31]. We used molecular modelling studies to rationalise thiabendazole's route of metabolism and possible binding modes in the CYP1A2 active site associated with the various mechanisms of inhibition (Fig. 3). Two of the drugs, artemisinin and thiabendazole where further evaluated for inhibitory effects on CYP1A2 in vivo in humans. This was done using caffeine as an in vivo probe for CYP1A2 activity and volunteers took caffeine alone or in combination with thiabendazole or artemisinin [4]. The in vitro data was in good

Table 2 Inhibitory effects of antiparasitic drugs on major drug metabolising enzymes

CYP and inhibitory compounds	Ki (μM)	Type of inhibition	Plasma concentrations (Cmax)	Predicted % inhibitory effects from in vitro data	Observed % inhibitory effects in vivo
CYP1A2					
Artemisinin	0.43	Competitive	1.38	76	66
Thiabendazole	1.54	Mixed	89	98	92
Primaquine	0.22	Competitive	0.44	67	ND
Dihydroartemisinin	3.67	Competitive	2.50	41	ND
CYP2D6					
Quinine	15.51	Competitive	15.41	50	
Chloroquine	12.68	Competitive	0.39	<10	ND
Amodiaquine	2.1	Competitive	0.074	<10	ND
Desethylamodaiquine	4.13	Mixed	0.444	<10	ND
Proguanil	6.76	Mixed	0.76	<10	ND
Cycloguanil	5.97	Competitive	0.21	<10	ND

ND - not determined

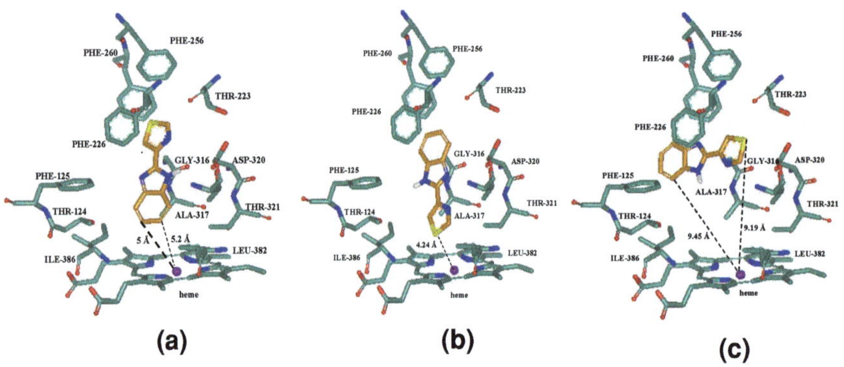

Fig. 3 Examples of different orientations in which thiabendazole docks into the active site of CYP1A2. The docking experiment was performed in GLUE. In 5 of the top 10 ranked solutions, the thiazole group was the group closest to the haem (**b**); in 3 solutions, the benzene ring in which hydroxylation occurs was closest (**a**); and in 2 solutions, both groups were further away (**c**). Interactions of the benzene or thiazole moiety of thiabendazole with phenylalanine 226 of CYP1A2 seems to be important in determining the orientation of thiabendazole in the enzyme active site

agreement with in vivo observations which gives credibility to the predictive power of the in vitro systems we use. Our data on the inhibitory effects of thiabendazole explains earlier clinical observations of an interaction between thiabendazole and theophylline and proposition of a 50 % theophylline dose reduction in asthma patients also taking thiabendazole [12]. Knowing the mechanistic basis for this interaction will now enable us to predict potential drug–drug interactions involving thiabendazole and other CYP1A2 substrate drugs.

3 Pharmacogenetics of Drug Metabolism

Based on our understanding that DNA is the blue print of life, the international community embarked on a project to sequence the whole human genome, a feat which was completed in 2003. The human genome is composed of 3 billion base pairs, about 25,000 genes, and approximately 10–30 million single nucleotide polymorphism (SNPs) (www.hugo.org). Genetic variability is thought to be the mechanism that facilitates the evolutionary process, where organisms with certain genetic status survive or succumb to some environmental selection pressures. For humans, this is supported by a number of known genetic variants that have been associated with disease susceptibility/resistance, with longevity, and with variable ability to tolerate potentially poisonous chemical exposures. Our work focuses on how genetic variability can affect our responses to medicines, a field of study referred to as pharmacogenetics. As indicated before, levels of drug exposure (PK) will determine the drug's pharmacological effect (PD). Since metabolism is the key determinant of the PK of most drugs, genetic variability of genes coding for drug metabolising enzymes could result in altered PK of a drug and subsequently affect the pharmacological effects of the affected drugs.

Genetic variation in the major drug metabolising enzymes such as CYP2D6, 2C9, 2C19, 2B6, FMO, and NAT-2 has been shown to influence the plasma concentrations of drug substrates and has been associated with increased incidences of adverse drug reactions in carriers of defect alleles. Most such studies have been done in Caucasian populations of Europe and North America and an increasing body of knowledge is being published for Asian populations. Our work, therefore sought to establish the status of some of these polymorphisms in African populations.

Following a workshop organised by AiBST on pharmacogenetics of drug metabolism in 2003 (the year the completion of the human genome was announced) in Nairobi, African scientists from six different countries (Nigeria, Kenya, Tanzania, Zimbabwe, Uganda, and South Africa) formed a consortium for biobanking and pharmacogenetic databasing in African populations. Samples from volunteers were subsequently collected from nine ethnic groups (Yoruba, Ibo, Hausa, Luo, Masai, Kikuyu, Shona, San, Venda and other mixed Bantu populations) and screened for genetic variants of key drug metabolising enzymes (Table 3). The data was subjected to multivariate analysis (Fig. 4) and clearly showed that the frequency of many genetic variants clusters Caucasian, Oriental and African populations into distinct groups [7, 8, 15–18, 21, 24, 28]. The general implication of this observation is that for drugs metabolised by these enzymes, the capacity to metabolise and eliminate them will differ significantly among these three major population clusters. This has implications for the use of medicines discovered and optimised for clinical use in Europe and then used in African populations [23].

The molecular epidemiological studies were followed by mechanistic investigations which lead to the discovery of a number of novel variants of CYPs and NAT, some of them unique to the African populations. Based on the phenotypic

Table 3 Genetic polymorphisms (% allele frequencies) of major drug metabolising enzymes in major African populations compared to Caucasian and Oriental populations

	CYP2C19		CYP2D6								NAT2				GST		CYP2B6	FMO		
	*2	*3	*2/2	*3	*4	*5	*10	*9	*17	*29	*5	*6	*7	*14	M1 del/del	T1 del/del	*6	K158	M257	G308
Orientals	30	10	2	0	1	6	51	0	0	0	5	25	13	0	55	65	18	–	–	–
Chinese	37	8	1	0	1	6	51	0	0	0	6	31	16	0	58	53	21	22.9	20.3	14.8
Japanese	35	11	1	0	1	3	43	0	0	0	2	19	10	0	44	44	16	22.7	14.5	21.0
Koreans	21	12	0	0	2	6	51	0	0	0	3	19	11	0	53	60	15	18.9	–	18.3
Caucasian	15	0	5	2	25	5	2	2	0	0	49	27	2	0	50	15	21	–	–	–
Swedes	17	0	1	3	23	5	1	0	0	0	51	28	2	0	51	20	–	44.3	7.1	22.4
Germans	18	0	2	2	20	2	2	0	0	0	46	27	4	0	51	21	–	42.6	6.9	22.5
American	14	0	2	–	–	–	–	–	0	–	45	28	2	0	54	15	–	39-53	6.7-22	15-22
Mix African	16	1	2	0	2	4	6	0	30	15	34	20	5	13	30	–	40	–	–	–
African American	25	0	1	0	7	6	4	1	15	5	30	22	2	9	28	24	47	41-52	5-7	0-5.2
Tanzanian	18	1	3	0	2	4	4	0	18	20	34	21	3	13	33	25	39	–	–	–
Shona	13	0	2	0	2	4	6	0	34	17	31	21	6	14	24	26	38	50	2	1.5
Venda	21	0	–	0	3	5	12	0	24	6	39	22	5	11	23	20	36	48.4	2.4	1.6
Ghanaian	–	–	2	0	7	6	3	0	28	–	–	–	–	–	39	–	49	–	–	–
Ethiopians	14	2	15	0	4	3	9	0	9	–	–	–	–	–	–	–	–	–	–	–
Kikuyu	16	0	–	0	1	–	–	0	33	14	58	24	–	–	28	25	34	49	4.6	1.0
Luo	18	0	–	0	4	–	6	0	23	16	34	22	3	14	29	22	37	50	4.5	0
Maasai	11	0	–	0	8	–	5	0	18	8	42	27	4	9	16	40	35	42	4.2	0
Igbo	29	0	–	0	8	–	10	0	14	20	28	29	4	11	23	36	38	43.9	0.5	0.5
Yoruba	10	0	–	0	3	–	7	0	22	10	33	27	3	8	31	35	42	52	1.0	0.5
Hausa	12	0	–	0	2	–	13	0	18	10	27	33	3	3	37	42	42	44	4.0	0.5
San	12	–	–	–	9	–	–	0	22	2	20	8	–	–	45	–	40	33.3	0.0	0.8

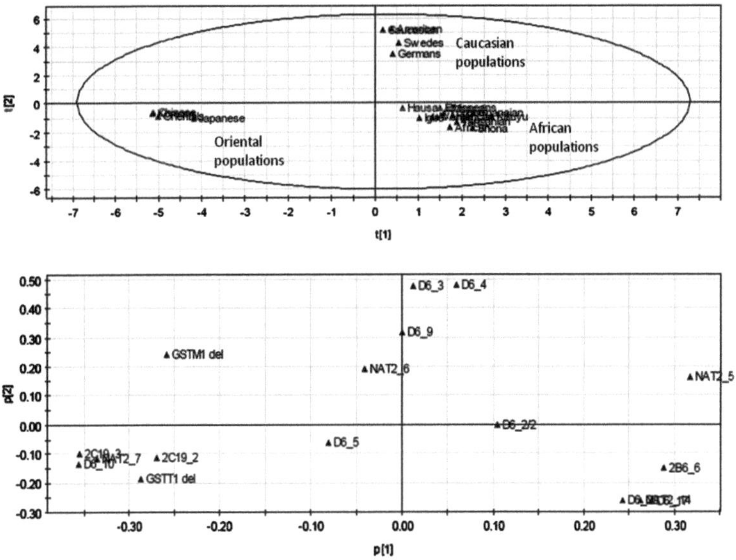

Fig. 4 The scores plot (*above*) showing correlations between populations. The loadings plot (*below*) show correlations between SNPs. Comparing the loadings plot to the scores plot enables one to understand how the variables (SNPs) relate to the observations (populations)

observations that African populations had reduced capacity to metabolise CYP2D6 substrate, sequence analysis led to the discovery of a novel variant, CYP2D6*17 (originally called CYP2D6*Z in recognition of its discovery in our work on Zimbabweans) [17, 18, 21]. The enzyme kinetic impact of key amino acid changes associated with this variant, T107I, R296C, and S486T was investigated using both in silico modelling and in vitro metabolism [3, 30] and shown to result in reduced affinity for CYP2D6 in a substrate dependent manner. Many publications have now demonstrated that CYP2D6*17 is the molecular basis of reduced CYP2D6 function in all populations of African origin [1]. Our group has also discovered variants of NAT-2 [6] and of CYP2C19, 2C5 and NAT-2 ([27], www.cypalleles.ki.se) whose functional significance is yet to be established.

Efavirenz, is a non-nucleoside analogue HIV reverse transcriptase inhibitor (NNRTI) which is part of the highly active antiretroviral therapy (HAART) being used to treat HIV/AIDS. It is normally used in patients who will have shown adverse drug reactions to nevirapine, a cheaper NNRTI and in patients who will be under treatment of both HIV/AIDS and TB. The latter use is done to avoid the drug–drug interactions between nevirapine, a mainly CYP3A4 substrate and rifampicin, a component of the TB treatment regimen which is a potent inducer of CYP3A4. Efavirenz on the other hand is mainly metabolised by CYP2B6 which is associated with reduced metabolic interaction with rifampicin. Genetic polymorphism of CYP2B6 has been associated with high plasma levels of efavirenz and increased incidence of CNS adverse drug reactions in patients homozygous for the

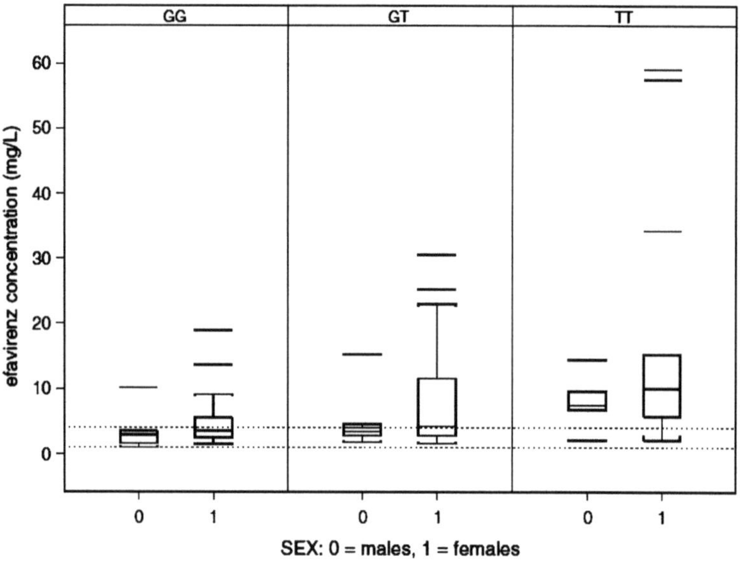

Fig. 5 Box and whiskers plot showing the observed efavirenz plasma concentrations at steady state in 71 HIV/AIDS patients of Zimbabwean origin. GG extensive metabolizer, GT intermediate metabolizer, TT poor metabolizer. The dotted horizontal lines show the optimum concentration interval (1–4 mg/L)

CYP2B6*6 variant. Following our observation that up to 20 % of African people (in contrast to <5 % in Caucasian and Oriental populations) were homozygous for CYP2B6*6 (Table 3), we conducted studies to evaluate the clinical impact of this polymorphism in HIV/AIDS patients.

Over 50 % of the 71 patients had concentrations above the minimum toxic concentration (MTC) of 4 μg/ml and none were below the minimum effective concentration (MEC) of 1 μg/ml. There was a clear gene dose concentration relationship with CYP2D6*1/*1, *1/*6, and *6/*6 patient groups showing increasing plasma concentrations. We also observed that in each genotype category, women had much higher concentrations than the men (Fig. 5). We conducted pharmacokinetic simulations to estimate the likely impact of CYP2B6 genotype on efavirenz dose requirements. The data indicated that patients homozygous for the CYP2B6*6 genotype might require 300 mg/day instead of the standard dose of 600 mg/day [29]. Based on this prediction, we are now conducting a large clinical study to evaluate the effect of demographical, physiological, and genetic factors on efavirenz dosing. If we observe similar results, it could be the basis to introduce a genetic pharmacodiagnsotic test and dosing algorithm to guide the use of efavirenz in Zimbabwe and other African countries.

4 Tools for Drug Discovery and Development Research

In the process of characterising the DMPK properties of antiparasitic drugs and demonstrating the utility of such information in the clinical use of these drugs, AiBST has also setup a platform for the integration of DMPK in the discovery and development of new chemical entities against infectious diseases. We now have in silico and in vitro methods for the prediction and measurement of physicochemical parameters such as solubility, lipophilicity (logP), ionisability (pKa), protein binding, blood/plasma partitioning and ADME parameters such as permeability, volume of distribution, metabolic stability, enzyme and metabolite identification, and enzyme inhibition. We have applied this platform in the characterisation of over 100 compounds from various drug discovery projects in Africa and for WHO–TDR funded discovery projects. The data is used to either guide the design of new chemical entities predicted to have better PK or towards the design of the first-time in men dose finding and drug–drug interactions studies (Masimirembwa et al. 2002) [19, 25, 26]. WHO–TDR has recognised our expertise and the utility of this platform and has nominated AiBST as DMPK Centre of Excellence.

The work on pharmacogenetics has evolved towards the development of a Biobank of African populations [28]. The Biobank will be used as a biomedical research tool towards healthcare solutions tailored for African populations. It will be useful in early stages of drug discovery in target identification and validation where the pharmaceutical industry needs to ensure that molecular drug targets against which hits, leads and candidate drugs are discovered do not have genetic variability that might result in the drug not working in some people. The tool will also be used to design clinical studies that evaluate exposure levels and possible adverse drug reactions profiles in subjects who might be carriers of genetic defects in ADME genes for a new chemical entity under investigation. Our work has also been a wake-up call to Research & Ethics committees in Africa to upgrade their basic and clinical research guidelines to address the potentially complex ethical and intellectual property issues that come with genomic research.

5 Conclusion

Work from our laboratory demonstrates that with modest but long term funding that focuses on human resource development (through postgraduate training) and research capacity strengthening (through purchase of equipment and support of effective South–North collaborations), Africa can achieve international standards in biomedical research. As this is being achieved, there is now a need to increase research collaborations among African institutions towards sharing expertise acquired over many years of North–South collaborations.

Acknowledgments We acknowledge funding from ISP, WHO–TDR, and EU that supports the work in our laboratory. We acknowledge the long term scientific and technical support from AstraZeneca that has resulted in AiBST having world class expertise in DMPK. We also acknowledge the many international collaborators, AiBST scientists, and students whose vision and hard work have resulted in the work presented in this chapter.

References

1. Akililu E, Dandara C, Bertilsson L et al (2007) Pharmacogenetics of cytochrome P450s in African populations: clinical and molecular evolutionary implications. In: Suarez-Kurtz G (ed) Pharmacogenomics in admixed populations, Ed. Landes Bioscience, Germany (ISBN 978-1-58706-130-9)
2. Bapiro TE, Egnell AC, Hasler JA et al (2001) Application of higher throughput screening (HTS) inhibition assays to evaluate the interaction of antiparasitic drugs with cytochrome P450s. Drug Metab Dispos 29:30–35
3. Bapiro T, Hasler JA, Ridderström M et al (2002) The molecular and enzyme kinetic basis for the diminished activity of the cytochrome P450 2D6*17 variant. Implications for CYP2D6 phenotyping studies and the clinical use of substrate drugs in some African populations. Biochem Pharmacol 7422:1–12
4. Bapiro TE, Sayi J, Hasler JA et al (2005) Artemisinin and thiabendazole are potent inhibitors of cytochrome P450 1A2 (CYP1A2) activity in humans. Eur J Clin Pharmacol 61:755–761
5. Castro N, Jung H, Medina R et al (2002) Interaction between Grapefruit Juice and Praziquantel in Humans. Antimicrob Agents Chemother 46:1614–1616
6. Dandara C, Masimirembwa C, Magimba A et al (2002) Arylamine N-acetyltransferase (NAT2) genotypes in Africans: the identification of a new allele with nucleotide changes 481C > T and 590 > A. Pharmacogenetics 12:1–4
7. Dandara C, Masimirembwa CM, Magimba A et al (2001) Genetic polymorphism of CYP2D6 and CYP2C19 in east- and southern African populations including psychiatric patients. Eur J Clin Pharmacol 57:11–17
8. Dandara C, Sayi J, Masimirembwa CM et al (2002) Genetic polymorphism of cytochrome P450 1A1 (Cyp1A1) and glutathione transferases (M1, T1 and P1) among Africans. Clin Chem Lab Med 40:952–957
9. Johansson T, Jurva U, Grönberg G et al (2009) Novel metabolites of amodiaquine formed by CYP1A1 and CYP1B1: structure elucidation using electrochemistry, mass spectrometry, and NMR. Drug Metab Dispos 37:571–579
10. Jurva U, Holmén A, Grönberg G et al (2008) Electrochemical generation of electrophilic drug metabolites: characterization of amodiaquine quinoneimine and cysteinyl conjugates by MS, IR, and NMR. Chem Res Toxicol 21:928–935
11. Keogh A, Spratt P, McCosker C et al (1995) Ketoconazole to reduce the need for cyclosporine after cardiac transplantation. N Engl J Med 333:628–663
12. Lew G, Murray WE, Lane JR et al (1989) Theophylline-thiabendazole drug interaction. Clin Pharm 8:225–227
13. Li XQ, Björkman A, Andersson TB et al (2002) Amodiaquine clearance and its metabolism to N-desethylamodiaquine is mediated by CYP2C8: a new high affinity and turnover enzyme-specific probe substrate. J Pharmacol Exp Ther 300:399–407
14. Li XQ, Björkman A, Andersson TB et al (2003) Identification of human cytochrome P(450)s that metabolise antiparasitic drugs and predictions of in vivo drug hepatic clearance from in vitro data. Eur J Clin Pharmacol 59:429–442
15. Mao M, Matimba A, Scordo MG et al (2009) Flavin-containing monooxygenase 3 polymorphisms in 13 ethnic populations from Europe, East Asia and sub-Saharan Africa: frequency and linkage analysis. Pharmacogenomics 10:1447–1455

16. Masimirembwa C, Bertilsson L, Johansson I et al (1995) Phenotyping and genotyping of S-mephenytoin hydroxylase (cytochrome P450 2C19) in a Shona population of Zimbabwe. Clin Pharmacol Ther 57:656–661
17. Masimirembwa C, Hasler JA, Bertilsson L et al (1996) Phenotype and genotype analysis of debrisoquine hydroxylase (CYP2D6) in a black Zimbabwean population: reduced enzyme activity and evaluation of metabolic correlation of CYP2D6 probe drugs. Eur J Clin Pharmacol 51:117–122
18. Masimirembwa C, Persson I, Bertilsson L et al (1996) A novel mutant variant of the CYP2D6 gene (CYP2D6*17) common in a black African population: association with diminished debrisoquine hydroxylase activity. Br J Clin Pharmacol 42:713–719
19. Masimirembwa CM, Bredberg U, Andersson TB (2003) Metabolic stability for drug discovery and development: pharmacokinetic and biochemical challenges. Clin Pharmacokinet 42:515–528
20. Masimirembwa CM, Dandara C, Sommers DK et al (1998) Genetic polymorphism of cytochrome P4501A1, microsomal epoxide hydrolase, and glutathione S-transferases M1 and T1 in Zimbabweans and Venda of southern Africa. Pharmacogenetics 18:83–85
21. Masimirembwa CM, Gustafsson LL, Dahl ML et al (1996) Lack of effect of chloroquine on the debrisoquine (CYP2D6 and S-mephenytoin (CYP2C19) hydroxylation phenotypes. Br J Clin Pharmacol 41:344–346
22. Masimirembwa CM, Hasler JA (1994) Characterisation of praziquantel metabolism by rat liver microsomes using cytochrome P450 inhibitors. Biochem Pharmacol 48:1779–1783
23. Masimirembwa CM, Hasler JA (1997) Genetic polymorphism of drug metabolising enzymes in African populations: implications for the use of neuroleptics and antidepressants. Brain Res Bull 44:561–571
24. Masimirembwa CM, Hasler JA, Johansson I (1995) Inhibitory effects of antiparasitic drugs on cytochrome P450 2D6. Eur J Clin Pharmacol 48:35–38
25. Masimirembwa CM, Ridderström M, Zamora I et al (2003) Combining pharmacophore and protein modeling to predict CYP450 inhibitors and substrates. Methods Enzymol 357:133–144
26. Masimirembwa CM, Thompson R, Andersson TB (2001) In vitro high throughput screening of compounds for favourable metabolic properties in drug discovery. Comb Chem High Throughput Screen 4:245–263
27. Matimba A, Del-Favero J, Van Broeckhoven C et al (2009) Novel variants of major drug-metabolising enzyme genes in diverse African populations and their predicted functional effects. Hum Genomics 3:169–190
28. Matimba A, Oluka MN, Ebeshi BU et al (2008) Establishment of a biobank and pharmacogenetics database of African populations. Eur J Hum Genet 16:780–783
29. Nyakutira C, Röshammar D, Chigutsa E et al (2008) High prevalence of the CYP2B6 516G–> T(*6) variant and effect on the population pharmacokinetics of efavirenz in HIV/AIDS outpatients in Zimbabwe. Eur J Clin Pharmacol 64:357–365
30. Oscason M, Hilderstrand M, Johansson I et al (1997) A combination of mutations in the CYP2D6*17 (CYP2D6Z) allele causes alteractions in enzyme function. Mol Pharmacol 52:1034–1040
31. Thelingwani RS, Zvada SP, Hugues D et al (2009) In vitro and in silico identification and characterisation of thiabendazole as a mechanism-based inhibitor of CYP1A2 and simulation of possible pharmacokinetic drug–drug interactions. Drug Metab Dispos 37:1286–1294
32. Vazquez ML, Jung H, Sotelo J (1987) Plasma levels of praziquantel decrease when dexamethasone is given simultaneously. Neurology 37:1561–1562
33. Zeldin RK, Petruschke RA (2004) Pharmacological and therapeutic properties of ritonavir-boosted protease inhibitor therapy in HIV-infected patients. J Antimicrob Chemother 53:4–9

Role of Flavonoid and Isoflavonoid Molecules in Symbiotic Functioning and Host-Plant Defence in the Leguminosae

Nyamande Mapope and Felix D. Dakora

Abstract Inoculating symbiotic legumes with infective rhizobial symbionts increases the nod-gene-inducing activity of root exudates, and alters the profile of nod gene inducers. The application of *Sinorhizobium meliloti* cells to the roots of alfalfa seedlings specifically causes the release of the aglycone and glycoside forms of the phytoalex in medicarpin, and a formononetin—O-(6″-O-malnylglycoside). Similarly, in the presence of *Rhizobium leguminosarum* biovar *phaseoli* bacteria, root exudates of common bean also contain more of the phytoalexin coumestrol, and its isoflavonoid precursor daidzein than exudates of uninoculated plants. This paper discusses the effects of root-nodule bacteria (hereafter called "rhizobia") on the synthesis and release of flavonoid and isoflavonoid signal compounds, and explores the biological significance of phytoalexin production in legume plant nodulation and defense against pathogens and insect pests.

1 Introduction

Nitrogen, phosphorus, and water are the most limiting factors to increased crop yields in Africa. With the high cost of chemical fertilizers, interest has increased in seeking new approaches to promote N nutrition in crops. In global terms, biological N_2 fixation contributes about 65 % of N used in agriculture today, and therefore provides an

N. Mapope
Department of Crop Sciences, Faculty of Science,
Tshwane University of Technology, Pretoria, South Africa

F. D. Dakora (✉)
Chemistry Department, Faculty of Science, Tshwane University
of Technology, Pretoria, South Africa
e-mail: dakorafd@tut.ac.za

A. Gurib-Fakim and J. N. Eloff (eds.), *Chemistry for Sustainable Development in Africa*, DOI: 10.1007/978-3-642-29642-0_3,

alternative to chemical fertilizers. N_2 fixation in root nodules of symbiotic legumes represent the major source of N for food production in the traditional cropping systems in Africa. During nodule formation, some legumes release flavonoid and isoflavonoid signal molecules, which act as chemo-attractants to invading root-nodule bacteria or "rhizobia" [1]. Additionally, these phenolic molecules fulfill various other functions, including promotion of bacterial growth [2] and induction of nodulation (*nod*) genes in compatible bacterial strains [3, 4]. Different legumes release different profiles of flavonoids and isoflavonoids from their roots and germinating seeds. Alfalfa grown in the absence of its N_2-fixing symbiont, releases five distinct flavonoids [3, 5, 6] and two betaines [7] that induce transcription of *nod* genes in homologous *Sinorhizobium meliloti.* The common bean plant also releases nine compounds from seeds and roots that cause expression of *nod* genes in its microsymbiont [8, 9], while sterile *Vicia* roots release only two *nod* gene-inducing compounds [10].

The early steps in nodule formation involves a two-way molecular communication between the legume and rhizobial symbiont. Flavonoid and isoflavonoid signal molecules released by leguminous host induce expression of rhizobial *nod* genes followed by a coordinated synthesis of rhizobial signal compounds, such as lipo-chito-oligosaccharide nod factors. It is these nod factors that affect morphological changes in legume root hairs, leading to nodule formation. In compatible interactions, rhizobial signaling to host root hairs results in root hair deformation, root hair branching, root hair curling, cortical cell division, infection thread formation [11], and ultimately nodule development. It is in the mature nodules that N_2 fixation occurs through the activity of the enzyme nitrogenase.

2 Microsymbiont-Induced Biosynthesis of Flavonoid and Isoflavonoid Compounds in Legumes

Studies with white clover [12] and later *Vicia sativa* subsp.*nigra* [13] were the first to demonstrate that inoculating host legume with infective rhizobial strains increased *nod* gene-inducing activity. Because that finding implied increased production of new and/or existing compounds, Recourt et al. [10] found from analysis of root exudates that, with *Rhizobium* inoculation, *Vicia* released 8 new additional nod gene-inducing compounds. However, there was no evidence that the inoculated roots released isoflavonoid phytoalexins [10] even though most of the 5-deoxy flavonoids produced by inoculated seedings could easily be metabolized to medicarpin, a common phytoalexin in *Vicia faba* [14].

Like *Vicia,* alfalfa root exudates also showed increased *nod* gene-induction following seedling inoculation with *Rhizobium meliloti* [15]. Similarly, root exudates of the common bean exhibited greater nod gene-inducing activity when plant roots were inoculated with *Rhizobium leguminosarum bv phaseoli* compared to uninoculated controls [16]. HPLC analyses followed by NMR and MS studies revealed that, unlike *Vicia*, those two symbioses exuded isoflavonoid phytoalexins in the presence of rhizobial symbionts. *Rhizobium meliloti* elicited the exudation of medicarpin, medicarpin-

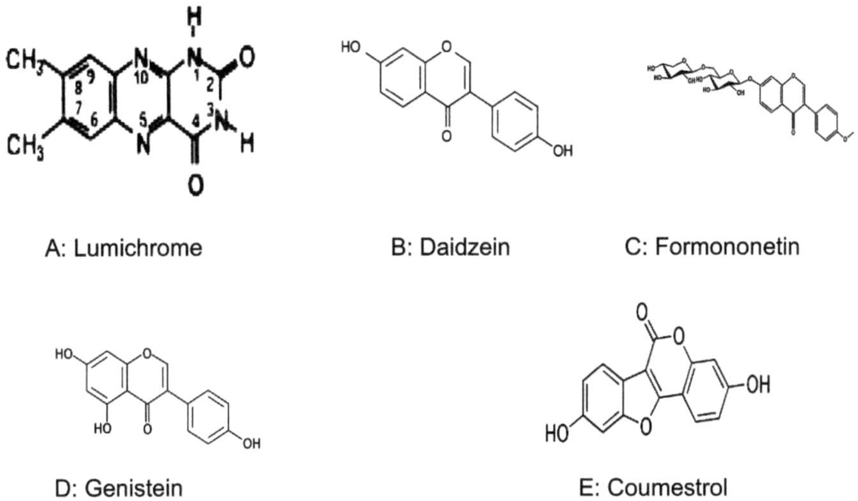

Fig. 1 Flavonoid compounds

3-Oglycoside, and formononetin-7-O-(6″-O-malonylglycoside) from the roots of alfalfa plants. In addition to genistein, eriodictyol, and naringenin which are normally present in root exudates of sterile-grown bean plants [9]; the phytoalexin coumestrol and its precursor daidzein (Fig. 1a) were released by roots of bean in response to *Rhizobium* inoculation [16]. Tests for biological activity showed that the two compounds were active *nod* gene inducers. Thus, coumestrol can function both as a nod gene inducer in bean rhizobia and as a phytoalexin for host plant defense.

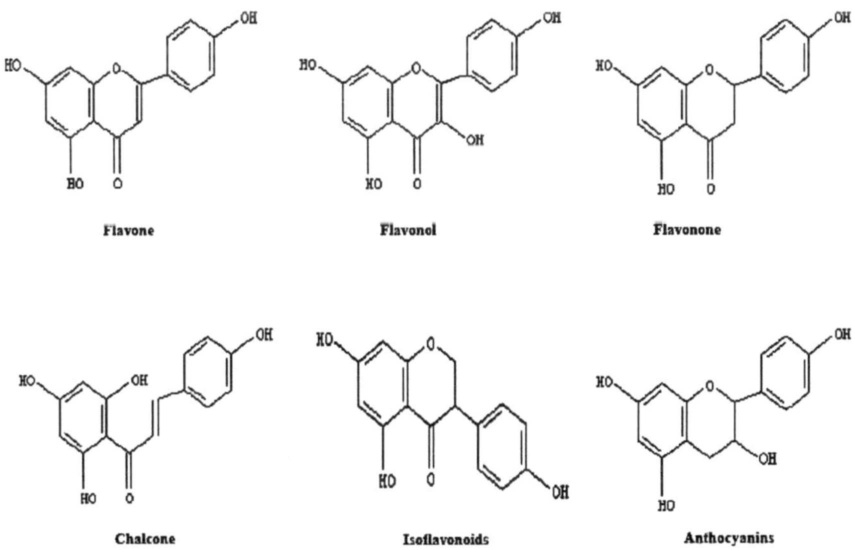

Other studies have shown that exposing soybean plants to infective strains of *Bradyrhizobium japonicum* led to the exudation of glyceollin, a major phytoalexin of soybean [17]. However, the amount of glyceollin released in the presence of *Bradyrhizobium japonicum* was less compared to levels exuded in the presence of *Phytophthora* pathogen [17]. With *Sinorhizobium fredii* (another soybean symbiont), however, the level of glyceollin exudation was comparable to that caused by pathogenic *Phytophthora* [17]. Those studies [15–17] clearly indicate that both pathogens and rhizobial symbionts can elicit phytoalexin exudation in N_2-fixing legumes. Even symbiotic fungi have been shown to trigger the formation of isoflavonoid phytoalexins in their host plants. In one study, analysis of soybean root extracts revealed increased tissue concentrations of glyceollin, coumestrol, and its precursor daidzein following *Glomus* infection of roots [18]. In alfalfa, *Glomus* infection resulted in accumulation of only formononetin in the roots [19], suggesting that phytoalexins are formed in both alfalfa and bean in response to fungal symbionts. These findings demonstrate that isoflavonoid phytoalexin production is caused by both pathogenic and symbiotic microbes, so to view these molecules simply as occurring in diseased tissue is misleading.

In contrast, some studies have found no accumulation of phytoalexins with fungal or rhizobial symbionts. Soybean cv. Maple Arrow, for example, showed no accumulation of glyceollin in response to *Glomus mosseae,* though with *Rhizoctonia solani* pathogen, glyceollin accumulation was very high [20]. The observed lack of phytoalexin accumulation in response to mycorrhizal fungus is comparable to *Bradyrhizobium japonicum* infection of soybean roots without inducing phytoalexin in accumulation [21]. However, with loss of symbiotic ability, rhizobial strains can be perceived as pathogens by host plants. For example, infection of soybean root by fix mutants of *Bradyrhizobium japonicum* led to rapid development of necrosis and instant phytoalexin accumulation [21], symptoms reminiscent of pathogen invasion.

Apart from these inconsistencies in overall experimental observations, similarities exist between rhizobial and fungal symbioses, especially with respect to host plant response to symbionts. *Sinorhizobium* inoculation of alfalfa seedlings stimulates the formation and release of conjugated formononetin into root exudates [15]. With mycorrhizal infection, the aglycone form of the same molecule is also found in alfalfa root extracts [19] and possibly exudates, suggesting that host plant recognition of symbiont, be it a fungus or rhizobium, is by the same mechanism. In that regard, mycorrhizal roots have been found to produce proteins which are immunologically related to nodulins in N_2-fixing nodules [22]. Another study has also shown that both pathogens and symbionts of soybean identify their hosts by recognizing the same chemical signals [23].

3 Phytoalexin Effects on Rhizobial Symbionts

Phytoalexin accumulation to higher concentrations in host plant tissue can have detrimental effects, especially where micro-symbiont cells are in contact with such metabolites. An early study by Cruickshank [24] showed that pisatin, (a classical

Table 1 Legume flavonoids and isoflavonoids involved in nod gene expression and N_2 fixation

Legume species	[1]Nod gene-inducing compounds	[2]Amount of N-fixed (kg. N ha^{-1})
Alfalfa	4,4′-dihydroxy-2′-methoxychalcone[a]	2–208[a]
	4′-7-Dihydroxyflavone liquiritigenin[a]	
Cowpea	Daidezein[b]	24–201[b]
	Genistein[b]	
	Coumestrol[b]	
Common bean	Genistein-3-O-glucoside[c]	0–165[c]
	Eridictyol[c]	
	Naringenin[c]	
	Daidzein[c]	
	Genestein[c]	
	Coumestrol[c]	
Soybean	Isoliquiritigenin[d]	26–197[d]
	Genestein[d]	
	Genestein-7-O-glucoside[d]	
	Coumestrol[d]	
	Daidzein[d]	
	Fomononetin[d]	
	Biochanin A[d]	
Bambara groundnut	Daidzein[e]	40–62[e]
	Genestein[e]	
	Coumestrol[e]	
Sesbania	Liquitigenin[f]	505–581[f]

Functional role [a] Ref. [6], [b] Ref. [72], [c] Ref. [73], [d] Ref. [29], [e] Ref. [74], [f] Ref. [75]
N-fixed [a] Ref. [76], [b] Ref. [70], [c] Ref. [77], [d] Ref. [78], [e] Ref. [70], [f] Ref. [70]

phytoalexin from pea and vetch) significantly inhibited growth of rhizobia compared with other rhizosphere microorganisms. Interestingly, *Rhizobium leguminosarum* biovar *viciae*, the microsymbiont of pea and vetch, exhibited a much broader range of pisatin tolerance relative to other rhizobial species. A survey on the inhibitory effects of selected isoflavonoid phytoalexins revealed three levels of sensitivity: pisatin, coumestrol, formononetin, vestitol, biochanin A, genistein, and rotenone were inhibitory; phaseollin, and maackiain, moderately inhibitory; while medicarpin and kievitone were strongly inhibitory [25]. In that study, kievitone and medicarpin, the latter a common photoalexin of agriculturally important legumes, such as *Vigna unguiculata, Cicer arietinum, Canavalia ensiformis, Vicia faba, Medicago, Melilotus*, and *Trifolium* spp., showed considerably marked inhibition of growth in slow-growing root-nodule bacteria compared to fast-growing strains (Table 1). Phaseollin also exhibited significant inhibition of growth in bradyrhizobia but had little effect on fast-growing species.

While the concentrations used in in-vitro studies [24, 25] probably surpass those encountered in plant tissues, the data obtained nevertheless provide a useful indication of the spectrum of symbiont sensitivity to isoflavonoids. Classical phytoalexins produced and released in purely symbiotic systems (Table 2), include medicarpin from alfalfa-*Rhizobium meliloti* symbiosis [15], coumestrol from bean-

Table 2 Mechanisms of bioprotection in plants by symbiotic rhizobia

Interaction	Chemical molecule	Reference
Alfalfa-*Rhizobium melitoti*	Medicarpin	[15]
Bean-*Rhizobium leg.* bv. *viciae*	Coumerstrol	[16]
Soybean-*Bradyrhizobium japonicum*	Glyceolin	[17]
Groundnut-*Bradyrhizobium*	Stilbenes	[79]
Chickpea-*Rhizobium*	Formononetin and biochanin A	[55]
Rice-*Rhizobium leguminosarum bv. phaseoli*	Gallic, tannic, ferulic, and cinnamic acids.	[43]
Clover-*Rhizobium*	Biochanin A–7-O-glucoside–malonate	[80]

Rhizobium leguminosarium biovar *phaseoli* interaction [15], coumestrol from bean-*Rhizobium leguminosarum* biovar *phaseoli* interaction [16], and glyceollin from soybean-*Bradyrhizobium japonicum* association [17]. Although the in-vitro sensitivity of *Rhizobium meliloti* to medicarpin, the major phytoalexin of alfalfa, is low, it is difficult to imagine how the bacterial symbiont copes with increasing metabolite concentration in tissues and in the rhizosphere. Initial experiments revealed that at 150 micromolar concentration, glyceollin inhibited growth of *Bradyrhizobium japonicum* [26], suggesting that the microsymbiont must have a mechanism for overcoming glyceollin toxicity in soybean tissue and/or rhizosphere.

Various mechanisms have been suggested for phytoalexin tolerance in pathogens and symbionts. Phytoalexin detoxification is one such mechanism by which microbial symbionts and pathogens overcome plant defence response. The phytoalexins medicarpin and maackiain, which are produced by a number of legumes, can be degraded by *Nectria haematococca* and *Ascochyta rabiei,* (two common fungal pathogens of chickpea and other legumes) via NADPH-dependent reductase conversion of the molecule to less toxic isoflavan, isoflavanone, or 6a-hydroxyptercocarpan (see Van Etten et al. [27]). Whether *Rhizobium meliloti* employs such a mechanism to detoxify medicarpin and other phytoalexins, remains to be seen Table 3.

The bean pathogen, *Fusarium solani* f. sp. *Phaseoli,* can also detoxify four major bean phytoalexins, namely kievitone, phaseollin, phaseollidin, and phaseollinisoflavan; kievitone detoxification occurs through kievitone hydratase-catalyzed hydration of the isopentenyl side chain to yield the less toxic kievitone hydrate [27]. There is evidence that *Rhizobium* and *Bradyrhizobium* strains modify various flavonoid molecules within their vicinity [28], suggesting that bacterial and fungal symbionts probably overcome isoflavonoid phytoalexin toxicity in tissues and soil through biotransformation of these compounds.

Besides chemical modification, bacterial tolerance of phytoalexins may be flavonoid-induced. Growing cells of a pathogen or symbiont in low-level concentrations of a phytoalexin can confer resistance to the microbe against the toxic effects of phytoalexins. It has been shown recently that this induction of resistance against phytoalexins may be effected by low concentrations of a second,

Table 3 Isoflavonoid compounds from legume roots with potential for use as defense molecule against soil borne pests (insect larvae and pathogens) in cropping systems

Legume species	Isoflavonoid	Defense role	Reference
Lotus pedunculatus	Vestitol	Insect deterrent	[81]
Glycine max	Glyceollin	Insect deterrent phytoalexin	[71]
Vigna unguiculata	Medicarpin	Insect deterrent phytoalexin	[71]
Phaseolus vulgaris	Phaseolin	Insect deterrent phytoalexin	[71]
Phaseolus linatus	Coumestrol	Nematicide	[71, 82]
Cajanus cajan	Cajanin	Insect deterrent phytoalexin	[71]
Lonchocarpus nicou	Rotenone	Insecticide phytoalexin	[83]
Derris malaccensis	Rotenone Deguelin Sumatrol taxicarol	Insecticide phytoalexin	[83]
Mundulea serica	Munduserone	Insecticide phytoalexin	[83]
Pachyrrhizus erosus	Pachrrhizone	Insecticide phytoalexin	[83]
Neoratanenia pseudopachyrrhiza	Dolineone	Insecticide phytoalexin	[83]

structurally-related molecule other than the phytoalexin itself. In a classical study by Parniske et al. [26], it was demonstrated that preculturing *Bradyrhizobium japonicum* with 10 μmM concentration of genistein or daidzein induced resistance in the bacterium against the phytoalexin glyceollin. In that study, cell viability tests involving bacterial growth in geneistein-free medium followed by transfer to 300 μM glyceollin showed a strong bactericidal effect, while rhizobia precultured with genistein were unaffected. Kape et al. [29] have also found that isoliquiritigenin (2′,4′,4-trihydroxychalcone) is both a strong nod gene inducer and glyceollin resistance inducer. This confirms the role of flavonoids in phytoalexin resistance induction. While these findings may be the first for bacterial symbionts, induced phytoalexin resistance is not new in fungal pathogens. Denny van Etten [30, 31] have reported and induced resistance to pisatin in *Nectria haematococca.*

Although experimental results show that isoflavonoids can induce resistance in soybean microsymbionts against glyceollin phytoalexins [26], the mechanism by which this resistance is induced is still not properly understood. It is, however, clear that, unlike rhizobial *nod* gene induction which requires flavonoid interaction with *nod*D protein, induction of glyceollin resistance in *Bradyrhizobium* does not involve the common *nod* genes. This was evidenced by the successful induction of glyceollin resistance in a *nod*D_1D_2YABC deletion mutant [26]. Equally unknown is the mechanism by which soybean rhizobia and pathogens overcome the inhibitory effects of glyceollin. Even the mode of action of phytoalexins on pathogens and/or symbionts remains undefined. In the case of the glyceollins, however, the lethal effect of these phytoalexins on pathogens and symbionts is exerted via inhibition of plasma membrane and tonoplast H^+-transporting ATPases [32], and/or NADH-ubiquinone-oxidoreductase [33].

4 Nodulation Significance of Phytoalexins From Legumes

The release of phytoalexin compounds by legume roots, whether in response to pathogen invasion, symbiont infection, or both, is likely to lead to higher concentrations of phytoalexins in the rhizosphere. Root exudation of phytoalexins, such as medicarpin, coumestrol, and glyceollin is known to occur from challenge by symbionts [15–17] and pathogens [26]. This suggests that the rhizosphere of nodulating legumes must be the site of continuous accumulation of phytoalexins triggered by invading rhizobia, root-borne pathogens, and various environmental stimuli. Also, because phytoalexin exudation is a localised response in roots, rhizosphere soil associated with symbiotic field legumes is likely to consist of spatial, and temporal phytoalexin concentration gradients. The accumulation of *nod*-gene inducing isoflavonoid phytoalexins in the rhizosphere is likely to increase the chance of enhancing nodulation while warding off pathogens. In fact, differences in legume nodulation have been shown to relate to limitation in the synthesis and release of *nod* gene-inducing flavonoids [34, 35].

A number of studies [36] have shown that soybean plants release the isoflavones daidzein and genistein into root exudates and these induce nod genes in *Bradyrhizobium japonicum*. Coumestrol is another phytoalexin from soybean which moderately induces *nod* genes in the host's microsymbiont [37]. Thus, the nodulation potential of soybean depends largely on adequate availability of daidzein, genistein, and coumestrol in the rhizosphere to transcribe *nod* genes in *Bradyrhizobium japonicum*. Besides *nod* gene induction, daidzein, and coumestrol released into the rhizosphere also promote growth of the soybean bacterial symbiont [38]. The growth-promoting effect of these isoflavonoids on bradyrhizobial populations in the rhizosphere increases the chance for enhanced nodulation and N_2 fixation in soybean from increased rhizobial numbers. Furthermore, in soybean, daidzein, genistein, and their glycosyl conjugates act in concert against pathogen invasion of this legume [39], resulting in healthy plant growth for better nodulation.

Phytoalexins released by legume roots play other roles essential for nodule formation. As indicated before, the isoflavonoids daidzein and genistein can each independently induce glyceollin resistance in *Bradyrhizobium japonicum* [26], thus permitting selective development of bacterial strain population needed for host plant nodulation. This means that with soybean release of daidzein and genistein into the rhizosphere, bradyrhizobia are induced to develop resistance to glyceollin exuded by legume to ward-off pathogens. Nutritionally, this offers a competitive advantage to microsymbiont, and enhances its ability to survive in the rhizosphere of phytoalexin-producing host roots. Biochemically, daidzein and genistein are important precursors for glyceollin formation [40], indicating that the synthesis and release of glyceollin would depend on the pool size of isoflavones in tissues as well as their rates of exudation from roots. The ecological significance of induced glyceollin resistance is not only to increase the survival of the microsymbiont in the rhizosphere which is constantly challenged by microsymbionts, pathogens, and environmental stimuli, but also to provide bacteria with physiological plasticity for adaptation to changing phytoalexin concentrations within the rhizosphere.

A recent study has shown that alfalfa growing in iron-deficient medium releases the phytoalexin 2-(3′5′-dihydroxyphenyl)-5,6-dihydroxybenzofuran, which solubilises iron from insoluble ferric compounds for the plant, while controlling *Fusarium oxysporum* f. sp. *phaseoli*, a root pathogen of alfalfa [41]. It is, therefore, likely that many legumes growing in poor soils, use compounds such as isoflavonoids to enhance nutrient uptake and control pathogens.

Coumestrol phytoalexin inhibits growth of pathogens, and acts as *nod*-gene inducer in strains of *Rhizobium leguminosarium* bv *phaseoli*, the causal organism of nodulation in the common bean. The molecule also moderately transcribes *nod*-genes in *Bradyrhizobium japonicum*, the microsymbiont of soybean. In soybean, isoliquiritigenin is both a nod gene inducer and glyceollin resistance inducer [29]. Breeding for increased levels of such compounds in legumes for enhanced nodulation and improved disease control could be useful for agriculture. Alternatively, increased nodulation and pathogen control can be achieved directly and together by applying isoflavonoid phytoalexin inducer compounds to field legumes.

5 Symbiotic Rhizobia as Biopesticides Against Plant Pathogens

Novel findings from field and glasshouse studies have shown that inoculating legumes and non-legume plants with N_2-fixing rhizobia can provide protection against pathogens. In common bean the severity of *Fusarium* root rot was reduced by inoculating it with rhizobia. Similarly, living and heat-killed bacterial cells of *Rhizobium leguminosarum* provided total protection against pathogen infection of lentil plants [42]. That study further revealed that the culture filtrate and the killed bacterial cells contained signals able to induce plant resistance. However, those signals were suppressed once *Rhizobium* was in contact with the plant. Mishra et al. [43] showed that in *Rhizobium*-inoculated rice plants, synthesis of phenolic compounds was consistently more enhanced than in the control, and maximum accumulation of phenolic compounds was observed in plants co-inoculated with *Rhizobium leguminosarum* bv. *phaseoli* and *Rhizoctonia solani*.

Phenolic acids mediate induced systemic resistance and provide bioprotection to plants during pathogenic stresses. In a related study, Khaosaad et al. [44] showed that arbuscular mycorrhizal fungi (AMF) root colonization provides a bioprotectional effect against a broad range of soil-borne fungal pathogens, including take all disease caused by *Gaeumannomyces gramminis var. tritici*. Infection by AMF also enhances shoot and root growth in wheat plants infected with *Gaeumannomyces gramminis var. tritici* compared to non AMF colonized diseased plants. Similarly co-inoculation of *Rhizobium leguminosarum* and arbuscular aycorrhizal fungi conferred resistance to *Bothytis fabae* in *Vicia faba* as a result of elevated Na-uptake and phenolics concentration in the bean plant [45].

Rhizobia are major biocontrol agents in natural and agricultural ecosystems. There is evidence that a strain of *Bradyrhizobium japonicum* can cause up to 75 % decrease in sporulation of *Phytophthora megasperma*, 65 % in *Pythium ultimum*, 47 % in *Fusarium oxysporum*, and 35 % in *Ascochyta imperfecta* [46]. Antoun et al. [47] identified 49 strains of *Sinorhizobium meliloti* that inhibited growth of *F. oxysporum* by up to 50 %. Rhizobia isolated from root nodules of *Acacia pulchella* similarly decreased the survival of the zoospores of *Phytophthora cinnamoni* in vitro [48], thus potentially providing bioprotection for the host plant.

Field and glasshouse studies show that inoculating plants with rhizobia can be a cheap and effective method of controlling soil-borne pathogens in cropping systems. For example, inoculating soybean and common bean plants with their respective microsymbionts significantly decreased the severity of *Phytophthora* and *Fusarium* root rot in these species [46, 49]. As with most parasitic interactions, the level of root rot decreased with increasing rhizobial numbers in soil [46]. Whether applied as seed dressing or soil drench, different rhizobial strains successfully protected field-grown soybean, mungbean, sunflower, and okra plants from infection by the root-borne pathogens *Macrophomina phaseolina*, *Rhizoctonia solani*, and *Fusarium* species [34]. Although Tu [46] has suggested that rhizobia achieve this bioprotection by parasitizing the hyphal tips of the fungal pathogens and decreasing contact with the host plant cells, other mechanisms may exist. For example, the elicitation of isoflavonoid phytoalexins by rhizobial cells [15, 16] and/or by their *nod* factors [50] can indirectly control pathogens in legumes. However, it is still unclear whether the same protection can be achieved in non-legume hosts such as the cereals and vegetables in mixed cropping systems.

Similar observations of biocontrol of plant pathogens have been reported for mycorrhizae, the second most important mutualism after the rhizobial symbiosis. Following infection, the AM fungus *Glomus mosseae* is claimed to confer bioprotection against *Phytophthora parasitica* in roots of tomato plants [51]. This was shown by the presence of pathogenesis-related proteins in both mycorrhizal and non-mycorrhizal roots, suggesting that the tomato plant probably acquired both localized and systemic resistance against the pathogen. This observation parallels the finding that inoculating legumes with infective rhizobial cells [15, 16], or with their *nod* factors [50], induces the synthesis, and release of isoflavonoid phytoalexins that confer bioprotection to the plant. Interestingly, some rhizobia are themselves protected from the antimicrobial effects of induced phytoalexins by their nod gene inducers [26, 29]. Thus, pathogens are kept under control as rhizobia infect their host-plant roots. Although no specific studies have been done, it is possible that root hair infection by rhizobial bacteria also induces resistance against various pathogens of the host plant, as observed with *G. mosseae* [51]. Hopefully, future studies will provide direct evidence for this claim. Biological control of fungal pathogens can therefore be achieved through a range of mechanisms, including mycoparasitism, production of phytoalexins, the induction of host plant defense proteins and peptides, as well as competition for nutrients [52]. Tu [46] suggested parasitisation hyphal tips by rhizobia. Purified *nod* factors from rhizobia induced the biosythesis and increased exudation of phytoalexins (daidzein, genistein, and coumestrol) from soybean roots [53]. Cyclic β-glucans produced by bradyrhizobia elicit glyceollin, a phytoalexin that controls pathogens in soybean [53]. Inoculating alfalfa with *Sinorhizobium* induced the release of medicarpin, a classical phytoalexin that control fungal pathogens of the host plant [15]. Similarly inoculating *Phaseolus vulgaris* bean with *Rhizobium leguminosarum bv phaseoli* elicited the exudation of coumestrol, daidzein, and genistein, isoflavonoid phytoalexins that control many fungal and bacterial pathogens of beans and other legumes. *Bradyrhizobium japonicum* secretes rhizobitoxine which inhibits charcoal rot fungus, *Macrophomina phaseolina* [54].

The pre-treatment of chickpea seedlings with *Rhizobium* isolates before challenging them with *Fusarium spp.* significantly increased levels of total phenolics and the levels of constitutive isoflavonoids such as formononetin and biochanin A [55]. It has been suggested that pathogen invasion there is increased the mRNA of the defense gene encoding phenylalanine ammonia lyase (PAL) and leading to the biosythesis of high levels of secondary phenolic metabolites.

Limited data are currently available on flavonoids as signaling compounds for fungal pathogens of legumes. Information about possible effects of flavonoids in root exudates on root infecting/colonizing fungi of non-legume plants remains scanty. Ruan et al. [56] have, however, shown that certain flavonoids, including isoflavonoid phytoalexins, stimulate spore germination of *Fusarium solani formae specia-les* pathogenic on pea and bean. Thus flavonoids in legume root exudates may be perceived as signals in a number of plant–microbe interactions [56]. For example, Steinkellner et al. [57] have reported that the flavonoid compounds, myricetin and luteolin, exhibited a low stimulating activity on microconidia germination of *Fusarium oxysporum* f. sp. *lycopersici*.

6 Mechanisms for *Rhizobium* Control of Plant Diseases

Many studies have provided both glasshouse and field evidence for biocontrol of plant pathogens by symbiotic rhizobia. These include *Phytophthora* root rot control by *B. japonicum* [46, 58], *Fusarium* root rot by *Sinorhizobium melitoti* [59], *Fusarium* root rot, bean bacterial wilt, and *Fusarium* wilt control by *Rhizobium leguminosarum* ([42, 49, 60, 61]; see Table 1). However, except for studies by Tu [46, 58], there are no data from microscopy to support rhizobial entry and intercellular/extracellular localization within the host plant cells. Some evidence for rhizobial invasion of host plant cells, especially where the test plants were non-legumes, has been provided by Matiru and Dakora [62]. However, a huge gap still exists in our understanding of possible mechanisms inducing bioprotection with rhizobial inoculation of crop plants.

While it is understood that rhizobial invasion of homologous legume hosts can trigger de novo synthesis of isoflavonoids and/or enhance the production and release of constitutively-present phytoanticipins [63], the same cannot be said of non-legume plants such as sunflower or okra [54]. Even with symbiotic legumes there is evidence of non-rhizobial bacteria present in root nodules. In earlier studies, [64] showed the presence of many non-nodulating agrobacteria-like strains in root nodules of cowpea. [65] also found many non-nodule forming bacteria that cooccupied root nodules of peanut with legitimate peanut rhizobia. Many studies have since shown that cohabitation of legume root nodules by authentic N_2-fixing rhizobia and agrobacteria is a common phenomenon in the nodulation world. In fact, we have recently isolated over 700 bacteria from cowpea nodules and many of them have been found to be unable to incite nodulation in both cowpeas and siratro (F. Pule-Meulenburg and F. D. Dakora, unpublished data). The question that then arises is; could rhizobial protection of host plants at least in the case of homologous legumes be caused by induction of phenylpropanoid pathway by illegal bacterial coinhabitant of legume root nodules? The recent report of large numbers of non-rhizobial bacteria isolated from nodules of symbiotic legumes suggest that the so-called bioprotection by rhizobia actually comes from opportunistic rather than the direct effect of nodule forming rhizobia. New studies are needed to answer these questions.

Plant natural products derived from phenylalanine and the phenylpropanoid pathways are impressive in their chemical diversity and are the result of plant evolution, which has selected for the acquisition of large repertoires of pigments, structural, and defensive compounds, all derived from a phenylpropanoid backbone via plant-specific phenylpropanoid pathway [66]. The phenylpropanoid pathway is known to be easily induced by both biotic and abiotic factors [67]. Petunia flavonoids such as kaemferol and its glycosides accumulate with pollen tube development on stigmas [68]. While this could suggest induction of the phenlpropanoid pathway by penentration of the infection thread down legume root hair, no data currently exist on flavonoid formation in response to rhizobial entry into host plant cells. Assuming this was to happen, it would mean that even with "crack entry" by rhizobia [69] internal invasion of host plant cells could trigger the phenylpropanoid pathway leading to the synthesis and/or release of isoflavones that unintetendedly serve in host plant defense. But for the absence of microscopy data on rhizobial localization in root cells of test plants a similar mechanism could be advanced for possible accumulation of flavonoids with rhizobial application as biopesticide to field crops. There is evidence that root tips incur injury during growth in difficult highly compact soils [67]. Such a wounding can also induce the phenylpropanoid pathway leading to tissue accumulation of newly formed isoflavonoids and increase in phytoanticipins for plant defense [67]. A combination of these factors with rhizobial application could provide significant defense against pathogens because of the high levels of host plant phenylpropanoid compounds.

7 Conclusion

Research has shown that in addition to fixing atmospheric nitrogen, rhizobia can have not only a positive influence on plant growth but can also help in its protection. Research on the use of rhizobia as a protectant against plant diseases is scanty and mechanisms for bioprotection are not yet well understood. Future research should reveal these mechanisms for bioprotection by symbiotic rhizobia.

References

1. Caetano-Annoles G, Christ-Estes D, Bauer WD (1988) Chemotaxis of *Rhizobium meliloti* to the plant flavone luteolin requires functional nodulation genes. J Bacteriol 170:3164–3169
2. Harwig UA, Joseph CM, Phillips DA (1991) Flavonoids released naturally from alfalfa seeds enhance growth rate of *Rhizobium meliloti*. Plant Physiol 95:797–803
3. Peters NK, Frost JW, Long SR (1986) A plant flavone.luteolin, induces expression of *Rhizobium meliloti* nodulation genes. Science 233:977–980
4. Redmond JW, Batley M, Djordjevic MA, Innes RW, Kuempel PL, Rolfe BG (1986) Flavones induce expression of nodulation genes in Rhizobium. Nature 323:632–635
5. Harwig UA, Maxwell CA, Joseph CM, Phillips DA (1990) Chrysoeriol and luteolin released from alfalfa seeds induce nod genes in *Rhizobium meliloti*. Plant Physiol 92:116–122
6. Maxwell CA, Hartwig UA, Joseph CM, Phillips DA (1989) Achalcone and two related flavonoids released from alfalfa roots induce nod genes of *Rhizobium meliloti*. Plant Physiol 91:842–847
7. Phillips DA, Joseph CM, Maxwell CA (1992) Trigonelline and stachydrine released from alfalfa seeds activate NodD2 protein in *Rhizobium meliloti*. Plant Physiol 99:1526–1531
8. Hungria M, Joseph CM, Phillips DA (1991) Anthocyanidins and flavonols, major nod-gene inducers from seeds of a black-seeded common bean. Plant Physiol 97:751–758
9. Hungria M, Joseph CM, Phillips DA (1991) Rhizobium nod-gene inducers exuded naturally from roots of common bean (*Phaseolus vulgaris* L.). Plant Physiol 97:759–764
10. Recourt K, Schripsema J, Kinje JW, Van Brussel AAN, Lugtenberg BJJ (1991) Inoculation of *Vicia sativa* subsp. *nigra* roots with *Rhizobium leguminosarum* biovar viciae results in release of nod gene activating tlavanones and chalcones. Plant Mol Biol 16:841–852
11. Fisher RF, Long SR (1993) *Rhizobium*—plant signal exchange. Nature 357:655–659
12. Rolfe BG, Batley M, Redmond JW, Richardson AE, Simpson RJ, Bassam B, Sargent CL, Weinman JI, Djordjevic MA, Dasso FB (1988) Phenolic compounds secreted by legumes. In: Bothe H, de Bruijn FJ, Newton WE (eds) Nitrogen fixation: hundred years after. Gustav Fischer, Stuttgart, p 405
13. Van Brussel AAN, Recourt K, Pees E, Spaink H, Tak T, Wijffelman CA, Kijne JW, Lugtenberg BJ (1990) A biovar-specific signal of *Rhizobium leguminosarum* bv. viciae induces increased nodulation gene-inducing activity in root exudate of *Vicia sativa* subsp. *nigra*. J Bacteriol 172:5394–5401
14. Hargreaves JA, Mansfield JW, Coxon DT (1976) Identification of medicarpin as a phytoalexin in the broad bean plant (*Vicia faba* L.). Nature 262:318–319
15. Dakora FD, Joseph CM, Phillips DA (1993) Alfalfa (*Midicago sativum* L.) root exudates contain isoflavonoids in the presence of *Rhizobium meliloti*. Plant Physiol 101:819–824
16. Dakora FD, Joseph CM, Phillips DA (1993) Common bean root exudates contain elevated levels of diadzein and coumestrol in response to *Rhizobium* inoculation. Mol Plant Microbe Interact 6:665–668
17. Schmidt PE, Parniske M, Werner D (1992) Production of the phytoalexin glyceollin I by soyabean roots in response to symbiotic and pathogenic infection. Bot Acta 105:18–25

18. Morandi D, Bailey J, Gianinazzi-Pearson V (1984) Isoflavonoid accumulation in soyabean roots infected with vesicular-arbuscular fungi. Physiol Plant Pathol 24:357–364
19. VolPin H, Elking Y, Okon Y, Kapulnik Y (1994) A vesicular-arbuscular mycorrhizal fungus (*Glomus intraradix*) induces a defence response in alfalfa roots. Plant Physiol 104:683–689
20. Wyss P, Boller T, Wiemken A (1991) Phytoalexin response is elicited by a pathogen (*Rhizoctonia solani*) but not by a mycorrhizal fungus (*Glomus mosseae*) in soyabean roots. Experientia 47:395–399
21. Werner D, Mellor RB, Hahn MG, Grisebach H (1985) Soyabean root response to symbiotic infection. Glyceollin I accumulation in an ineffective type of soyabean nodules with an early loss of the peribacteroid membrane. Z Narurforsch 40:179–181
22. Wyss P, Mellor RB, Wiemken A (1990) Vesicular-arbuscular mycorrhizas of wild-type soyabean and non-nodulating mutants with *Glomus mosseae* contain symbiosis-specific polypeptides (mycorrizins). Immunologically cross-reactive with nodulins. Planta 182: 22–26
23. Morris PF, Ward EWB (1992) Chemoattraction of zoospores of the soyabean pathogen, *Phytophthora sojae*, by isoflavones. Physiol Mol Plant Pathol 40:17–22
24. Cruickshank LAM (1962) Studies on phytoalexins. IV. The antimicrobial spectrum of pisatin. Aust J Biol Sci 15:147–159
25. Pankhurst CE, Biggs DR (1980) Sensitivity of *Rhizobium* to selected isoflavonoids. Can J Microbiol 26:542–545
26. Parsnike M, Ahlborn B, Werner D (1991) Isoflavonoid-inducible resistance to phytoalexin glyceollin in soyabean rhizobia. J Bacteriol 173:3432–3439
27. Vanetten HD, Mathews DE, Mathews PS (1989) Phytoalexin detoxification: importance for pathogenicity and practical implications. Ann Rev Phytopathol 27:143–164
28. Rao JR, Sharma ND, Hamilton JTG, Boyd DR, Cooper IE (1991) Biotransformation of the pentahydroxy flavone quercetin by *Rhizobium loti* and *Bradyrhizobium* strains (lotus). Appl Environ Microbiol 57:1563–1565
29. Kape R, Parniske M, Brandt S, Werner D (1992) Isoliquiritigenin, a strong nod gene- and glyceollin resistance-inducing flavonoid from soyabean root exudate. Appl Environ Microbiol 58:1705–1710
30. Denny TP, Vanetten HD (1983) Tolerance of *Nectria haematococca* MP VI to the phytoalexin pisatin in the absence of detoxification. J Gen Microbiol 129:2893–2901
31. Denny TP, Vanetten HD (1983) Characterisation of an inducible, non-degradative tolerance of *Netria haematococca* MP VI to phytoalexins. J Gen Microbiol 129:2903–2913
32. Giannini Jl, Briskin DP, Holt JS, Paxton JD (1988) Inhibition of plasma membrane and tonoplast H+-transporting ATPases by glyceollin. Phytopathology 70:894–896
33. Boydstron R, Paxton JD, Koeppe DE (1983) Glyceollin: a site-specific inhibitor of electron transport in isolated soyabean mitochondrion. Plant Physiol 72:151–155
34. Hungria M, Phillips DA (1993) Effects of a seed colour mutation on rhizobial nod-gene-inducing flavonoids and nodulation in common bean. Mol Plant Microbe Interact 6:418–422
35. Kapulnik Y, Joseph CM, Phillips DA (1987) Flavone limitations to root nodulation and symbiotic nitrogen fixation in alfalfa. Plant Physiol 84:1193–1196
36. Smit G, Puvanesarajah V, Carlson RW, Barbour WM, Stacey G (1992) *Bradyrhizobium japonicum* nodD I can be specifically induced by soyabean tlavonoids that do not induce the nodY ABCSUIJ operon. J Biol Chem 267:310–318
37. Kosslak RM, Bookland R, Barkei J, Paaren HE, Appelbaum ER (1987) Induction of bradyrhizobium japonicum common nod genes by isoflavones isolated from glycine max. Proc Natl Acad Sci U S A 84:7428–7432
38. Kosslak RM, Joshi RS, Bowen BA, Paaren HE, Appelbaum ER (1990) Strain-specific inhibition of nod-gene induction in *Bradyrhizobium japonicum* by flavonoid compounds. Appl Environ Microbiol 56:1333–1341
39. Morris PF, Savard ME, Ward EWB (1991) Identification and accumulation of isoflavonoids and isoflavone glucosides in soyabean leaves and hypocotyls in resistance responses to *Phytophthora megasperma* f.sp. *glycinea*. Physiol Mol Plant Pathol 39:229–244

40. Hahlbrock K, Grisebach H (1975) Biosynthesis offlavonoids. In: Harbome JB, Mabry TJ, Mabry H (eds) The flavonoids Part 2. Academic Press, New York, pp 866–915
41. Masaoka Y, Kojima M, Suguhara S, Yoshihara T, Koshino M, Ichihara A (1993) Dissolution of ferric phosphate by alfalfa (*Medicago satim* L.) root exudates. Plant Soil 155(156):75–78
42. Essalmani H, Lahlou H (2003) Bioprotection mechanisms of the lentil plant by *Rhizobium leguminosarum* against *Fusarium oxysporum* f. sp. *lentis*. Curr Biol 326(12):1163–1173
43. Mishra RPN, Singh RK, Jaiswal HK, Kumar V, Maurya S (2006) *Rhizobium*-mediated induction of phenolics and plant growth promotion in rice (*Oryza sativa* L.). Curr Microbiol 52:383–389
44. Khaosaad T, Garcia-Garrido JM, Steinkellner S, Vierheilig H (2007) Take-all disease is systematically reduced in roots of mycorhizal barley plants. Soil Biol Biochem 39:727–734
45. Rabie GH (1998) Induction of fungal disease resistance in vicia faba by dual inoculation with *Rhizobium leguminosarum* and vesicular-arbuscular mycorrhizal fungi. Mycopathologia 141:3
46. Tu JC (1979) Evidence of deferential tolerance among some root rot rhizobial parasitism in vitro. Physiol Plant Pathol 14:171–177
47. Antoun H, Beauchamp CJ, Goussard N, Chabot R, Lalande CR (1998) Potential of *Rhizobium* and *Bradyrhizobium* species as plant growth promoting rhizobacteria on non-legumes: effects on radishes (*Raphanus sativus* L.). Plant Soil 204:57–67
48. Malajczuk N, Pearce M, Lichfield RT (1984) Interaction between *Phytopythora cinnamoni* and *Rhizobium* isolates. Trans British Mycol Soc 82:491–500
49. Buonassissi AJ, Copeman RJ, Pepin HS, Eaton GW (1986) Effect of *Rhizobium* spp. on *Fusarium solani* f.sp. *phaseoli*. Can J Plant Pathol 8:140–146
50. Savoure A, Maygar Z, Perre M, Brown S, Schultze M, Dudits D, Kondorosi A, Kondorosi E (1994) Activation of the cell cycle machinery and the isoflavonoid biosynthesis by active *Rhizobium melitoti* Nod signal molecules in *Midicago microcallus* suspensions. EMBO J 13:1093–1102
51. Cordier C, Pozo MJ, Barea JM, Gianianazzi-Pearson V (1998) Cell defence response associated with localised and systemic resistance to *Phytophthora parasitica* induced in tomato by arbuscula mycorrhizal fungus. Mol Plant Microbe Interact 11:1017–1028
52. Brimner TA, Boland GJ (2003) A review of the non-target effects of fungi used to biologically control plant diseases. Agric Ecosyst Environ 100:3–16
53. Phillips DA, Kapulnik Y (1995) Plant isoflavonoids, pathogens and symbionts. Trends Microbiol 3:58–64
54. Eheshamul-Haque S, Ghaffar A (1993) Use of rhizobia in the control of root rot disease of sunflower, okra, soyabean and mungbean. J Phytopathol 138:157–163
55. Arfaoui A, Sifi B, Boudabaous A, Hadrami IE, Cherif M (2006) Identification of *Rhizobium* isolates possessing antagonistic activity against *Fusarium oxysporium* f.sp. *ciceris*, the causal agent of *Fusarium* wilt of chickpea. J Plant Pathol 88:67–75
56. Ruan Y, Kotraiah V, Straney DC (1995) Flavonoids stimulate spore germination in *Fusarium solani* pathogenic on legumes in a manner sensitive to inhibitors of cAMP-dependent protein kinase. Mol Plant Microbe Interact 8:929–938
57. Steinkellner S, Lendzemo V, Langer I, Schweiger P, Khaosaad T, Toussaint JP, Vierheilig H (2007) Flavonoids and strigolactones in root exudates as signals in symbiotic and pathogenic plant-fungus interactions. Molecules 12:1290–1306
58. Tu JC (1978) Protection from severe phytophthora root rot by *Rhizobium*. Physiol Plant Pathol 12:233–240
59. Husain S, Ghaffar A, Aslam M (1990) Biological control of *Macrophomina phaseolina* charcoal rot sunflower and mungbean. J Phytopathol 130:157–160
60. Chao WL (1990) Antagonistic activity of *Rhizobium* spp. against beneficial and plant pathogenic fungi. Lett Appl Microbiol 10:213–215
61. Huang HC, Kodama F, Akashi K, Konno K (2002) Impact of crop rotation on soilborne diseases of kidney bean: A case study in northern Japan. Plant pathol Bull 11:87–96
62. Matiru VN, Dakora FD (2004) Potential use of rhizobial bacteria as plant growth for increased yield in landraces of African cereal crops. Afr J Biotechnol 3(1):1–7

63. Curir P, Marchesini A, Danielit B, Mariani F (1996) 3-Hydroxyacetophenone in Carnations is a Phytoanticipin active against *Fusarium oxysporum* f. sp. *Dianthi*. Phytocheraistr 41(2):447–450
64. Dakora FD, Vincent JM (1984) Fast-growing bacteria from nodules of cowpea (*Vigna unguiculata* (L.) Walp.). J Appl Bacteriol 56:327–330
65. van Rensburg J, Strijdom H, Kriel MM (1976) Necessity for seed inoculation of soybeans in South Africa. Phytophylactica 8:91–96
66. Ververidis F, Trantas E, Douglas C, Vollmer G, Kretzschmar G, Panopoulos N (2007) Biotechnology of flavonoids and other phenylpropanoid-derived natural products. Part II: Reconstruction of multienzyme pathways in plants and microbes. Biotechnol J 2:1235–1249
67. Dixon RA, Paiva NL (1995) Stress-Induced Phenylpropanoid Metabolism. Plant Cell 7(7):1085–1097
68. Vogt T, Pollak P, Tarlyn N, Taylor LP (1994) Pollination-or Wound-lnduced Kaempferol Accumulation in Petunia Stigmas Enhances Seed Production. The Plant Cell 6:11–23
69. Sprent JI (2008) 60 Ma of legume nodulation. What's new? What's changing? J Exp Bot 59(5):1081–1084
70. Dakora FD, Keya SO (1997) Contribution of legume nitrogen fixation to sustainable agriculture in sub-Saharan Africa. Soil and Biochem 29:809–817
71. Dakora FD, Phillips DA (1996) Diverse functions of isoflavonoids in legumes transcend anti-microbial definitions of phytoalexins. Physiol Mol Plant Pathol 49:1–20
72. Dakora FD (2000) Commonality of root nodulation signals and nitrogen assimilation in tropical grain legumes belonging to the tribe Phaseoleae. Aust J Plant Physiol 27:885–892
73. Hungria M, Joseph CM, Phillips DA (1991) Rhizobium nod-gene inducers exuded naturally from roots of common bean (Phaseolus vulgaris L.). Plant Physiol 97:759–764
74. Dakora FD and Muofhe ML (1996) Molecular signals involved in nodulation of the African Bambara groundnut. In: Proceedings of the International Bambara Groundnut Symposium, University of Nottingham, UK pp 171–179
75. Messens E, Geelen D, van Montagu M, Holsters M (1991) 7,4-Dihydroxyflavanone is the major Azorhizobium nod-gene-inducing factor present in Sesbania rostrata seedling exudate. Mol Plant-Microbe Interact 4:262–267
76. Haby VA, Stout SA, Hons FM, Leonard AT (2006) Nitrogen fixation and transfer in a mixed stand of alfalfa and Bermuda grass. Agron J 98:890
77. Unkovich MJ, Pate JS (2000) An appraisal of recent field measurements of symbiotic N2 fixation by annual legumes. Field Crops Res 65:211–228
78. Maskey SL, Bhattarai S, Peoples MB, Herridge DF (2001) On-farm measurements of nitrogen fixation by winter and summer legumes in the Hill and Terai regions of Nepal. Field Crops Res 70:209–221
79. Azpilicueta CE, Zawoznik MS, Tomaro ML (2004) Phytoalexins synthesis is enhanced in groundnut plants inoculated with Bradyrhizobium sp.(Arachis). Crop Prot 23:1069–1074
80. De Rijke E. Aardenburg L, Van Dijk J, Ariese F, Ernst WHO, Gooijer C, Brinkman UA (2005). Changed isoflavone levels in red clover (Trifolium pratense L) leaves with disturbed root nodulation in response to water logging. J Chem Ecol 31(6):1285–1298
81. Russell GB, Sutherland ORW, Huchins RFN, Christmas RE (1978) Vestitol: A phytoalexin with insect feeding-deterrent activity. J Chem Ecol 4:571–579
82. Rich JR, Keen NT, Thomason IJ (1977) Association of coumestans with the hypersensitivity of Lima bean roots to Pratylenchus scribneri. Physiol Plant Pathol 10:105–116
83. Fukami H, Nakajima M (1971) Rotenone and the rotenoids. In: Jacobson M, Crosby DG (eds) Naturally Occurring Insecticides. Dekker, New York, pp 71–97

Sustainable Biodiesel Production Using Wastewater Streams and Microalgae in South Africa

T. Mutanda, D. Ramesh, A. Anandraj and F. Bux

Abstract The diminishing petroleum reserves in the world call for sustainable use of cheaply and readily available substrates such as wastewater streams for biomass and lipid production by microalgae. Treated wastewater is rich in macronutrients, such as nitrates and phosphates, and can therefore be used as a substrate for microalgal cultivation in open raceway ponds. The chemistry and composition of treated wastewater is of significance since it is made up of a wide range of compounds that support microalgal growth. The use of raceway pond technology utilizing wastewater streams feed is a new phenomenon that provides much needed phytoremediation of the wastewater as well as facilitating microalgal mass production. Macronutrient utilization by the microalgae justifies the application of treated wastewater as a sustainable raw material for renewable bioenergy production. The operational parameters in the raceway pond such as light intensity, photoperiod, pH, nutrients, salinity, and temperature are carefully optimized for maximal biomass and lipid yield. The biomass and lipid produced using the raceway pond system undergoes downstream processing in order to get the final product. The lipids are converted via transesterification to produce algae biodiesel. Other biologically active compounds and novel phytochemicals can also be derived from microalgae.

T. Mutanda (✉) · D. Ramesh · F. Bux
Department of Biotechnology and Food Technology,
Centre for Water and Wastewater Technology, Durban University of Technology,
Steve Biko Campus, P.O. Box 1334, 4001 Durban, South Africa
e-mail: taurai7@yahoo.com

F. Bux
e-mail: faizalb@dut.ac.za

A. Anandraj
Department of Nature Conservation, Mangosuthu University of Technology,
PO Box 12363, Jacobs, 4026 Durban, South Africa

A. Gurib-Fakim and J. N. Eloff (eds.), *Chemistry for Sustainable Development in Africa*, DOI: 10.1007/978-3-642-29642-0_4,

1 Introduction

Large-scale venture for commercial production of biomass and lipids from microalgae requires sustainable and cost-effective materials such as wastewater as feed for biomass and lipid production. There are basically two ways that can be used to grow microalgae at large scale viz. using photobioreactors and open raceway ponds [5]. Photobioreactors are reported to be more efficient than open raceway ponds [9, 5] though they are capital and energy intensive. Open raceway pond system for growing microalgae is economically viable since they are cheap to construct and maintain. However, one drawback of using raceway ponds is that they are easily contaminated by non-target microorganisms since they are open systems that are greatly affected by inclement weather.

The use of artificial media for growing microalgae at large scale is economically nonviable, hence the use domestic municipal wastewater as feed. Three models can be postulated for commercial use of post-chlorinated wastewater for growing microalgae in open raceway ponds. First, the treated wastewater can be introduced intermittently into the raceway pond to replenish water lost due to evaporation. Second, the wastewater can be used under batch conditions in supplementation with trace metals and macronutrients whenever it is necessary. Third, another approach is to continuously introduce the wastewater streams into the raceway pond and also taking the filtrate back into the pond after harvesting the biomass.

There are few reports on the optimization of the operational parameters in the open raceway ponds though to date a wide range of microalgal strains have been successfully cultured [5]. The main challenge using the open raceway pond system is harvesting the biomass since the microalgal cells will be in suspension. However, cost-effective methods such as settling by gravity can be employed for the dewatering exercise as compared to centrifugation which is expensive and energy intensive.

The main aim of this book chapter is to clarify the sustainable use of wastewater streams as feed for microalgal growth for biomass and lipid production. The lipids are transesterified to biodiesel. The concepts of raceway pond technology, macronutrient utilization, lipid synthesis, and downstream processing are described and discussed in detail.

2 Municipal Wastewater Treatment

Sustainability strives for the maintenance of economic well being, protection of the environment, and prudent use of natural resources for the benefit of mankind [58]. Copious amounts of industrial and domestic municipal wastewater are produced everyday in South Africa [85]. Domestic municipal wastewater presents a number of challenges to local government municipalities mainly due to its heterogeneous composition. Domestic municipal wastewater is of environmental concern; therefore it should meet certain stringent requirements before being discharged into receiving water bodies.

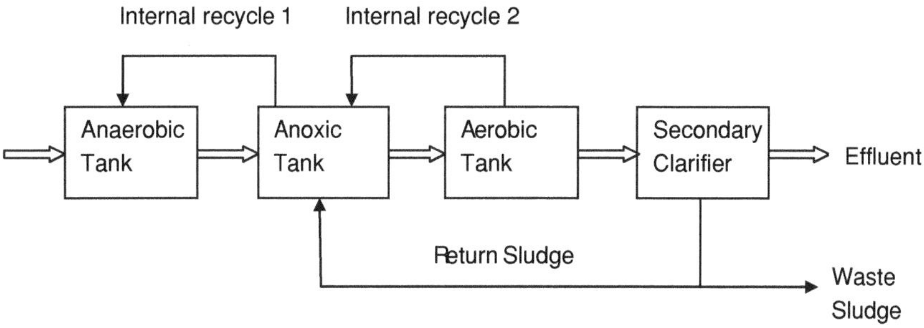

Fig. 1 The activated sludge method for primary sewage treatment using the University of Cape Town WWTP layout [25]

The wastewater requires treatment before discharging to the environment to avoid killing aquatic life as a result of eutrophication [42]. Eutrophication is the proliferation of aquatic plants and/or algal blooms in rivers and other water bodies in response to the high organic material load of the wastewater resultantly starving aquatic animals such as fish of dissolved oxygen. The excessive proliferation of algal blooms will consequently lead to the death of aquatic animals, cause water quality problems, including toxin production, odors, scums, and unsafe drinking water [60]. There are several methods that can be employed to treat domestic municipal wastewater to reduce the levels of organic material load to the acceptable regulatory levels involving aerobic and anaerobic systems [42].

The widely used method for the treatment of domestic municipal wastewater is mechanically by the activated sludge with secondary treatment method. This method involves a series of steps as shown in a flow chart in Fig. 1. In short, after primary treatment to remove mainly suspended solids, the wastewater is then treated microbiologically by either using stabilization ponds or activated sludge to further eliminate solids and organic matter [89]. The process of degrading the wastewater further produces a lot of nutrients in the wastewater which can be treated chemically before discharging. However, other biological, cheap, and environmentally friendly methods can be used to treat these nutrients such as the use of microalgae in maturation ponds or open raceway ponds [61]. Recent studies have demonstrated that microalgae have the potential for removing nitrogen and phosphorus from wastewater and incorporate these nutrients into algal cell biomass [42].

The utilization of wastewater is a cost-effective option for growing microalgae for biomass and lipid production for subsequent conversion to biodiesel. Domestic municipal wastewater is valuable as a cheap substrate that is readily available and amenable to microalgal growth. It is crucial to know the composition and chemistry of the wastewater before using it.

2.1 Wastewater Chemistry and Nutrient Profiles

The chemical composition of domestic wastewater is complex and varies greatly in response to the time of the day and also urban population density. The main nutrient profiles and chemical compounds levels permitted/required for treated wastewater or effluent before disposal according to the Water Act of South Africa [76] include both organic and inorganic materials. Inorganic materials, such as nitrates and phosphates, are found in treated effluents since they are not completely eliminated by the activated sludge treatment process [63]. The permitted concentrations of nitrates and phosphates are ≤ 1.5 and ≤ 1.0 mg/L, respectively, according to the Water Act of South Africa [76]. These macronutrients in such low dosages serve as good nitrogen and phosphorus sources, and therefore are required for microalgal propagation in maturation ponds resulting in further wastewater treatment and microalgal biomass production which can be used as a biofuel feedstock [42, 60]. A high concentration of nitrates and phosphates has been shown to inhibit the growth of phytoplankton due to the toxic nature of the intracellular side product nitrite [8]. The treated effluent has some recalcitrant aromatic pollutants which cannot be easily degraded biologically. These recalcitrant aromatic compounds have been demonstrated to be degraded by microalgae [51].

Treated wastewater that meets the regulatory requirements is crucial if the effluent is to be used as nutrient feed for growing microalgae such as *Chlorella* sp. Most of the recalcitrant compounds such as aromatic dyes, phenolics, etc., inhibit microalgal growth, therefore their removal can positively impact on microalgal growth since wastewaters are rich in nutrients, such as nitrates and phosphates, which are the predominant nutrients for microalgal propagation. Municipal domestic wastewater consists of a multiplicity of organic and inorganic materials which can either support or inhibit microalgal growth. The most critical macronutrients are nitrates, ammonium, and phosphates which are the same compounds as those found in artificial media though in varying proportions [8]. The main inorganic materials required for microalgal growth are selenium, manganese, iron, soluble orthophosphate, arsenic, and sodium.

2.2 Wastewater Bioremediation Using Microalgae

The discharge of residual nitrates and phosphates in wastewater effluents into receiving environments can result in negative environmental impacts and potential eutrophication of water bodies [72]. Nutrients in these wastewater streams can be used for beneficial purposes. Wastewater treatment offers economic benefits as cheap and readily available nutrient rich substrate that can be used for microalgal growth for biomass and lipid production for further conversion into biodiesel and other spinoff products. The mechanism of bioremediation of domestic municipal wastewater using microalgae is through absorption and active uptake of nitrates and phosphates through the algal cell wall. Furthermore, the microalgae release dissolved oxygen which is used by a consortium of microorganisms for the decomposition of the organic wastes. Microalgae can be used for the tertiary treatment of wastewater and *Chlorella* sp is the best candidate for this biological activity. In addition, *Chlorella* sp can also remove various nitrogen and phosphorous compounds, heavy metals, and toxic residues from the wastewater [12].

2.3 Post-Chlorinated Wastewater as Substrate

Various strategies are used to increase microalgal biomass yield. Media supplemented with varying concentrations of urea as a nitrogen source were investigated and this resulted in an increase in biomass and lipid content of *Chlorella* sp. [35]. Li et al. [50] supplemented $NaNO_3$ to the soil extract medium for the growth of the green alga *Neochloris oleobundans* and found that it enhanced biomass and lipid accumulation. The effect of chlorinated compounds on microalgal growth has also been reported [67]. These researchers established that *Chlorella* VT-1 showed some tolerance to all the chlorophenols tested at a concentration of 10 mg/L except for 2,4,5-trichlorophenol, which was toxic at all concentrations investigated. The use of post-chlorinated wastewater for microalgal growth has not been fully investigated though some reports indicate that chloride ions act as microalgal micronutrients [20]. Previous reports focused on the use of chlorinated organic compounds in wastewater as a pollution monitoring strategy. Simmons and Sivaborvorn [84] carried out a study to investigate the effects of chlorine containing organic compounds formed from the chlorination of domestic wastewater on phytoplankton growth. In an effort to prevent nutrient depletion and enhance biomass yield, Costa et al. [10], supplemented nutrients (carbon as sodium bicarbonate, nitrogen as urea, phosphate, sulfate, ferric iron, magnesium, and potassium) to their race way pond.

2.4 Macronutrient Utilization

Macronutrients contribute to molecules which make up the algal structure and therefore required in large quantities. Algae require carbon, nitrogen, phosphorus, oxygen, hydrogen, and also calcium, magnesium, sulfur, and potassium in "macro" quantities. Micronutrients are required in milligrams per liter or lower concentrations and are utilized as components of essential molecules such as growth factors or enzymes or pigments [38]. After carbon, nitrogen is the most important element contributing 1–10% of the dry matter of algal cells. It is generally low in diatoms where silica is the major element in the cell wall and in nitrogen deficient organisms that have produced large amounts of oils and polysaccharides [9]. Both inorganic and organic nitrogen compounds, such as nitrate (NO_3^-), nitrite (NO_2^-), or ammonia (NH_4^+), can serve as a nitrogen source for growth of various microalgae. When N is assimilated in an oxidized form as nitrate

(NO_3^-) or nitrite (NO_2^-), it must be reduced to ammonium (NH_4^+) by the enzyme nitrate reductase before it can be incorporated into organic molecules [38, 92]. Ammonium may be an alternative nitrogen source and may be added as NH_4Cl. At the typical pH of seawater (8.2), there is about 90 % NH_4 and 10 % NH_3 (ammonia). As the pH of the culture medium increases during algal growth, the ratio of NH_4: NH_3 increases and reaches 1:1 at a pH 9.3. Therefore, substantial amounts of ammonium may be lost from the culture if the algal culture is kept mixed by bubbling with air. Ammonium, at concentrations of 100–250 mM, may be inhibitory to some coastal species, but most coastal species tolerate concentrations as high as 1 M [34].

Phosphorus is another major nutrient element required for normal algal growth as it plays a significant role in most cellular processes such as those involved in energy transfer and in nucleic acid synthesis. The form in which microalgal cells assimilate phosphorus is as inorganic phosphate ($H_2PO_4^- + HPO_4^{2-}$) or collectively P_i. For organic phosphate compounds to be utilized as a primary source of phosphorus, they must be hydrolyzed by extracellular enzymes such as phosphoesterases or phosphatases and the product P_i is assimilated. Three major processes are used to incorporate orthophosphate into ATP: (1) substrate phosphorylation, (2) oxidative phosphorylation, and [3] photophosphorylation [38].

Optimum concentration of nutrients for a given algal strain in a commercial raceway pond may vary considerably as it is dependent on many factors such as population density, growth rate, PAR, temperature and pH, which varies daily and seasonally. Raceway ponds supplemented with essential macro nutrients (dissolved inorganic nitrogen and phosphate) from treated wastewater promotes algal blooms which, when harvested, provides an economically feasible biodiesel feedstock.

Irradiance is the key factor for maximum productivity of natural systems [2] and commercial raceway ponds. Irradiance is the input of light energy per unit surface area and time. Best light conditions for algal growth are normally determined from photosynthesis versus irradiance (PE) curves, by the application of a range of actinic light for long time periods (15–60 min). However, more recent techniques involve the use of rapid photosynthetic light curves (RPLC) generated from a PAM Fluorometer (1–2 min), by incubating the algal culture at a sequence of increasing actinic irradiance levels. Light curves indicate the minimum, optimum and maximum PAR levels required for photosynthesis and the level of PAR inducing photoinhibition, which results in loss of photosynthetic capacity due to excessive photon flux densities [80]. From a typical PE curve, photosynthesis increases linearly with irradiance until saturation is reached and the curve levels off, reaching maximum (P_{max}). The initial slope of the curve (β) is an indicator of photosynthetic efficiency and the declining slope (β) indicates photoinhibition. The light saturation constant is the intensity of light at which the specific biomass growth rate is half its maximum value, μmax, which tend to be much lower than the maximum midday irradiance. For example, the light saturation constants for microalgae *Phaeodactylum tricornutum* and *Porphyridium cruentum* are 185 and 200 μE m^{-2} s^{-1} [29], respectively [9]. In comparison with these values, the typical midday outdoor light intensity in southern African regions is about 2,000 μE m^{-2} s^{-1}. Light saturation reduces the biomass growth rate and consequently the total pond productivity [7, 9]. Algal strains with a higher threshold of photoinhibition generally have a higher average daily growth rate in raceway ponds, increasing the overall yield. Irradiance is therefore one of the most significant factors influencing algal growth [53].

The second important factor is temperature, which regulates all metabolic activity of microalgae occurring naturally [1] and in artificial ponds. It affects nutrient availability and uptake, the properties of the cell structure, enzyme activity, and biomass composition. Temperature of raceway ponds vary seasonally and during the diurnal period. Analysis of chlorophyll-*a* fluorescence using PAM fluorometry can once again be used to determine the optimum temperature range for growth of a specific strain of microalgae. Algal cultures are subjected to a range of temperatures and the corresponding photosynthetic activity recorded. Such experiments provide temperature tolerance levels of the strain and information on potential climatic regions which would encourage optimum growth. Thermal stress in algal cultures is indicated by a decrease in photosynthetic efficiency. Temperature can have a significant influence on the chemical composition of algae, especially fatty acid composition. A decrease in temperature can reduce lipid synthesis and alter the species of lipids. Maximizing algal biomass and cellular neutral lipids may not necessarily occur under the same nutrient regime, and therefore requires a rapid assessment of the chlorophyll-a and nutrient concentrations [77].

2.5 Determining Optimal Conditions for Growth and Lipid Synthesis

Analysis of chlorophyll fluorescence is a rapid, non-invasive technique of monitoring the photosynthetic activity, and the overall physiological health of microalgae. It is also a valuable tool to assess the optimum growth conditions required to maximize the biomass feedstock and to quantify the effect of nutrient or other extreme environmental stresses (salinity, temperature PAR and pH) on the algal culture. Instruments commonly used to analyze chlorophyll-*a* fluorescence are either a Pulse Amplitude Modulated (PAM) (Waltz, Germany) [79], and a Fast Repetition Rate (FRR) fluorometer [21, 41, 45]. Other instruments include, Plant Stress Meter (PSM), BioMonitor AB, Plant Efficiency Analyser (PEA) (Hansatech LTD) and PhytoFlash (Turner Designs). The quantum yield (F_v/F_m) of photosystem II (PS II), is the ratio of the variable fluorescence (F_v) and maximum fluorescence (F_m) and is known to decrease under nutrient limitation and other environmental stresses. This parameter has been used to assess the nutrient status of natural algal populations [4, 46] and cultures in raceway ponds [88, 92]. The quantum yield generally ranges from 0.6 to 0.8 in dark adapted green algae [55]. Nutrient stressed cultures in raceway ponds exhibit values closer to the lower end or below the range and indicate potential performance.

2.6 Lipid Synthesis

Neutral lipid synthesis is stimulated under nutrient depleted or limited conditions. Many microalgae have the ability to produce substantial amounts (15–80 % dry cell weight) of triacylglycerols (TAG) as a storage lipid [9] under nutrient or other environmental stress. Fatty acids, the building blocks for TAGs and all other cellular lipids, are synthesized in the chloroplast using a single set of enzymes, of which acetyl CoA carboxylase (ACCase) is key in regulating fatty acid synthesis rates. Synthesis and sequestration of TAG into cytosolic lipid bodies appear to be a protective mechanism by which algal cells cope with stressful conditions, but little is known about regulation of TAG formation at the molecular and cellular level. Unlike the glycerolipids found in membranes, TAGs do not perform a structural role but instead serve primarily as a storage form of carbon and energy [36]. After being synthesized, TAGs are deposited in densely packed lipid bodies generally located in the cytoplasm of the algal cell. As many algal species have been found to grow rapidly and produce substantial amounts of TAG or oil, and are thus referred to as oleaginous algae, it has long been postulated that algae could be employed as cell factories to produce oils and other lipids for biofuels and other biomaterials [36]. The productivity of oil, that is the mass of oil produced per unit volume of the microalgal medium per day, depends on the algal growth rate and the oil content of the biomass. Microalgae with high oil productivities are desired for producing biodiesel [9].

2.7 Raceway Pond Technology and Operational Parameters

Raceway pond technology is a cost-effective method for growing microalgae for biomass and lipid production and biomass productivities of 0.5 g/L can be achieved in some commercial raceway ponds [30, 95]. The operation of this outdoor culture open system is easy and requires minimal maintenance [5, 50]. The system has few operating costs, minimal power consumption, and little overheads as compared to photobioreactors [71]. The advantage of this method is that readily available wastewater can be used as media for cultivation with the added benefit of bioremediation. In addition, if the system is located near a power plant, cheaply available flue gas can be used to speedup the photosynthetic rates in the pond or pure carbon dioxide can be bubbled into the pond [37].

In order to avoid contamination from debris and rainfall, the roof of the raceway pond can be covered using a material that allows maximum sunlight to penetrate. The depth of the water in the pond should be at most 30 cm to allow maximum light penetration. A paddle wheel is installed to allow for water circulation at low speed of 9–30 m/min to avoid microalgae from settling [5]. Ambient carbon dioxide is frequently used but however when there is a need to speedup photosynthetic rates in the pond, pure carbon dioxide can be bubbled into the pond as required according to the size of the pond.

As the substrate is circulated in the raceway pond by the paddle wheel, the microalgae will be using the nutrients for their growth. As the nutrients in the substrate become limiting, the microalgae will start to store lipids and this is common with a cascading raceway pond system. The following operational parameters will be recorded daily: cell density, temperature, pH, light intensity, humidity, dissolved CO_2, dissolved O_2, conductivity, salinity, evaporation levels, suspension flow rate, nitrate levels, phosphate levels, and nitrite levels [30]. Since it is an open system, some of the water is lost due to evaporation and some fresh wastewater is added to replenish the lost water and to balance the salinity in the pond.

3 Downstream Processing of Algae Biomass

The downstream processes involved in biodiesel production from microalgae biomass are harvesting, algal oil extraction, and transesterification of the oil (Fig. 2). Algae such as *Chlorella* (3–15 μm) and *Scenedesmus* (30 μm) are difficult and costly to remove from wastewater because of their small size and low specific gravity [3, 11, 68]. Algae are poorly compacted by gravity because of negative surface charges and dilute concentrations within the liquid medium [22, 43, 59, 87]. Microalgae harvesting involves the concentration of dilute microalgal suspensions, typically 0.02–0.06 % total suspended solids (TSS) into a slurry or paste with concentrations of 5–25 % TSS or more depending on the target process objective or more depending on the target process objective [82]. The recovery of microalgal biomass which generally requires one or more solid–liquid separation steps is a challenging phase of the algal biomass production process [93]. After mass production of algae biomass in wastewater based raceway pond, the produced biomass could be harvested from the pond at regular intervals. In algae harvesting, the algae is separated from the huge quantities of wastewater in the raceway pond by using the suitable technology. The selection of harvesting technology is crucial to economic production of microalgal biomass and it may contribute to 20–30 % of total biomass production cost [28, 32]. Low cell densities (typically in the range of 0.3–5 g l^{-1}) when there is limited light penetration, and the small size of some algal cells (typically in the range of 2–40 mm), make the recovery of biomass difficult [49]. The selection of technology for algae harvesting is depends on microalgae size, quantity to be treated, biomass concentration in wastewater, biomass production per day, quality of end products, capital investment, cost of operation, less periodical maintenance of the plants, moisture content of algae paste after harvesting, and etc.

The harvesting of alga biomass can be achieved by physical, chemical, biological methods, or combination of any two these methods. The conventional methods employed for harvest the microalgae from wastewater are sedimentation, centrifugation, filtration, membrane filtration, air flotation, and flocculation. Ryll et al. [78] mentioned that drawback of membrane filtration was membrane fouling and clogging due to the small size of the microalgae. Mohn [57] reported

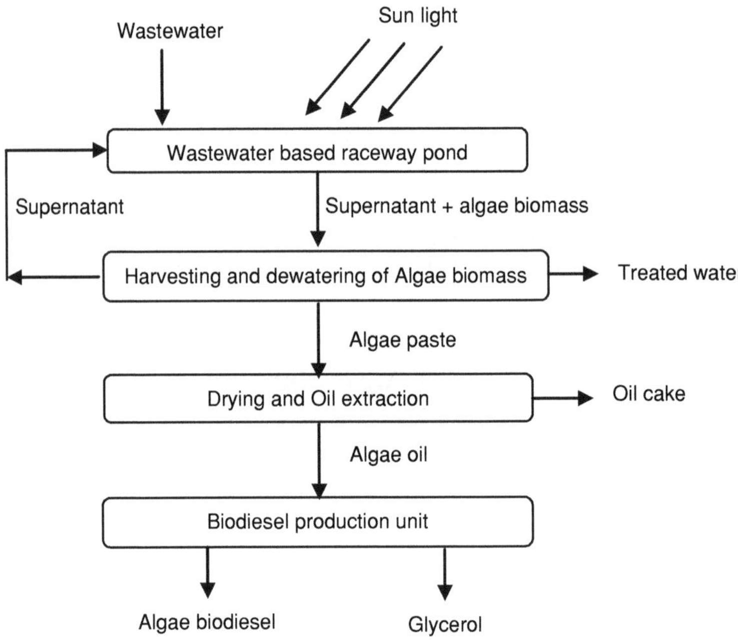

Fig. 2 Process involved in treatment of wastewater by microalgae for biodiesel production

that centrifugation and filtration are expensive and only economic in special cases. The high investment costs and additional energy requirements for pumps and the generation of compressed air are disadvantageous to flotation. Centrifugal recovery of the biomass is feasible for high-value products [28].

For extremely low value products, gravity sedimentation, possibly enhanced by flocculation, may be the method of choice [28]. Sedimentation tanks or settling ponds are generally used in biomass recovery from sewage-based processes [91]. Sedimentation requires special sedimentation tanks and occasionally the addition of flocculating agents. For an efficient production of algal biomass, the process should run maintenance and trouble free with minimized technical costs and expenditure on personnel. Gravity sedimentation is the most common harvesting technique for algae biomass in wastewater treatment because of the large volumes treated and the low value of the biomass generated. Sedimentation tanks or settling ponds are also possible, e.g., to recover algae biomass from sewage-based processes. [65]. The algae treated wastewater in raceway pond is pumped into the harvesting and dewatering unit. After removal of the algae biomass, the treated wastewater can be disposed into the streams, rivers or seas. Before disposal of treated wastewater into the water bodies, the quality of wastewater must be checked and it should meet the local standards. The slurry contains 80–90 % moisture content has to be sent to dewatering unit. The algae paste containing less than 15 % moisture content can be achieved after dewatering process. The algae paste to be sent to the drying unit.

3.1 Drying of Algae Biomass

Before the oil extraction, the moisture present in the algae paste has to be removed by suitable method. The harvested biomass paste (typical 5–15 % dry solid content) is perishable and must be processed rapidly after harvest; dehydration or drying is commonly used to extend the viability depending on the final product required. Methods that have been used include sun drying, low-pressure shelf drying, drum drying [70] spray drying [18], fluidised bed drying [48], freeze drying [27], and Refractance Window^{TM} technology drying [62].

Spray drying is commonly used for extraction of high-value products, but it is relatively expensive and can cause significant deterioration of some algal pigments [18]. Freeze drying is equally expensive, especially for large-scale operations, but it eases extraction of oils. Intracellular elements such as oils are difficult to extract from wet biomass with solvents without cell disruption, but are extracted more easily from freeze-dried biomass [27, 28]. Sun drying is the cheapest dehydration method; but main disadvantages include long drying times, the requirement for large drying surfaces, and the risk of material loss [70].

Drying temperature during lipid extraction affects both the lipid composition and lipid yield from the algal biomass. For example, drying at 60 °C still retains a high concentration of TAG in the lipids and only decreases slightly the lipid yield, with higher temperatures decreasing both the concentration of TAG and lipid yield [94].

3.2 Oil Extraction

For biodiesel production, lipids and fatty acids have to be extracted from the microalgal biomass. The three methods to extract the oil from algae biomass are expeller/press method, solvent extraction, and supercritical fluid extraction. A simple process is to use a press to extract a large percentage (70–75 %) of the oils out of algae [17]. Several solvents can be used such as hexane, ethanol (96 %), or a hexane–ethanol (96 %) mixture, being possible to obtain up to 98 % quantitative extraction of purified fatty acids [77]. Mostly, hexane was used for extracting oil from algae in solvent extraction method. Supercritical fluid extraction is far more efficient than traditional solvent separation methods. Supercritical fluids are selective, thus providing the high purity and product concentrations [69]. This can extract almost 100 % of the oils all by itself. Chemical solvents are often used in the extraction of the oils from dried algae biomass materials. In the supercritical fluid carbon dioxide (CO_2) extraction, CO_2 is liquefied under pressure and heated to the point that it has the properties of both a liquid and gas. This liquefied fluid then acts as the solvent in extracting the oil [17].

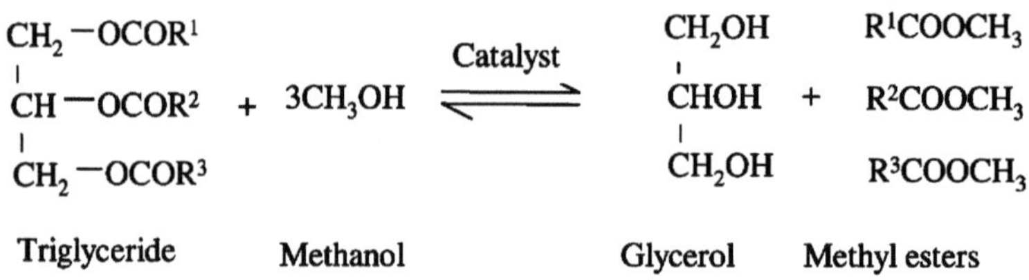

Fig. 3 General equation for transesterification of triglycerides

3.3 Biodiesel Production

Vegetable oils have chemical structures different than that of petroleum-based diesel fuels. Vegetable oils containing up to three fatty acids linked to a glycerin molecule with ester linkages are called triglycerides. The fatty acids are characterized by their carbon chain length and in numbers of double bonds [13]. Algal oils contain a high degree of polyunsaturated fatty acids when compared to vegetable oils, which makes it susceptible to oxidation in storage and therefore limits utilization [9].

The high kinematic viscosities of vegetable oils and animal fats ultimately lead to operational problems such as engine deposits when used directly as fuels [39]. Glycerol is suspected of contributing to engine deposit formation during combustion [40]. To overcome these problems in diesel engines, the microalgae oil has to improve the combustion properties. Four different methods like blending, transesterification, pyrolysis, and emulsification can be used to produce biodiesel from microalgae oils. Among these, the transesterification is the key and foremost important step to produce the cleaner and environmentally safe fuel from vegetable oils or animal fats [56]. The transesterification is a chemical reaction, which can be used lower the viscosity by removing of glycerol from oils. The end products of the transesterification reaction are fatty acid (m)ethyl esters and glycerol (Fig. 3). The name "biodiesel" has been given to transesterified vegetable oil to describe its use as a diesel fuel [14].

Biodiesel is produced by transesterification of large branched triglycerides of oil into smaller straight chain molecules of methyl esters (biodiesel), in presence of the catalyst. A catalyst is usually used to improve the reaction rate and yield. Because the reaction is reversible, excess alcohol is used to shift the equilibrium to the product side. The conversion of triacylglycerols (TAG) to biodiesel is a stepwise process whereby the alcohol initially reacts with TAG as the alkoxide anion to produce Fatty Acid Alkyl Ester (FAAE) and diacylglycerols (DAG), (reaction [1], Fig. 4), which react further with alcohol (alkoxide) to liberate another molecule of FAAE and generate monoacylglyerols (MAG), (reaction [2], Fig. 4). Lastly, MAG undergo alcoholysis to yield glycerol and FAAE (reaction [3], Fig. 4), with the combined FAAE collectively known as biodiesel. Three moles of biodiesel and one mole of glycerol are produced for every mole of TAG that undergoes complete conversion. The transesterification reaction is reversible,

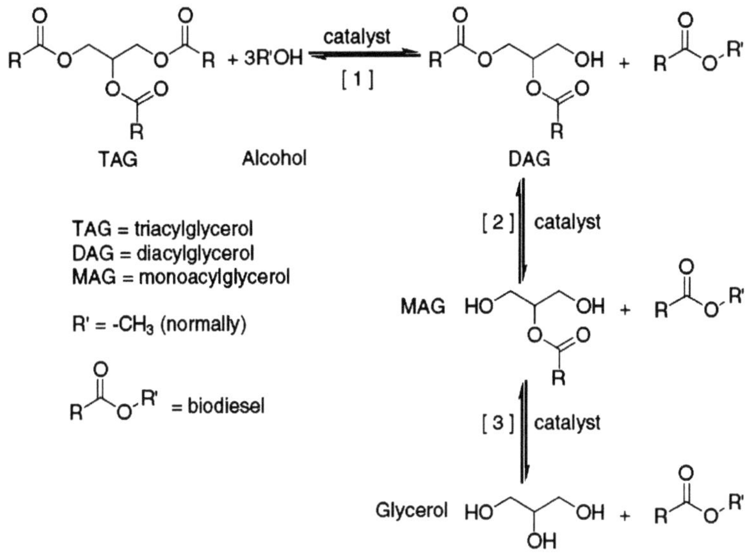

Fig. 4 Transesterification of triacylglycerols to yield fatty acid alkyl esters (biodiesel)

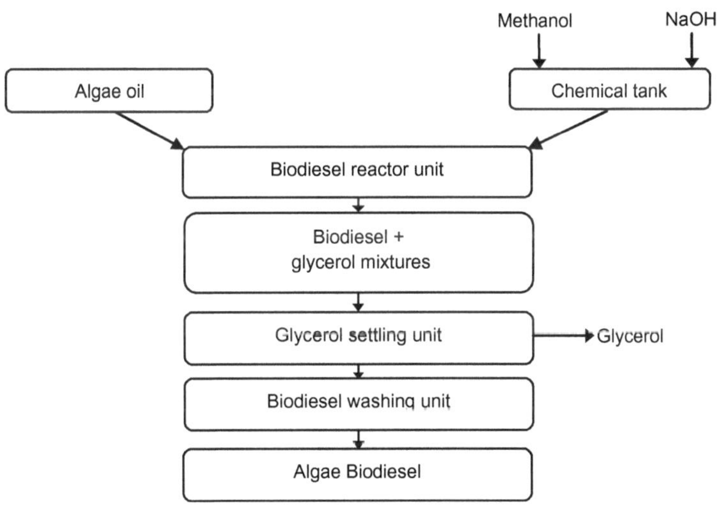

Fig. 5 Process flowchart for production of biodiesel production from microalgae oil

although the reverse reaction (production of MAG from FAAE and glycerol, for instance) is negligible largely because glycerol is not miscible with FAAE, especially fatty acid methyl esters (FAME) when using methanol as the alcohol component. The reaction system is biphasic at the beginning and the end of biodiesel production, as methanol and vegetable oil and glycerol and FAME are not

Table 1 Requirements for automotive biodiesel fuel—SANS 1935

S. no.	Property of the biodiesel	Requirements
1	Ester content, % mass fraction, min	96.5
2	Density at 15 °C, kg/m^3	860–900
3	Kinematic viscosity at 40 °C, mm^2/s	3.5–5.0
4	Flash point, °C, min	120
5	Sulfur content, mg/kg, max	10.0
6	Carbon residue (on 10% distillation residue), % max fraction, max	0.3
7	Cetane number, min	51.0
8	Sulfated ash content, % mass fraction, max	0.02
9	Water content, % mass fraction, max	0.05
10	Total contamination, mg/kg, max	24
11	Copper strip corrosion (3 h at 50 °C), rating, max	Class 1
12	Oxidation stability, at 110 °C, h, min	6
13	Acid value, mg KOH/g, max	0.5
14	Iodine value, g of iodine/100 g of FAME, max	140
15	Linolenic acid methyl ester, % mass fraction, max	12
16	Polyunsaturated (>= double bonds) methyl esters, % mass fraction, max	1
17	Methanol content, % mass fraction, max	0.2
18	Monoglyceride content, % mass fraction, max	0.8
19	Diglyceride content, % mass fraction, max	0.2
20	Triglyceride content, % mass fraction, max	0.2
21	Free glycerol, % mass fraction, max	0.02

Source [26]

miscible [6]. These are three stepwise reactions with intermediate formation of diglycerides and monoglycerides resulting in the production of three mol of biodiesel and one mol of glycerol.

Vegetable oils can be transesterified by heating them with a large excess of anhydrous alcohol and a catalyst. The transesterification reaction can be catalyzed by alkalis [31, 47, 73, 74, 98], acids [24, [96], enzymes [19, 33, 64, 66, 83], or supercritical methanol. Various studies have been carried out using different oils as raw material, different alcohols (methanol, ethanol, butanol), as well as different catalysts, including homogeneous ones such as sodium hydroxide, potassium hydroxide, sulfuric acid, and supercritical fluids and heterogeneous ones such as lipases [54].

Base-catalyzed reactions are performed at generally lower temperatures, pressures, and reaction times and are less corrosive to industrial equipment than acid-catalyzed methods. Therefore, fewer capital and operating costs are incurred by biodiesel production facilities in the case of the base-catalyzed transesterification method [16, 23]. The homogenous base-catalyzed transesterification reaction is about 4,000 times faster than the corresponding acid-catalyzed process [75, 86]. Among various base catalysts, sodium hydroxide is very well accepted and widely used because of its low cost and high product yield [15].

The biodiesel industry currently uses sodium methoxide, since methoxide cannot form water upon reaction with alcohol such as with hydroxides [99]. The sodium methoxide catalyst is dissolved in alcohol using a standard agitator or mixer. The methyl alcohol and catalyst mix is then charged into a closed reactor and the oil or fat is added. The reaction mix is kept just above the boiling point of the methyl alcohol (around 60 °C) to speedup the reaction, and the transesterification reaction takes place. Recommended reaction time varies from 1 to 8 h, and optimal reaction time is about 2 h [90]. Excess alcohol is normally used to ensure total conversion of the fat or oil into its esters. After the reaction is complete, two major products form: glycerin and biodiesel. The biodiesel product is sometimes purified by washing gently with warm water to remove residual catalyst or soaps, dried, and sent to storage [14, 52]. The most abundant composition of microalgal oil transesterified with methanol is $C_{19}H_{36}O_2$, which is suggested to accord with the standard of biodiesel [97] (Fig. 5).

The biodiesel is biodegradable, non-toxic, has low emission profiles, and it is environmentally beneficial [44]. Biodiesel has very low sulfur content and releases lower CO during combustion in diesel engines. Biodiesel contains oxygen, which leads to complete combustion in internal combustion (IC) engines. Biodiesel can be used in the existing diesel engines without any modification and can mix readily with diesel at any blending level [44, 86]. The major advantage of biodiesel is in reduced CO_2 emissions of up to 78 % compared to emissions from petroleum diesel [81]. For algal biodiesel to be an accepted substitution fuel for fossil fuels, its properties must meet out the International Biodiesel Standard for diesel vehicles developed different countries. The biodiesel standards developed in South Africa (SANS 1935) is furnished in the Table 1.

References

1. Anandraj A, Perissinotto R, Nozais C (2008) The recovery of microalgal production and biomass in a South African temporarily—open/closed estuary, following mouth breaching. Estuar Coast Shelf Sci 79:599–606
2. Anandraj A, Perissinotto R, Nozais C (2007) A comparative study of microalgal production in a marine versus a river-dominated temporarily open/closed estuary, South Africa. Estuar Coast Shelf Sci 73:768–780
3. Bare WFR, Jones NB, Middlebrooks EJ (1975) Algae removal using dissolved air flotation. J Water Pollut Control 47(1):153–169
4. Beardall J, Young E, Roberts S (2001) Approaches for determining phytoplankton nutrient limitation. Aquat Sci 63:44–69
5. Borowitzka MA (2005) Culturing microalgae in outdoor ponds. In: Andersen RA (ed) Algal culturing techniques. Elsevier Academic Press, UK
6. Moser Bryan R (2009) Biodiesel production, properties, and feedstocks. In Vitro Cell. Dev Biol Plant 45:229–266. doi:10.1007/s11627-009-9204-z
7. Camacho Rubio F, García Camacho F, Fernández Sevilla JM, Chisti Y, Molina Grima E (2003) A mechanistic model of photosynthesis in microalgae. Biotechnol Bioeng 81:459–473
8. Chen W, Zhang Q, Dai S (2009) Effects of nitrate on intracellular nitrite and growth of *Microcystis aeruginosa*. J Appl Phycol 21:701–706
9. Chisti Y (2007) Biodiesel from microalgae. Biotechnol Adv 25(3):294–306

10. Costa JAV, Colla LM, Filho PD (2003) *Spirulina platensis* growth in open raceway ponds using fresh water supplemented with carbon, nitrogen and metal ions. Z Naturforsch 58:76–80
11. Craggs RJ, McAuley PJ, Smith VJ (1997) Wastewater nutrient removal by marine microalgae grown on a corrugated raceway. Water Res 31(7):1701–1707
12. de-Bashan LE, Moreno M, Hernandez JP, Bashaan Y (2002) Removal of ammonium and phosphorus ions from synthetic wastewater by the microalgae *Chlorella vulgaris* coimmobilised in alginate beads with the microalgae growth-promoting bacterium *Azospirillum brasilense*. Water Res 36:2941–2948
13. Demirbas A (1998) Fuel properties and calculation of higher heating values of vegetable oils. Fuel 77:1117–1120
14. Demirbas A (2002) Biodiesel from vegetable oils via transesterification in supercritical methanol. Energy Convers Manage 43:2349–2356
15. Demirbas A (2003) Biodiesel fuels from vegetable oils via catalytic and non catalytic supercritical alcohol transesterifications and other methods: a survey. Energ Convers Manage 44:2093–2109. doi:10.1016/S0196-8904(02)00234-0
16. Demirbas A (2008) Production of biodiesel from tall oil. Energy Sour Part A 30:1896–1902
17. Demirbas A (2009) Production of biodiesel from algae oils. Energy Sour Part A 31:163–168. doi:10.1080/15567030701521775
18. Desmorieux H, Decaen N (2006) Convective drying of *spirulina* in thin layer. J Food Eng 66(4):497–503
19. Du W, Xu Y, Liu D, Zeng J (2004) Comparative study on lipase-catalyzed transformation of soybean oil for biodiesel production with different acyl acceptors. J Mol Catal B Enzymat 30:125–129
20. Eyster C (1958) Chloride Effect on the Growth of *Chlorella pyrenoidosa*. Nature 181:1141–1142
21. Falkowski PG, Wyman K, Ley AC, Mauzerall DC (1986) Relationship of steady-state photosynthesis to fluorescence in eucaryotic algae. Biochim Biophys Acta 849:183–192
22. Folkman Y, Wachs AM (1970) Filtration of *Chlorella* through Dune-Sand. Proc Am Soc Civil Eng, J Sanit Eng Div 96:675–690
23. Freedman B, Butterfield RO, Pryde EH (1986) Transesterification kinetics of soybean oil. JAOCS 63:1375–1380
24. Furuta S, Matsuhashi H, Arata K (2004) Biodiesel fuel production with solid super acid catalysis in fixed bed reactor under atmospheric pressure. Catal Commun 5:721–723
25. Gernaey KV, van Loosdrecht MCM, Henze M, Lind M, Jorgensen SB (2004) Activated sludge wastewater treatment plant modelling and simulation: state of the art. Environ Model Softw 19:763–783
26. Green diesel (2009) http://www.green-diesel.co.za/info_standards.htm. Accessed 20 Oct 2009
27. Grima (1994) Comparison between extraction of lipids and fatty acids from microalgal biomass. JAOCS 71(9):955–959
28. Grima ME, Belarbi EH, Fernandez FGA, Medina AR, Chisti Y (2003) Recovery of microalgal biomass and metabolites: process options and economics. Biotechnol Adv 20(7–8):491–515
29. Grima ME, Acién Fernández FG, García Camacho F, Camacho Rubio F, Chisti Y (2000) Scale-up of tubular photobioreactors. J Appl Phycol 12:355–368
30. Grobbelaar JU (2007) Photosynthetic characteristics of *Spirulina platensis* grown in commercial-scale open outdoor raceway ponds: what do the organisms tell us? J Appl Phycol 19:591–598
31. Gryglewicz S (1999) Rapeseed oil methyl esters preparation using heterogeneous catalysts. Bioresour Technol 70:249–253
32. Gudin C, Therpenier C (1986) Bioconversion of solar energy into organic chemicals by microalgae. Adv Biotechnol Process 6:73–110
33. Hama S, Yamaji H, Kaieda M, Oda M, Kondo A, Fukuda H (2004) Effect of fatty acid membrane composition on whole-cell biocatalysts for biodiesel-fuel production. Biochem Eng J 21:155–160
34. Harrison PJ, Berges JA (2005) Marine culture media. In: Andersen RA (ed) Algal culturing techniques. Elservier Academic press, London

35. Hsieh CH, Wu WT (2009) Cultivation of microalgae for oil production with a cultivation strategy of urea limitation. Bioresour Technol 100:3921–3926
36. Hu Q, Sommerfeld M, Jarvis E, Ghirardi M, Posewitz M, Seibert M, Darzins A (2008) Microalgal triacylglycerols as feedstocks for biofuel production: perspectives and advances. Plant J 54:621–639
37. Huber GW, Iborra S, Corma A (2006) Synthesis of transportation fuels from biomass: chemistry, catalysts, and engineering. Chem Rev 106:4044–4098
38. Kaplan D, Richmond AE, Dubinsky Z, Aaronson S (1986) Algal Nutrition. In: Richmond A (ed) Handbook of microalgal mass culture. CRC Press Inc, USA
39. Knothe G, Steidley KR (2005) Kinematic viscosity of biodiesel fuel components and related compounds. Influence of compound structure and comparison to petrodiesel fuel components. Fuel 84:1059–1065. doi:10.1016/j.fuel.2005.01.016
40. Knothe G, Van Gerpen J, Krahl J (2005) The biodiesel handbook. AOCS, Urbana
41. Kolber Z, Falkowski PG (1993) Use of active fluorescence to estimate phytoplankton photosynthesis in situ. Limnol Oceanogr 38:1646–1665
42. Kong Q, Li L, Martinez B, Chen P, Ruan R (2010) Culture of microalgae *Chlamydomonas reinhardtii* in wastewater for biomass feedstock production. Appl Biochem Biotechnol 160:9–18
43. Koopman B, Lincoln EP (1983) Autoflotation of algae from high-rate pond effluents. Agric Wastes 5(4):231–246
44. Krawczyk T (1996) Biodiesel—alternative fuel makes inroads but hurdles remain. Inform 7:801–829
45. Kromkamp JC, Forster RM (2003) The use of variable fluorescence measurements in aquatic ecosystems: differences between multiple and single turnover measuring protocols and suggested terminology. Eur J Phycol 38:103–112
46. Kromkamp J, Peene J (1999) Estimation of phytoplankton photosynthesis and nutrient limitation in the Eastern Scheldt estuary using variable fluorescence. Aquat Ecol 33:101–104
47. Chung Kyong Hwan, Kim Jin, Lee Ki-Young (2009) Biodiesel production by transesterification of duck tallow with methanol on alkali catalysts. Biomass Bioenergy 33(1):155–158
48. Leach G, Oliveira G, Morais R (1998) Spray-drying of *Dunaliella salina* to produce abcarotene rich powder. J Ind Microbiol Biotechnol 20(2):82–85
49. Li Y, Horsman M, Wu N, Lan CQ, Dubois-Calero N (2008) Biofuels from microalgae. Biotech Prog 24(4):815–820
50. Li Y, Horsman M, Wang B, Wu N, Lan CQ (2008) Effects of nitrogen sources on cell growth and lipid accumulation of green alga *Neochloris oleoabundans*. Appl Microbiol Biotechnol 81:629–636
51. Lima SAC, Raposo MFJ, Castro PML, Morais RM (2004) Biodegradation of p-chlorophenol by a microalgae consortium. Water Res 38:97–102
52. Ma F, Hanna MA (1999) Biodiesel production: a review. Bioresour Technol 70:1–15
53. MacIntyre HL, Cullen JJ (1996) Primary production by suspended and benthic microalgae in a turbid estuary: time-scales of variability in San Antonio Bay. Tex Mar Ecol Prog Ser 145:245–268
54. Marchetti JM, Miguel VU, Errazu AF (2007) Possible methods for biodiesel production. Renew Sustain Energy Rev 11:1300–1311
55. Masojidek J, Koblizek M, Torzillo G (2004) Photosynthesis in microalgae. In: Richmond A (ed) Handbook of microalgal culture: biotechnology and applied phycology. Blackwell Science Ltd, Oxford, pp 20–39
56. Meher LC, Vidya SD, Naik SN (2006) Technical aspects of biodiesel production by transesterification—a review. Renew Sust Energy Rev 10:248–268
57. Mohn FH (1988) Harvesting of micro-algal biomass. In: Borowitzka LJ, Borowitzka MA (eds) Micro-algal biotechnology. Cambridge University Press, Cambridge, pp 395–414
58. Muga HE, Mihelcic JR (2008) Sustainability of wastewater treatment technologies. J Environ Manage 88:437–447
59. Mulaku WO, Nyanchanga EN (2004) Dissolved air flotation process for algae removal in surface water treatment in Kenya. J Civil Eng Res Pract 1(2):27–38

60. Mulbry W, Kondrad S, Buyer J (2008) Treatment of dairy and swine manure effluents using freshwater algae: fatty acid content and composition of algal biomass at different manure loading rates. J Appl Phycol 20:1079–1085
61. Mutanda T, Karthikeyan S, Mustapha S, Bux F (2010) The utilisation of post-chlorinated municipal domestic wastewater for biomass and lipid production by *Chlorella* sp. under batch conditions. Biomass Bioenergy (in press)
62. Nindo CI, Tang J (2007) Refractance window dehydration technology: a novel contact drying method. Dry Technol 25:37–48
63. Noue J, Laliberte G, Proulx D (1992) Algae and wastewater. J Appl Phycol 4:247–254
64. Noureddini H, Gao X, Philkana RS (2005) Immobilized *Pseudomonas cepacia* lipase for biodiesel fuel production from soybean oil. Bioresour Technol 96:769–777
65. Nurdogan Y, Oswald WJ (1996) Tube settling rate of high-rate pond algae. Water Sci Technol 33:229–241
66. Oda M, Kaieda M, Hama S, Yamaji H, Kondo A, Izumoto E, Fukuda H (2004) Facilitatory effect of immobilized lipase-producing Rhizopus oryzae cells on acyl migration in biodiesel-fuel production. Biochem Eng J 23:45–51
67. Olivier S, Scragg AH, Morrison J (2003) The effect of chlorophenols on the growth of *Chlorella* VT-1. Enzym Microb Technol 32:837–842
68. Oswald WJ, Lee EW, Adan B, Yao KH (1978) New wastewater treatment method yields a harvest of saleable algae. WHO Chron 32:348–350
69. Paul PFM, Wise WS (1971) The principle of gas extraction. Mills Boon, London
70. Prakash J, Pushparaj B, Carlozzi P, Torzillo G, Montaini E, Materassi R (1997) Microalgal biomass drying by a simple solar device. Int J Solar Energy 18(4):303–311
71. Pushparaj B, Pelosi E, Tredici MR, Pinzani E, Materassi R (1997) An integrated culture system for outdoor production of microalgae and cyanobacteria. J Appl Phycol 9:113–119
72. Ramdhani N, Bux F (2007) Functional characterization of heterotrophic denitrifying bacteria in activated sludge. S Afr J Sci 103:113–116
73. Ramesh D, Samapathrajan A, Venkatachalam P (2005) Pilot biodiesel plant for vegetable oils. Periyar J Res Dev 3(1):15–19
74. Ramesh D, Samapathrajan A, Joshua Davidson S (2005) Fuel properties of palm oil and its biodiesel production. Periyar J Res Dev 2(3):25–29
75. Reid EE (1911) Studies in esterification. IV. The interdependence of limits as exemplified in the transformation of esters. Am Chem J 45:479–516
76. Republic of South Africa (1998) National Water Act. Act No 36 of 1998
77. Richmond A (2004) Handbook of microalgal culture: biotechnology and applied phycology. Blackwell Science Ltd, Oxford
78. Ryll T, Dutina G, Reyes A, Gunson J, Krummen L, Etcheverry T (2000) Performance of small-scale CHO perfusion cultures using an acoustic cell filtration device for cell retention: Characterization of separation efficiency and impact of perfusion on product quality. Biotechnol Bioengng 69:440–449
79. Schreiber U (1986) Detection of rapid induction kinetics with a new type of high-frequency modulated chlorophyll fluorometer. Photosynth Res 9:261–272
80. Serodio J, Vieira S, Cruz S, Barroso F (2005) Short-term variability in the photosynthetic activity of microphytobenthos as detected by measuring rapid light curves using variable fluorescence. Mar Biol 146:903–914
81. Sheehan J, Camobreco V, Duffield J, Graboski M, Shapouri H (1998) Life cycle inventory of biodiesel and petroleum diesel for use in an urban Bus. Final Report NREL/SR-580-24089. National Renewable Energy Laboratory, Golden, Colorado
82. Shelef GA, Sukenik A, Green M (1984) Microalgae harvesting and processing: a literature review report. Solar Energy Research Institute, Golden Colorado, SERI/STR-231-2396
83. Shieh CJ, Liao HF, Lee CC (2003) Optimization of lipase-catalyzed biodiesel by response surface methodology. Bioresour Technol 88:103–106
84. Simmons MS, Sivaborvorn K (1979) Effects of chlorinated organics from wastewater treatment on algal growth. Bull Environ Contam Toxicol 23:766–773

85. Sidat M, Kasan HC, Bux F (1999) Laboratory scale investigation of biological phosphate removal from municipal wastewater. Water SA 25(4):459–462
86. Srivastava A, Prasad R (2000) Triglycerides-based diesel fuels. J Renew Sustain Energy Rev 4:111–133
87. Teixeira MR, Rosa MJ (2006) Comparing dissolved air flotation and conventional sedimentation to remove cyanobacterial cells of *Microcystis aeruginosa*. Part 1: the key operating conditions. Sep Purif Technol 52(1):84–94
88. Torzillo G, Bernardini P, Masojidek J (1998) On-line monitoring of chlorophyll fluorescence to assess the extent of photoinhibition of photosynthesis induced by high oxygen concentration and low temperature and its effects on the productivity of outdoor cultures of *Spirulina platensis* (*Cyanobacteria*). J Phycol 34:504–510
89. Valderramma LT, Campo CMD, Rodriguez CM, de-Bashan LE, Bashan Y (2002) Treatment of recalcitrant wastewater from ethanol and citric acid production using the microalga *Chlorella vulgaris* and the macrophyte *Lemma minuscule*. Water Res 36:4185–4192
90. Van Gerpen J, Shanks B, Pruszko R, Clements D and Knothe G (2004) Biodiesel production technology. National Renewable Energy Laboratory. 1617 Cole Boulevard, Golden, CO. Paper contract No. DE-AC36-99-GO10337
91. Venkataraman LV (1978) New possibility for microalgae production and utilisation in India. Arch Hydrobiol Beih 11:199–210
92. Vonshak A (1997) *Spirulina*: growth, physiology and biochemisty. In: Vonshak A (ed) *Spirulina platensis* (*Arthrospira*): physiology, cell-biology and biochemistry. Taylor & Francis, London, pp 43–65
93. Wang B, Li Y, Wu N, Lan CQ (2008) CO_2 bio-mitigation using microalgae. Appl Microbiol Biotechnol 79(5):707–718
94. Widjaja A, Chien CC, Ju YH (2009) Study of increasing lipid production from fresh water microalgae *Chlorella vulgaris*. J Taiwan Inst Chem Eng 40(1):13–20
95. Wijffels RH (2007) Potential of sponges and microalgae for marine biotechnology. Trends Biotechnol 26(1):26–31
96. Miao Xiaoling, Li Rongxiu, Yao Hongyan (2009) Effective acid-catalyzed transesterification for biodiesel production. Energy Convers Manage 50(10):2680–2684
97. Xu H, Miao XL, Wu QY (2006) High quality biodiesel production from a microalga *Chlorella protothecoides* by heterotrophic growth in fermenters. J Biotechnol 126:499–507
98. Zhang Y, Dub MA, McLean DD, Kates M (2003) Biodiesel production from waste cooking oil: 2. Economic assessment and sensitivity analysis. Bioresour Technol 90:229–240
99. Zhou W, Boocock DBG (2006) Phase behavior of the base-catalyzed transesterification of soybean oil. JAOCS 83:1041–1045. doi:10.1007/s11746-006-5160-5

Antifungal Properties of Plant Extract and Density on Some Fungal Diseases and Yield of Cowpea

Gabriel Onyengecha Ihejirika

Abstract Cowpea (*Vigna unguiculata* (L) Walp) is one of the most important food legume crops grown in many tropical and subtropical countries. An experiment was carried out during the planting seasons of 2006 and 2007 respectively, to determine the antifungal properties of plant extract and plant density on some fungal diseases and yield of Ife brown 825–124 (erect type) cowpea. Analysis of variance indicated that plant density was highly significant on leafspot disease at 6 weeks after planting (0.71, 0.82) and at 8 weeks after planting (0.74, 0.82) at 5 % probability level in 2006 and 2007, respectively. Plant extract was significant on leafspot disease severity at 4 weeks after planting (0.43, 0.51) as well as 6 weeks after planting (0.051, 0.22) in 2006 and 2007, respectively. *Gongronema latifolium* treated plots recorded the lowest of all the diseases and seasons investigated. No sprayed plots (control), recorded the highest leafspot disease severity (10.3, 13.3) while, *G. latifolium* had very high antifungal (7.7, 6.10). *Vernonia amygdalina* recorded the lowest severity of blight (4.3, 5.5). 50 × 100 cm recorded the highest leafspot disease severity (10.3, 15.0) as well as blight (5.7, 5.1), while 100 × 100 cm had the lowest (7.0, 6.2) in 2006 and 2007, respectively. Similarly, 50 × 100 cm recorded the highest severity of blight (5.7, 5.1) with 100 × 100 cm recording lowest (4.0, 4.6). Investigation revealed that the severity of leafspot and blight diseases had a direct relation with plants' age. Interaction of *G. latifolium* and 100 × 100 cm, as well as interaction of no spray with 50 × 100 cm, recorded the lowest leafspot disease severity. Interaction of no spray with 50 × 100 cm recorded the lowest severity of leafspot disease (3.0, 4.2). *Gongronema latifolium* interaction with 100 × 100 cm, recorded the lowest occurrence of blight (3.0, 0.5). Similarly, interaction of *V. amygdalina* with 75 × 100 cm recorded the lowest severity of blight (0.3, 0.1), while that of no spray

G. O. Ihejirika (✉)
Department of Crop Science and Technology, Federal University of Technology, PMB 1526, Owerri, Nigeria
e-mail: ihejirikagabriel@yahoo.co.uk

A. Gurib-Fakim and J. N. Eloff (eds.), *Chemistry for Sustainable Development in Africa*, DOI: 10.1007/978-3-642-29642-0_5,

with 50 × 100 cm recorded the highest severity of blight (1.0, 1.5) in 2006 and 2007, respectively. *Vernonia amygdalina* interaction with spacing 50 × 100 cm recorded the highest stalk yield (119.7, 132.0). Investigation showed that closer spacing and no treatment increased the severity of the diseases on yield and performance of cowpea. The microorganisms identified with infected cowpea were *Aspergillus* species, *Penicillium* species, and *Rhizoctonia solani.*

1 Introduction

Cowpea [*Vigna unguiculata* (L) Walp] popularly known as bean is a tropical herbaceous, short stem annual grain legume which is cultivated in many tropical and subtropical countries. It belongs to the family *Leguminoseae* (Allen 1981). In Nigeria, cowpea is the most important indigenous grain legume. A total of 1,412,500 tons of cowpea were produced in the country for the 1989–1990 cropping season [4].

Cowpea is the most important food legume, especially for low income farmers in less developed tropical countries. It is used as food for man and feed for livestock, as cash crop, as well as for restoration of soil fertility [5]. The grain is a major source of protein in the diet and it helps to balance the diet for the majority of people. In addition, because the grain is widely treated out of the major production area, it is a cheap and nutritious food for relatively poor urban communities.

In fresh form, the young leaves, immature pods, and peas are used as vegetables, while several snacks and main meal dishes are prepared from the grain. All the edible parts of the plant are nutritious providing proteins, vitamins, and minerals. Petty trading of fresh produce and processed food provides both rural and urban opportunities for earning cash, particularly for women. They constitute a significant proportion of the total dietary protein and energy intake of Nigeria [11].

The cowpea in symbiotic association with the bacteria *Rhizobium japonica* infects the roots giving rise to a membranous structure called a nodule that houses the bacteria. This symbiotic association fixes a substantial amount of nitrogen to the soil and plant take up to 200–500 kg of N/ha/year which is vital for maintaining soil productivity over a long period.

Cowpea can also serve as a cover crop, smothering weeds, protecting the soil from raindrop impacts that would have caused surface sealing, crusting, compaction, and subsequent soil erosion. However, the call for self-sufficiency in cowpea production today requires that not only land area be brought under arable land but also that low input technologies affordable by resource poor farmers be evolved.

The production of cowpea is treated by biotic and abiotic factors, especially the menace of diseases and pests. Fungal and bacterial diseases are capable of causing a loss in grain yield of up to 100 % [9]. Crop losses of varying degrees due to specific fungal and bacterial diseases have been reported by various workers. [1, 18, 24]. Crop loss of up to 70 % was observed in the savannah region of Nigeria due to *Sphaceloma* Scab. Also, *Cercospora* and *Pseudocercospora* leaf-spot was listed as the major disease of cowpea [12]. Oputa [21] observed that plant population and nitrogen fertilizer could be adjusted to reduce the severity of leafspot disease and yield of groundnut.

Cowpea grown in the tropics are attacked by diseases such as bacteria blight (*Xanthomonas compestris*), bacteria wilt (*Pseudomonas syringae*), Anthracnose, spot, or leaf smut, Brown blotch, Brown rust, *Cercospora* leafspot, *Pseudocercospora* leafspot, powdery mildew, *Pythium* soft stem rot, *Septoria* leafspot, *Sphaceloma* scab, web blight, and Basal stem rot. The leafspot symptoms appear initially on the lower surface; dark red to black lesions occur along the veins on larger leaf veins with these lesions expanding into sunken cankers, within which *acervauli* bearing conidia are produced. Lesions also commonly develop on cotyledons, petioles, branches, stems, and pods. Seedlings that develop from infected seeds show severe symptoms [3].

Rust (*Uromyces appendiculatus*) occur worldwide wherever bean are grown. The extent of crop loss depends on the growth stage at which infection occurs, the susceptibility of the cultivars, and the environmental conditions dissemination occurs principally by means of wind borne *urediniospores*. Secondary dispersal of urediniospores is favored by humid, cloudy weather with heavy dew, and temperatures of 21–27 °C. Minute yellow raised spots appear on both surfaces of infected leaves as well as on petals and pods. The spots enlarge and rupture the epidermis to form reddish brown *uredial* pustules, which may be surrounded by yellow haloes and then by rings of smaller secondary pustules. Dry powdery spores are typical of rust fungi. As the infection ages, much of the leaves become chlorotic while the tissues colonized by the fungus remain green. While *Ascochyta* blight causes severe defoliation and lesions on stems and pods. The fungus is seed borne and also survives on bean straw. Secondary spread occurs through rain splash and infection depends on high relative humidity and cool temperatures (21–24 °C). Dark gray to black lesions appear on leaves and later become concentric ringed. These lesions develop also on petioles, stem nodes, peduncles, and pods and they can girdle stems and thus skill the plant. Flower infection can lead to stem-end rot of the pods, which can also be infected directly. Pod lesions expand and merge, causing significant damage [3]. Therefore, disease control strategies especially those that are effective, cheap, and environmentally nonhazardous, are needed. The presence of anti-fungal compounds in higher plants had been recognized long ago, as an important factor to disease management [19]. Similarly, Kuruchve et al. [17] reported that the extracts from plant parts were recommended for the control of disease.

Alabi [2] reported that the extracts from four botanical plants namely, *V. amygdalina*, *Bryophllum pinnatum*, *Eucalyptus globulus*, and *Ocimum gratissimum* increased plant height, plant shelf life, relative water content, chlorophyll contents of extract treated plants significantly ($P = 0.05$), when compared to their control during the planting seasons of 2000–2001.

O'Donovan [20] reported that crop density had a major effect on crop/weed interactions. Blumenthal and Ison [8] noted plant population density as a major determinant of agronomic success in annual crops in the sense that it determines the total dry matter yield. While Jagtap et al. [16] showed that optimum plant density is important for the maximum biological yield of crops.

Hence, the objectives of the research are to determine the antifungal properties of plant extract and plant density on some fungal diseases and yield of cowpea, as well as to assess the microorganisms associated with diseased cowpea.

2 Materials and Methods

The experiment was conducted in the research farm as well as the laboratory of the School of Agriculture and Agricultural Technology and the Federal University of Technology, Owerri, during the growing seasons of 2006 and 2007, respectively. Ife brown 825–124 (erect type) variety of cowpea was used for the experiment, while plant extracts used were extracts from Onugbu (*V. amygdalina*) and Utazi (*G. latifolium*). Land preparation was done by slashing using cutlass, followed by packing of debris, mapping out of plot, and making of ridges. Plots measuring 4 by 4 m were mapped out and ridges measuring 1.0 by 1.0 m were made, with 0.4 m as main plot gap and 0.2 m as subplot gap. Three levels of spacing were investigated upon and they included 50 $\times$ 100 cm, 75 $\times$ 100 cm, and 100 $\times$ 100 cm, respectively.

Preparation of the extract: Each of the plant materials was obtained and dried till they appeared crispy. Then they were separately ground with mortar and pestle to a fine powder and sieved. Ten grams of each of the ground plant powder were mixed in 1 litre of boiled water and the crude extract obtained was spread over a plot of 4 x 4 meters.

The experiment was laid out in a split plot design with plant extract as the main plot treatment while plant density constituted the subplot. The experiments consisted of $3 \times 3 = 9$ combinations. At 3 replications $= 3 \times 3 \times 3 = 27$ plots.

Data collection: Data were collected based on the following parameters: Number of leaves, leaf area and severity of leafspot, and blight diseases.

Number of leaves produced: This was obtained by counting the number of leaves produced per treatment one after the other and recorded.

Leaf area: The length and width of each sampled leaf were measured with a ruler and recorded at 2 weeks' interval and the area was determined by multiplying each sampled plant by a common factor.

Disease severity: The severity of leafspot and blight diseases was estimated using visual observation and scoring as proposed by Ford and Hewitt [13].

2.1 *Identification of Microorganism*

Preparation of Potato Dextrose Agar (PDA) medium: 250 g of Irish potato was washed and peeled with a knife. It was cut into small pieces and placed in a large beaker. 10 ml of water was added to it and the content was gently boiled for 30 minutes and allowed to cool and settle. Two hundred and fifty millilitres of the supernatant (potato broth) was poured into a 1000 ml conical flask. Then, 20 g of Agar powder was poured into the flask and 20 g of dextrose was added and made up to 1000 ml with distilled water. It was then steamed in an autoclave to melt the Agar resulting in the PDA.

Table 1 Analysis of variance for leafspot and blight diseases of cowpea on weeks after planting in 2006 and 2007

	Blight disease severity			Leafspot disease severity		
	Weeks after planting			Weeks after planting		
	4	6	8	4	6	8
2006						
Factor A	0.1	6.8*	6.7*	0.43	0.05*	0.51*
Block/plot SSK	0.2	0.48	0.7	0.28	1.15	0.6
Error (SSAK)	0.13	6.95	1.83	0.1	0.33	0.85
Factor B	6.24	2.25*	2.95*	0.03	0.71*	0.75*
Factor A and B interaction	0.16	0.36	0.7	0.21	0.38*	0.3
Error	0.18	6.77	0.61	0.09	0.1	0.51
TSS	0.15	0.81	1.1	0.11	0.75	0.63
2007						
Factor A	1.2	5.4*	7.51	0.51	0.22*	0.63*
Block/plot SSK	1.0	0.6	0.88	0.26	1.40	0.81
Error (SSAK)	0.42	7.77	2.14	0.2	0.30	0.92
Factor B	7.5	2.68*	3.62	0.08	0.82	0.82
Factor A and B interaction	0.24	0.52	0.91	0.3	0.48*	0.36*
Error	0.26	8.0	0.8	0.12	0.16	0.66
TSS	0.18	0.94	0.96	0.15	0.71	0.55

Key Factor A: Plant Extract; Factor B: Plant Density
*significant at 5 % probability level

Fifteen millilitres of the PDA was dispensed into each of the MacCartney bottles plugged with cotton wool and covered with aluminium foil and kept in the autoclave for 15 minutes at 121 °C for sterilization. The contents were allowed to cool to 45 – 50 °C and poured into sterile Petri dishes and allowed to solidify.

An advancing edge of leaf and stem was inoculated on the PDA plates and incubated at 25 °C for 4 days. The culture was then sub-cultured to obtain pure cultures of the different organisms.

Preparation of growth for microscopy: The colonies were merely touched with a fine wet inoculating needle and a wet mount made on clean microscope slides. The slides were viewed under a X40 high power microscope. Then the microorganisms were identified using laboratory guidelines according to Barnett and Hunter [7].

Data were analyzed using the methods of Statistical Analysis System [23] and means were separated using the Fisher's Protected Least Significant Difference (LSD).

3 Results

The result of the investigation revealed that plant extract was highly significant on leafspot disease severity at 6 weeks after planting (0.71, 0.82) and at 8 weeks after planting (0.74, 0.82). Plant density also was significant on leafspot disease severity

Table 2 Analysis of variance for number of leaves of cowpea on weeks after planting in 2006 and 2007

	2006			2007		
	Weeks after planting			Weeks after planting		
	2	4	6	2	4	6
Factor A	7.3	18.8*	974.4*	7.8	20.0*	966.2*
Block/plot SSK	22.5	19.2	95.5	24.5	22.5	98
Error (SSAK)	1.28	12.1	143.3	1.36	14.3	140
Factor B	1.15	16.5*	838.8*	1.22	15.8*	845.2*
Factor A and B interaction	2.9	13.6*	410.1*	3.5	14.0*	428.5*
Error	4.29	34.9	92.8	4.68	36.1	87.5
TSS	4.9	24.3	274.8	5.5	24.8	280.4

Key *significant at 5 % probability level

Table 3 Analysis of variance for leaf area on weeks after planting and stalk yield in 2006 and 2007

	Leaf area						Stalk yield	
	2006			2007			2006	2007
	Weeks after planting			Weeks after planting				
	2	4	6	2	4	6		
Factor A	2.79	11.5*	5.1	2.5	12.6*	7.2	1,235.1*	1,250.0*
Block/plot SSK	0.79	179.3	403.9	0.82	188	388.5	725.7	730.1
Error (SSAK)	27.5	56.8	461.6	24.8	54.5	464.1	1,431.2	1,442
Factor B	75.4	83.1*	103.7*	77	80.0*	110.5*	970.9*	966.2*
Factor A and B interaction	3.5	86.6*	243.9	3.85	84.5*	245.2	1,240.9	1,260.8
Error	29.2	1,531.6	269.6	27.5	1,510.6	277	2,181.6	2,200
TSS	24.4	746.2	272.4	25.1	755	280.4	1,643.5	1,652.1

Key *significant at 5 % probability level

at 6 weeks after planting (2.25, 2.68) and at 8 weeks after planting (2.95, 3.62) in 2006 and 2007 respectively, at 5 % probability level (Table 1).

Analysis of variance revealed that plant extract recorded the highest number of leaves at 4 weeks after planting (18.8, 20.0) and at 6 weeks after planting (974.4, 966.2). Also, plant density had a highly significant influence on the number of leaves produced by cowpea at 4 weeks after planting (16.5, 15.8) and at 6 weeks after planting (838.8, 845.2) at 5 % probability level in 2006 and 2007 respectively (Table 2).

Table 2 also shows that interaction of plant extract and plant density had influence on the number of leaves produced at 4 and 6 weeks after planting (13.6, 14.0); (410.1, 428.5) in 2006 and 2007, respectively. Plant extract recorded significant difference on leaf area at 4 weeks after planting (11.5, 12.6) as well as plant density at 4 weeks (83.1, 80.0) and at 6 weeks (103.7, 110.5) after planting, respectively. Also, interaction of plant extract and plant density was significant on leaf area at

Table 4 Mean of main effects of leafspot, blight diseases, and yield in 2006 and 2007

	Leafspot disease		Blight disease		Stalk yield	
	2006	2007	2006	2007	2006	2007
No extract (control)	10.3	13.3	8.7	14.8	220.5	240.3
V. amygdalina	8.3	6.5	4.3	5.5	268.5	288.2
G. latifolium	7.7	6.1	5.7	6.2	200.1	246.5
$LSD_{0.05}$	10.07	8.25	1.51	0.98	25.46	28.11
(50 × 100) cm	10.3	25.2	5.7	5.1	257.9	280.2
(75 × 100) cm	8.1	10.4	5.2	5.6	196.2	220.1
(100 × 100) cm	7.2	6.2	4.1	4.6	235.1	238.5
$LSD_{0.05}$	6.28	5.25	0.92	0.55	38.82	30.14

Table 5 Mean of main effects of plant extract and density on yield components of cowpea in 2006 and 2007

	Number of leaves produced		Leaf area produced	
	2006	2007	2006	2007
No extract (control)	112.1	108.4	108.6	104.5
V. amygdalina	153.1	166.2	157.1	175.5
G. latifolium	141.2	153.6	154.2	162.3
$LSD_{0.05}$	25.5	30.1	82.1	63.1
(50 × 100) cm	164.3	188.5	168.2	192.5
(75 × 100) cm	59.7	80.1	148.3	138.2
(100 × 100) cm	112.2	128.1	153.3	156.1
$LSD_{0.05}$	16.5	22.6	47.9	36.8

Table 6 Mean of interaction effects of plant extract and density on leafspot, blight. and stalk yield in 2006 and 2007

	Leafspot disease		Blight disease		Stalk yield	
	2006	2007	2006	2007	2006	2007
Control and 50 × 100 cm	3.1	4.2	1.1	1.5	63.5	65.5
Control and 75 × 100 cm	2.3	2.6	0.7	0.5	80.8	92.5
Control and 100 × 100 cm	1.3	1.1	0.7	0.6	76.2	86.2
V. amygdalina and 50 × 100 cm	2.2	2.6	1.3	1.2	119.7	132.1
V. amygdalina and 75 × 100 cm	1.7	1.9	0.3	0.1	94.1	100.5
V. amygdalina and 100 × 100 cm	1.3	1.5	1.4	1.1	66.8	80.2
G. latifolium and 50 × 100 cm	2.3	2.1	0.7	0.8	75.7	78.1
G. latifolium and 75 × 100 cm	1.7	1.8	1.2	1.3	59.6	62.5
G. latifolium and 100 × 100	1.3	1.1	0.3	0.5	64.8	62.8
$LSD_{0.05}$Factor A	1.69	1.75	0.59	0.66	25.46	28.14
$LSD_{0.05}$Factor B	1.37	1.16	1.94	2.11	38.82	35.12

Key V.a = *Vernonia amygdalina*

4 weeks after planting (86.6, 84.5) and at 6 weeks after planting (243.9, 245.2). Results showed that plant extract and density interaction were significant on stalk yield (1240.9, 1260.8) in 2007 and 2007, respectively (Table 3).

50 × 100 cm recorded highest number of leaves (164.3, 188.5) and the same was observed on leaf area (168.0, 192.5) followed by 100 × 100 cm (112.0, 128.1) as well as leaf area (153.3, 156.0) in 2006 and 2007, respectively. The mean value of main effect indicated that *V. amygdalina* recorded higher leaf area (157.1, 175.5) than *G. latifolium* (154.2, 162.3) when no spray (control), recorded the lowest (108.6, 104.5) (Table 4).

No spray (control), recorded the highest leafspot disease (10.3, 13.3), as well as the highest blight (8.7, 14.8), when *G. latifolium* had the lowest leafspot disease (7.7, 6.1). *Vernonia amygdalina* recorded the lowest severity of blight disease (4.3, 5.5) in 2006 and 2007, respectively (Table 5).

50 × 100 cm recorded highest leafspot disease severity (10.3, 25.0), while 100 × 100 cm was lowest (7.0, 6.2). Similarly, 50 × 100 cm recorded highest blight (5.7, 5.1), while 100 × 100 cm was lowest (4.0, 4.6) (Table 5).

Results indicated that interaction of *G. latifolium* with 100 × 100 cm recorded the lowest severity of leafspot disease while interaction of no spray and 50 × 100 cm recorded high leafspot disease severity (3.0, 4.2). Also, *G. latifolium* interaction with 100 × 100 cm recorded the lowest blight severity (0.3, 0.5). No spray interaction with 50 × 100 cm recorded highest blight disease severity also, while *V. amygdalina* interaction with 50 × 100 cm had the highest stalk yield (119.7, 132.0) (Table 6).

Results showed that closer spacing and no extract increased the effect of all the diseases investigated on the yield and performance of cowpea. The microorganisms identified were *Aspergillus, Penicillium, and Rhizoctonia* species respectively.

4 Discussion

Plant spacing was inversely related to leafspot and blight disease development. This is because, the more the crowded nature of the plants, the less the nutrients available per plant and hence the lower the biochemical processes and the ability to withstand disease attack reduced, hence high disease severity in agreement with FAO [10], which proposed that disease development is formed by warm humid conditions caused by closed spacing.

Gongronema latifolium recorded the least severity of leafspot; this may be as a result of the fact that *G. latifolium* contains higher chemical composition than *V. amygdalina*. This is in line with Alabi [2], who reported that extracts from four botanical plants significantly reduced disease infection rate, transpiration rate, and stomatal aperture of cowpea. No spray plots (control) was the most affected with leafspot and its interaction with 50 × 100 cm. This is in line with Hume [14], who proposed that crop density had a major effect on crop/weed interaction. Thus, interaction development *G. latifolium* with 100 × 100 cm recorded the highest resistance to leafspot and blight disease infection, because of induced resistance through high soil nutrient and low humid conditions.

The direct relation between the number of leaves produced as well as the leaf area of cowpea within weeks after planting may be attributed to the fact that increase in physiological and metabolic activities of the plant led to cell and tissue development, thereby resulting in increase in size as well as more photosynthetic activities, hence leaf production. This is in line with Ihejirika and Nwufo [15], who proposed that plant height and leaf production increased with plant age. Also, direct relation of leafspot and blight disease severity with plant age may be attributed to the fact that as a plant ages, the tissues become weak and the plant's ability to withstand the attack of pathogens is reduced, leading to increased disease penetration and symptom manifestation and spread.

The significant difference observed in plant extract on the severity of these diseases investigated may be attributed to the effectiveness of the plant extracts as fungicides. This might be due to the chemical constituents of the plants, which disrupt the normal metabolic activities of the pathogen as well as act as antifeedants, repellence, and insecticidal properties on insect vectors of plant pathogen, in agreement with Graigne et. al. (1985).

Aspergillus, Penicillium, and Rhizoctonia species were identified to be associated with diseased cowpea. This is due to the fact that these microorganisms have been identified to be responsible for the disease development and spoilage of tropical crops, in line with Bankole and Adebanjo [6], and Richardson [22].

In conclusion, G. *latifolium* treated plots recorded the lowest of all the diseases while no sprayed plots (control), recorded the highest. Also, 50 × 100 cm recorded the highest in all the diseases while 100 × 100 cm had the lowest in all the seasons investigated. *Aspergillus* species, *Penicillium* species, and *Rhizoctonia solani* were associated with diseased cowpea.

References

1. Alabi O (1994) Epidemiology of cowpea blown blotch induced by *Colletotricum capsicis* and assessment crop losses due to disease. J Sustain Agric Environ 4:5–9
2. Alabi DA (1999) Effect of acetyl salicylic acid on distribution and total photosynthates in three cultivars of tomato. Niger J Sci 2(1):101–103
3. Allen DJ, Ampofo JKO, Wortman CS (1996) Pests, diseases and nutritional disorders of the common bean in Africa. Field Guide 4:49–59
4. Anon O (1990) Annual report of the Agricultural Products Monitoring and Evaluation Unit (APRAEU) of the Federal Ministry of Agriculture, Lagos, Nigeria, 125 pp
5. Awurum AM, Emechebe AM, Amadioha AC (2001) Effect of cropping system on disease development in cowpea in Umudike, South Western Nigeria. J Sustain Agric Environ 2:25–28
6. Bankole SA, Adebanjo A (2003) Aflatoxin contamination of dried yam chips marketed in Nigeria. Trop Sci 43:201–203
7. Barnett HL, Hunter BB (1998) Descriptions and illustrations of genera. Illustrated genera of imperfect fungi. The American Phytopathological Society press, St. Paul, pp 59–218

8. Blumenthal MJ, Ison RL (1994) Plant population dynamics. Insuerranean Colver and Murex Medic Swards, 60 pp
9. Emechebe AM, Shoyinko SA (1985) Fungal and bacterial diseases of cowpea in Africa. In Cowpea research, production and utilization, pp 173
10. FAO (1990) Food and Agricultural Organization of the United Nations. Bot Maize Level Crop Corn 10:2–5
11. Fetuga BL, Ologhbo AD (1987) Energy values in differently processed cowpeas. Nig Food J 5:18–23
12. Florini DA (1997) Nematodes and other soil-borne pathogens of cowpea. In: Singh BB, Mohana Raj, Dashiell KE, Jackai LEN (eds) Advances in cowpea research. Center for Agricultural Science (JIRCAS). IITA, Ibadan, pp 193–206
13. Ford JE, Hewitt D (1980) In: Bond DA (ed) Vicia Faba feeding values, processing and viruses. L Martinus Nihoff, The Hague, pp 125–139
14. Hume 1 (1985) Crop losses in wheat (*Triticum aestivum*). As determined using weeded and non-weeded quadrats. Weed Sci 33:734–740
15. Ihejirika GO, Nwufo MI (2001) Effects of plant population and nitrogen application on severity of leafspot disease, growth and yield of Benniseed. (*Sesamum indicum*). Afr J Sci Technol 2(1, 2):203–207
16. Jagtap SS, Alabi RT, Adelege O (1998) The influence of maize density or resources use and productivity. Afr Crop Sci J 6:259–272
17. Kuruchve V, Ezhilan JG, Jayam J (1997) Screening of higher plants for fungi toxicity against Rhizoctonia solani in vitro. Indian Phytopathol 50(2):235–241
18. Ltunde-Dada AO (1990) Assessmant of Antracnose disease in some cultivars of cowpea. J Phytopathol 130(2):147–156
19. Mahadeum A (1982) Biochemical aspects of plant disease resistance. Annu Plant Rev 10:125–129
20. O'Donovan JT (1994) Canola (*Brassica rapa*) plant density influence tartary buck wheat (*Fagoprum tatarium*) interference, biomass, and seed yield. Weed Sci 42:385–389
21. Oputa EE (2004) Effect of NPK fertilizer and plant population on severity of leafspot disease and yield of groundnut. Department of Crop Science & Technology, Federal University of Technology, Owerri, Nigeria, 45 pp
22. Richardson MJ (1990) An annotated list of seed—borne diseases, 4th edn. Th International Seed Testing Association, Zurich
23. SAS (1999) Statistical analysis system: user's guide. Statistics SAS Institute Inc., Cary
24. Stoffella PJ, Bullock RC, Sonoda KM (1990) Influence of pesticide spray schedules on the yield of cowpea. Trop Hortic 34:83–87

Part II
Emerging Areas and Technologies

6. Mammino: Promoting the development of computational chemistry research: Motivations, challenges, options and perspectives (31 MS pages)
7. Meck: Geochemistry for sustainable development in Africa: Zimbabwe case study (22 MS pages)
8. Musee: Relevance of nanotechnology to Africa: Synthesis, applications and safety (62 pages)
9. Simate: Biotechnology and nanotechnology—a means for sustainable development in Africa (40 MS pages)

Promoting the Development of Computational Chemistry Research: Motivations, Challenges, Options and Perspectives

L. Mammino

Abstract Computational chemistry is a fast growing area of modern chemistry, capable of interfacing with the other research areas in chemistry and with other sciences involving consideration of substances and materials, and enjoying increasing industrial relevance. Its presence in Sub-Sahara African tertiary institutions is still scarce, mostly because of scarcity of experts. This chapter analyses the current situation, discusses the importance of developing it and the relevance of such development for research and education, outlines its relevance for sustainable development, offers reflections for possible development pathways and a feasibility assessment based on the concrete experience of its recent development, *ex novo*, in an underprivileged university in South Africa.

1 Introduction

The significance and roles of theoretical chemistry and computational chemistry (its applied component) are easily evident from a brief consideration of the nature and roles of chemistry. Chemistry is the science of substances. It studies everything concerning substances: their composition, properties and behaviour; the pathways for their production; their possible utilizations. It is at the basis of the chemical industry—the huge industrial sector devoted to the design and production of substances. Lots of substances are utilised in everyday life for a variety of purposes. The objectives and the ways in which different substances are utilised depend closely on their properties. The properties of a substance determine for

L. Mammino (✉)
Department of Chemistry, University of Venda, Thohoyandou, South Africa
e-mail: Liliana.Mammino@univen.ac.za

A. Gurib-Fakim and J. N. Eloff (eds.), *Chemistry for Sustainable Development in Africa*, DOI: 10.1007/978-3-642-29642-0_6,

what it can be used, in which ways it has to be used, how much of it can be utilised for a given purpose or at a given moment, which precautions one might need to adopt on using it, etc. The search for new substances with specific properties is continuous, to respond to acknowledged needs: the need for new drugs to treat diseases; the need for new materials to store energy, or to store information, increasingly more effectively; the need for ingredients giving desired properties to the materials used in the construction industry; the need for health-friendly and environmental-friendly substances for a variety of uses, from paints to agro-chemicals; and many more.

The properties of a substance depend on the properties of its molecules. Therefore, the more we know and understand about the properties of molecules, the more we understand about the properties of existing substances, or we are able to predict the properties of substances that have not yet been synthesised. Computational chemistry studies the properties of molecules, utilising the models developed within theoretical chemistry. The fast increase of computers' powers in recent decades has enabled increasing sophistication of the computational approaches, thus continuously enhancing the interpretation and prediction abilities of computational results. This, in turn, has expanded the roles of computational chemistry beyond the domain of academic research, making its results important for industrial applications, from the design of new substances (drug design, design of new catalysts, etc.) to novel areas such as nanotechnologies.

Computational chemistry research is currently active in most countries, including developing countries in Asia, Latin America and Northern Africa. The recognition of its importance for a better understanding of the properties of substances and materials has made it enter typically technological institutes (e.g., the Central Leather Research Institute in Chennai, India, under the auspices of the Council for Scientific and Industrial Research) as well as the research sections of a number of industrial companies. However, computational chemistry research is not yet active in many countries in Sub-Sahara African. This chapter aims at presenting the motivations for an effective development of computational chemistry, spreading to as many institutions as possible. After a brief review of the nature and significance of computational chemistry (Sect. 2), the next sections:

- outline the importance and the expected benefits of fostering its development in Sub-Sahara African institutions;
- discuss the challenges to be addressed for effective development;
- recall expressions of concerns and initiatives aimed at fostering the development;
- and suggest options apt to combine feasibility within current contextual situations with the obtainment of desirable results.

The case of the development, in recent years, of computational chemistry research at the University of Venda (UNIVEN, an underprivileged institution in South Africa) is utilised as illustrative example in the discussion, as it can be viewed as a successful story from a "typically African" institution.

2 The Properties of Substances and the Properties of Their Molecules

As mentioned earlier, the properties of substances depend on the properties of their molecules. The major challenge of chemistry theory and interpretation is that of understanding the relationships between the two sets of properties. Pursuing this objective requires sufficiently reliable knowledge of the properties of molecules as a prerequisite stage.

The objective of relating the properties of substances to the properties of their constituting particles dates back to the first particulate models of matter, as far away in time as the models proposed by ancient Greek atomistic philosophers. Throughout many centuries, only mechanical properties and actions were known and, therefore, attempts to imagine relationships between the properties of substances and the properties of their particles were developed in terms of hypothesised shapes and mutual shape adaptabilities. Shapes were hypothesised from known macroscopic properties: e.g. needle-like shapes for the particles of acids, to account for their ability to attack and "demolish" other substances and hollow shapes for the particles of bases, into which the needle-like acid particles could fit, to account for acid–base neutralization. It was however clear that, without any possibility of actual verification, hypotheses about particle shapes did not have much probability of corresponding to reality. As the role of mathematics in physics increased up to becoming a fundamental tool both for the description of systems and phenomena and for the generation of new information, scientists started perceiving that mathematics (a new mathematics, still to be developed) will be the key to the description of the microscopic world. As early as 1765, Buffon was confident that:

> Posterity, with the aid of calculations, will be able to open this new field of knowledge and ascertain the shapes of the constituents of bodies with considerable precision.

Two decades later, Macquer expressed the same hope with explicit reference to the integration of mathematical and chemical knowledge [1]:

> Maybe time, experience, the enhancement of chemical knowledge, and the dedicated activity of persons who are expert in mathematics and in chemistry will, in future, bring much more light on these issues, of which now we have only confused notions.

Mathematics developed and, when classical physics encountered the "crisis" of not being able to account for the behaviour of the microscopic world, a new physics—quantum mechanics—was born and developed in the first decades of the twentieth century. However, there was still a long way before the study of molecules became a realistic possibility, because molecules are complex systems and their complexity poses great demands on the mathematical approaches. It became necessary to develop approximation options with sufficiently realistic physical meaning to enable adequate reliability of the description of molecular systems, while simplifying the computational part sufficiently to make it feasible. The necessity of finding appropriate approximation methods had been clearly expressed by Dirac [2, 3]:

> It therefore becomes desirable that approximate practical methods of applying quantum mechanics should be developed, which can lead to an explanation.

Modern computational chemistry constitutes the current state-of-the-art progress for the realisation of these objectives.

Thinking in terms of relationships between molecular features and substance behaviour is actually an all-permeating feature of the entire chemical approach. It is embedded in the content of any chemistry course, including the most basic ones. Simple examples are the expectation of acidic behaviour when an H atom is bonded to a VII group nonmetal or to an O atom bonded to a nonmetal, or the expectation of Lewis-base behaviour when an atom has a lone electron pair. Other relationships are encountered within more specific branches of chemistry: for instance, several aspects of the behaviour of an organic compound can be predicted on the basis of the presence of specific groups of atoms in its molecule (functional groups). All these predictions are of a rather "gross" character, since they consider only some aspects of the molecular structure and, therefore, refer to categories of compounds and properties that are common to a whole category, rather than to the specificities of each individual compound. The behaviour of individual substances depends on the details of the molecular structure. An impressive example is offered by enantiomers, where the fact that two molecules are mirror images of each other may generate significant differences in some properties: these may concern just the smell, like in the case of limonene; but it may also concern something so important as the biological activity, like in the case of thalidomide, where one enantiomer is effective against the morning nausea in pregnancy, while the other enantiomer is teratogenic, i.e. causes malformations in a developing foetus.

When important properties such as the biological activity are of interest, it becomes necessary to consider all the finest details of the molecular structure and of the properties associated with it, such as the charge distribution, the dipole moment, and others. For instance, the anticancer activity of anthracyclines depends on the dipole moment of the molecule and decreases as the dipole moment increases [4]; therefore, only molecular structures with dipole moment in the range favourable for anticancer activity would be considered interesting for synthesis and testing. As the number of known molecular properties increases, the probability that predictions of the substance behaviour may be realistic also increases. Refinements in the calculation of molecular properties enable refinements in the predictions, to better approach the details of the actual behaviour of individual compounds. When a substance already exists, the comparison between the computational predictions of its properties and the experimental observations is part of the interpretation activity: it contributes to validate both the prediction ability of a given computational approach and the interpretation of experimental results. When a substance has not yet been synthesised, predicting its properties is important to decide whether the actual synthesis is interesting or not—an issue that has conspicuous economic implications, because of the high costs of synthesis. The computational study of molecules is fundamental for the prediction of the

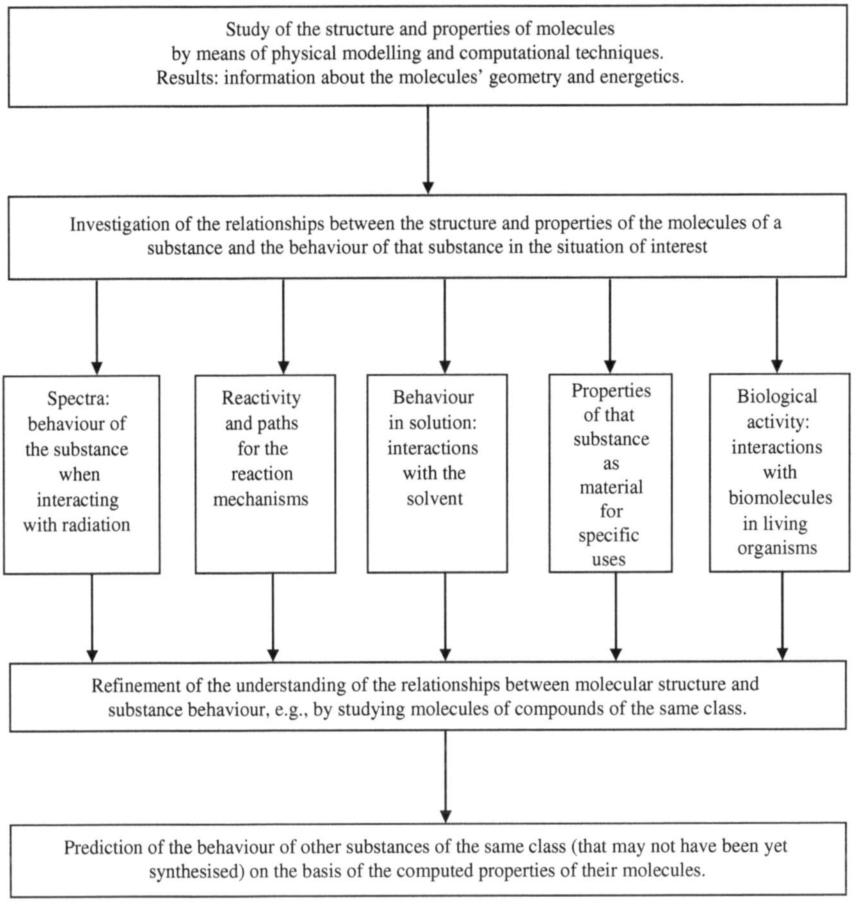

Fig. 1 A simple outline of the computational approach from the study of the properties of substances to the prediction of the properties of new substances

behaviour of not yet synthesised substances and, therefore, for the design of new substances responding to specific needs. Figure 1 provides a simple overview of what computational chemistry can do for the study of the properties of substances.

3 Developing Computational Chemistry Research in African Universities

3.1 The Current Situation: A Gap With the Other Continents

In the other continents, computational chemistry research is active and computational chemistry is becoming an integral component of the preparation of chemists in various core chemistry areas (physical chemistry, organic chemistry) as well as for biochemistry, pharmaceutical chemistry, chemistry of materials and other areas with industrial relevance. The increasing role of molecular studies in modern chemistry is clearly shown by the high number of published studies on molecular calculations and their applications (including the frequent mention of computationally obtained information in the literature devoted to material sciences, pharmacology, biochemistry and life sciences), by the growth of journals specifically devoted to individual application fields (e.g., QSAR, Computer-Aided Drug Design, etc.), by the presence of computational chemistry sessions in major chemistry conferences and by the fast increase of industrial applications in areas such as drug design, the development of efficient catalysts, the prediction of thermodynamic and kinetic data relevant to process design, the study and prediction of the properties of materials [5].

In many Sub-Sahara African institutions, the presence of computational chemistry is not yet adequate for a variety of reasons, first of all the scarcity of specialists who can engage in research and familiarise the younger generation with this field. In a number of institutions, and even in a number of countries, it is still totally absent, including countries with prestigious research records in many other fields, such as Kenya, Tanzania and several others. As computational chemistry continues to grow in the other continents, the gap between Sub-Sahara African and the other continents continues widening, also in comparison with other developing countries (e.g. the number of research outputs in molecular studies in Latin American and Asian countries has increased rapidly, and molecular calculations constitute an established component of research activities in most institutions). The gap concerns both the research level and the educational level. The extent of the gap at educational level is better perceived by considering that, in other contexts, the introduction of computational approaches into the university undergraduate curriculum started in the late 1980s and early 1990s of the twentieth century [6–10] and that a basic introduction within secondary schools is currently the object of pioneering activities [11, 12].

Besides the scarcity of specialists (by far the major cause behind the gap), other aspects may have contributed to the so far inadequate development of computational chemistry. Financial constraints are obviously an important cause. The difficulties toward funds granting were enhanced by inadequate general information on the role of molecular studies. For instance, the fact that, until not so long ago, the term *computational chemistry* was not in general use, and molecular studies went altogether under the *theoretical chemistry* term, generated some misperceptions—the perception that these areas of investigation cannot be relevant for Africa, where priority must be given to applications, to things that give immediately visible results, and not to the development of theory. This contributes to highlight the importance of disseminating information about the roles of computational chemistry, and also about the gap with other developing contexts (Asia, Latin America, Northern Africa), to help dispel the perception that this type of investigation is not suitable for developing institutions, but only for "first world" contexts.

3.2 *The Expression of Concern and the Theoretical Chemistry Workshops in Africa*

The awareness of the gap, and of the need to try and address the problem, has increased since the early 1990s. Concern about the situation of theoretical/computational chemistry, and physical chemistry in general, in African universities was formally expressed at the *Fifth International Chemistry Conference in Africa (ICCA)* in Gaborone (Botswana) in July 1992, on initiative by Prof. Mjojo (then at the University of Malawi), promptly joined by other interested participants. On that occasion, three participants (Prof. Geoffrey Kamau, of the University of Nairobi, Prof. Pierre Claver Karenzi, of the University of Rwanda (who later remained a victim of the Rwanda genocide) and Prof. Liliana Mammino, then at the University of Zambia) decided to try and establish some initiatives to disseminate information about computational chemistry and its potentialities for African universities, and to explore ways of fostering research initialisation. The decision marked the birth of the *Theoretical Chemistry Workshops in Africa (TCWA)*. It is interesting to note that, while for Prof. P. C. Karenzi and Prof. L. Mammino, theoretical/computational chemistry was the area of expertise, Prof. G. Kamau is a specialist in a different area of chemistry, and his enthusiastic support and leading organisation roles show a recognition of the importance of developing computational chemistry that overcomes the boundaries of personal expertise to think and act in favour of the development of modern chemistry as a whole in African universities.

Thanks to Prof. Kamau's leading organisation role and the prompt support by his colleagues from the University of Nairobi, the first three workshops were held in Nairobi (Kenya) in the years 1995 (February 22–26), 1996 (August 25–29) and 1998 (November 2–6), respectively. The growth of an international group of African chemists supporting the initiative prompted a rotation of the venue for the subsequent workshops. The fourth workshop was held in Addis Ababa (Ethiopia), 5–9 November 2001, the fifth in Dar-es-Salaam (Tanzania), 1–5 December 2003, the sixth in Windhoek (Namibia), 5–9 December 2005. In 2007, the initiative had sufficiently "grown" for the workshop to become a *Conference* (TCCA) that was held in Victoria Falls (Zimbabwe), 3–7 December. In October 2009, it was held jointly with the Conference of the Kenyan Chemical Society in Mombasa and in October 2011 it was held jointly with the Second Tanzania Chemical Society International Conference the in Dar es Salaam. From their third edition, the TCWA were held jointly with the Eastern and Southern Africa Environmental Chemistry Workshops (ESAECW), what fostered exchanges of views beyond individual research areas, taking into account the perspectives of the overall development of chemistry and of the role of chemistry for sustainable development, and also favoured some cross-discipline explorations [13].

Participation in the workshops has involved chemists from several countries and from various expertise backgrounds (not only the few theoretical/computational chemists available), and has enabled valuable exchanges of views and mutual updating. However, the establishing of research activities and students' training has not yet gained sufficient momentum. An analysis of this situation during the Conference in Victoria Falls suggested the opportunity to expand the activity to other initiatives, to be conducted in the periods between conferences and to be more specifically focused on training and research initialisation. Explorations in this regard are currently in progress (the main obstacle to implementation being the usual financial constrains).

3.3 Significance of Developing Computational Chemistry

3.3.1 Significance for Research Capacity Building

Developing computational chemistry activities can be considered one of the important tasks facing chemistry and chemists in Sub-Sahara African. It is not only a question of reducing the gap with the other continents. The most important aspect is the benefits that can derive from the development. The development obviously needs to take into account both educational aspects and the establishing of research. The two components (research and education) are interdependent: the essential basis for the development of a research area is the preparation of specialists, and the presence of postgraduate students is essential for carrying out research activities in the given area.

The potential issues of interest for the initiation or expansion of research activities in theoretical and computational chemistry are many and diverse, and the selection can be linked to the more relevant needs in a given community, e.g. by relating it to the study of perspective drugs for the treatment of endemic diseases, or to the requirements of established or taking-off industrial activities (substances/materials design). The study of biologically active compounds with potentialities for drug development is given particular attention in the current discussion, as a suitable example for illustrations and as one of the areas whose development can be considered particularly important and prospective in the African context.

The study of drugs for the treatment of endemic diseases can ideally integrate with the study of indigenous natural products and traditionally utilised medicines, bringing a wealth of benefits [13, 14]. The investigation of natural products to discover new lead compounds for the development of drugs is a major endeavour of pharmaceutical research, because of the challenges posed both by new endemic diseases (such as HIV/AIDS) and by the fast-developing resistance to commonly utilised drugs for *older* diseases such as malaria or tuberculosis. Natural products are the richest and most prospective sources of lead compounds for drug development, and the knowledge accumulated through many centuries by traditional medicine is rightly expected to provide precious indications on optimal sources. For this reason, the study of natural products and traditional remedies is developing at a fast rate in all continents [15–17] and is producing valuable results; an important example is offered by artemisinin, currently the most effective antimalarial drug, derived from *Artemisia annua*, a plant utilised in Chinese traditional medicine. Moreover, the interest in lead compounds of natural origin by the pharmaceutical industry is expected to further increase in the next years, after the lower-than-expectations performance of approaches such as combinatorial chemistry or high-throughput screening contributed to emphasise the realisation that compounds of natural origin have a higher probability of being compatible with living organisms and, therefore, of reaching their target within the organism and exerting the expected activity.

On the other hand, too often the part of the study of products from natural origin performed in developing countries is limited to the isolation and identification of the active compounds, whereas further development is continued elsewhere, in contexts with more advanced facilities and more extensive support for research. Because of historical reasons, this problem concerns Africa to a higher extent than other developing areas. The development of theoretical and computational chemistry can play important roles in retaining additional components of the drug development research in the African continent, above all when the lead compounds are isolated from indigenous natural products. This, in turn, can contribute to foster other benefits, like:

- emphasising the importance of the indigenous biodiversity (an aspect with relevant links to overall sustainable development perspectives);
- expanding the opportunities to focus drug development research on potential drugs for the treatment of endemic diseases (some of which are still "neglected diseases");
- playing a driving role for the enhancement of research in other areas of chemistry. By focusing on the study of molecules and their properties, theoretical/computational chemistry is at the core of chemical thought and can interface with all the other areas of chemical research [14, 18, 19], providing interactions that are valuable both for the computational chemists and for the experimental ones. Figure 2 highlights interfacing pathways between computational chemistry and experimental research for the study of biologically active compounds of natural origin.

The timing to actively engage in the development of computational chemistry appears particularly ripe. The interest toward it is rising in several African countries, as became evident at the TCCA in Victoria Falls (December 2007). Young chemists are increasingly realising that publications on the investigation of new compounds, from institutions in other continents, include the computational investigation and, therefore, they realise the growing relevance of computational chemistry for all the branches of chemical research, and their interest in getting opportunities to learn more about it increases.

Finally, it may be appropriate to devote some deeper reflections to the issue of developing highly specialised research areas in institutions with limited resources (as is the case for several institutions in the continent). For institutions with limited resources, it may not be easy or affordable to develop all research areas; moreover, it may not be interesting to duplicate some of the activities that are already present in other institutions with more facilities and funds. Under such conditions, development plans might follow two major directions: research areas that are more

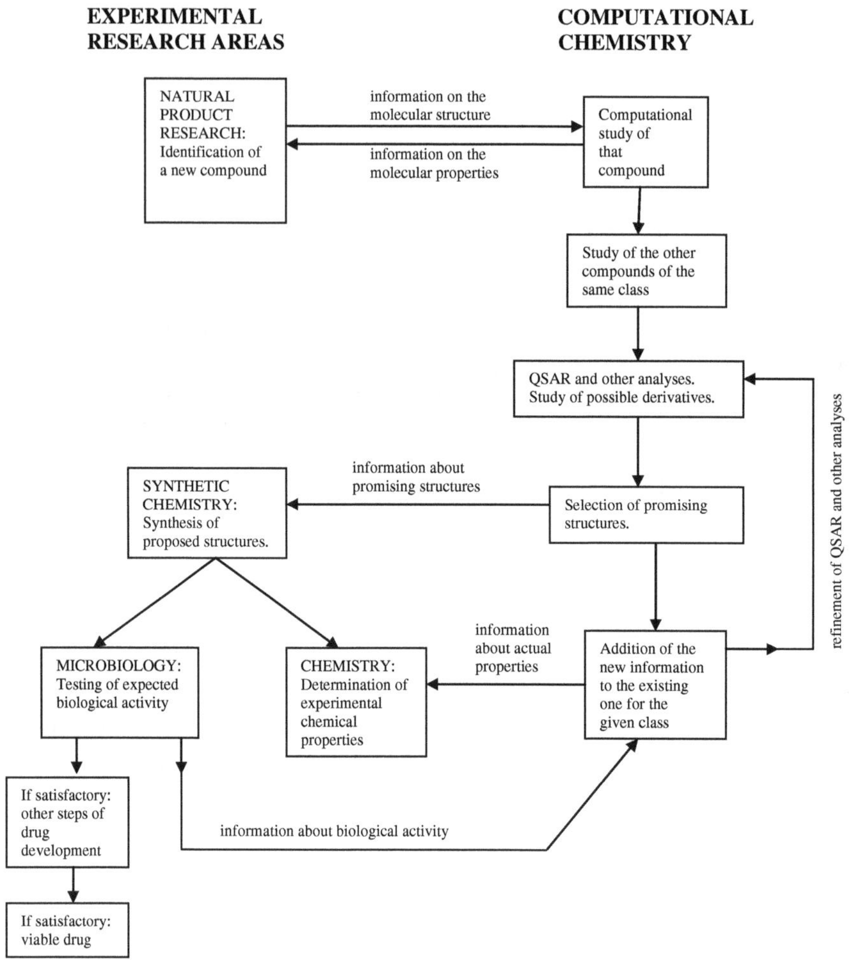

Fig. 2 Outline of possible interfaces between computational chemistry, other branches of chemistry and other sciences involved in drug development, for the investigation of biologically active compounds of natural origin

closely related to the needs of the surrounding community and some highly specialised areas that can interface with the former and for which the initialisation and running costs are comparatively low and the quality of the outputs depends more on human factors (on the expertise available). Computational chemistry is ideal for this role, for the reasons already outlined in previous considerations:

- It is a fast growing and advanced area of modern chemistry.
- It can interface with many other areas of chemistry and also with other sciences that utilise chemistry as part of their investigation and interpretation approaches. In particular, it can interface with other research areas of local interest

(e.g. natural products or ethnomedicine) as well as with areas of growing industrial interest (drug design, substances design, nanotechnologies).
- It is a highly specialised area, in which carrying out research and training new specialists depends mainly on human factors (expertise available).
- It is a research area where state-of-the-art results can be obtained also with comparatively low-cost facilities.

3.3.2 Significance for Chemical Education

Adequate incorporation of both the theoretical chemistry and the computational chemistry components into the educational curriculum is fundamental for the preparation of specialists. It is also fundamental to expose chemistry students to an important component of modern chemistry, with which students from other continents are getting increasingly sophisticated familiarisation since the undergraduate level or, sometimes, even earlier. The question of designing apt curricula and approaches, taking into account the average preparation level of chemistry students and trying to maximise their acquisition of information and skills, still requires major attention.

Quantum chemistry courses are often considered difficult by students. The extensive presence of mathematics is viewed as a major deterrent [20]. The difficulties inherent in studying through a second language makes it much more arduous to understand concepts that cannot be expressed only through syntactically simple one to two-clause sentences, but often require the ability to follow rather complex logical frameworks [20–22]. The objective of developing computational chemistry research demands that students' exposure to theoretical chemistry goes beyond basic literacy, to pursue sufficient insight into its motivations, methods and research questions to make it possible for a student to consider including it in the range of potential options for his/her future career. Therefore, it is necessary to design approaches that can help overcome—at least, to a significant extent—the difficulties experienced by students, so that they can attain sufficient insight into the nature of computational chemistry.

The issue of the teaching of theoretical/computational chemistry in African universities has been given specific attention in the TCWA, trying to identify the main existing problems [23], to optimise its position in the chemistry curriculum through maximisation of the interfaces with the contents of the other courses [24, 25] and to explore approaches that can be better tuned to the students' needs [26]. Interactive teaching [27, 28] appears to be particularly important for a subject that is perceived as difficult, or even very difficult, and the issue of overcoming students' "fear" of mathematics requires *ad hoc* attention and efforts [29]. A detailed discussion of educational and curricular approaches for a fruitful incorporation of computational chemistry in the chemistry curriculum is included in [15], taking into account both the need to attain adequate students' familiarisation with the content and the importance to attract their interest, so that some of them may include computational chemistry among the areas that they are ready to consider for their professional career. Some unconventional approaches to broaden students' views on the conceptual frameworks and mathematical methods of theoretical chemistry have also been explored [30–33]; although they may appear to border on even more difficult conceptual issues, some of them have proved interesting with small groups of students taking the quantum chemistry postgraduate course at UNIVEN some years ago (when the background preparation of incoming students was generally higher); for example, it was at UNIVEN that the attempt to broaden the view on quantization, and to decrease the abstractness perception so common at its first introduction, through comparisons with a simple issue of structural engineering encountered the unexpected response of students getting really involved with the engineering case [30]. This shows that it is possible, even in underprivileged contexts, to find ways of engaging students' attention in *difficult* issues, provided the passive attitude too many students still retain has been somehow overcome.

Passive attitudes, and the equalization of learning to memorization [20], are probably the major obstacle to students' engagement in conceptual explorations like the ones inherent in a quantum chemistry course and in any molecular study. At the same UNIVEN, in recent years fewer and fewer students appear to reach the awareness that *studying* is much more than passive memorization and requires personal engagement to pursue understanding; this complicates their performance in quantum chemistry courses, because of the conceptual demands of the content. When a student develops or attains that awareness, his/her performance increases sharply and rapidly. It is easy to infer that the stimulation of the awareness of need for personal intellectual engagement on *studying* is a fundamental educational challenge—a challenge that should be taken through all the chemistry courses, as it would be too limited (and, therefore, limitedly effective) to take it only in the quantum chemistry course.

Finally, it may be particularly important to recall that the last decade has witnessed an enormous development of user-friendly software that has made molecular calculations accessible to many users, so that "molecular modelling can now be performed in any laboratory or classroom" [34]. This is having important educational implications, as concepts and theories relevant to the study of molecules can be introduced in a concrete, visualised way even at secondary school level. The continuous rapid increase in the power of individual PCs is offering comparatively low-cost options. At introductory level, non-time-demanding computational options, such as semiempirical methods, can be utilised to familiarise students with important concepts like that of convergence and important practical abilities such as preparing inputs, analysing and comparing outputs and making inferences from the comparisons, always keeping chemical perspectives into account [35]. Simple case studies can be selected in such a way as to emphasise the interfaces with the material of the other chemistry courses. For example, at UNIVEN the study of selected molecules through simplest molecular mechanics methods is utilised as practical work for the quantum chemistry course (that pertains to the Honours postgraduate course): the selection of molecules is individual (different for different students) and includes molecules that the students study in other courses (e.g. compounds on which they are working for an organic chemistry project); for a preliminary basic training, the study focuses on the consideration of geometrical features and charge distributions and on their relationships with general chemistry concepts (size of atoms, hybridization, electronegativity, etc.), thus aiming at basic familiarisation with the 3D structure of molecules, the parameters describing it, and a way of analysing it in terms of already known chemistry concepts.

3.3.3 Significance for Sustainable Development

The development of computational chemistry research has considerable significance for sustainable development perspectives in the continent. The most important envisaged contributions to sustainable development can be summarised as follows:

- It would constitute an expansion and enrichment of the overall research capacity, with perspectives of producing state-of-the-art research outputs. This can contribute to the overall international status of African science and to linkages with those research activities ongoing in other contexts and focusing on issues relevant for sustainable development.
- It can contribute to retain in the continent important components of the development into marketable products of compounds isolated from indigenous natural sources (e.g. biologically active compounds with pharmacological potential), thus contributing to ensure some benefits for the communities (what is important for development in general) and to increase the awareness of the value of natural resources (what is important for sustainable development).
- It can contribute to the sustainability of industrial development through the design of new substances, such as catalysts or other substances needed for green industrial chemistry processes, or environmentally benign substances to replace some of the less environmentally benign currently in use (including some agrochemicals and other wide-usage substances). Developing substance design capacities up to standards adequate to meet the needs of emerging industries would enable African chemists to fully join the efforts aimed at making the use of green processes economically viable/attractive—one of the major and more urgent challenges facing chemists worldwide. Moreover, it would greatly enhance the possibility of designing substances that respond specifically to identify regional or continental needs (e.g. agrochemicals selectively targeting types of pests that are present in certain regions), and to design them taking into account their possible fate in the specific environmental conditions in which they are going to be used.

Pursuing the maximisation of the interfaces between the development of computational chemistry and the contextual needs of sustainable development can lead to novel perspectives whose interest would go beyond continental borders, to bring contributions to the overall search for sustainability options.

3.4 The Major Difficulty: The Scarcity of Experts

The major problem is currently the scarcity of experts. The number of computational chemistry experts in Sub-Sahara African is sorely inadequate. This affects all levels of activities, from offering exposure to chemistry students to the possibility of preparing new specialists and to the possibility of establishing active research. It is not easy to design realistic measures to address this problem. The first-priority objectives, to function as prerequisites to further developments, could include:

- searching for innovative ways to lower the impact of the scarcity of specialists;
- encouraging students to consider theoretical and computational chemistry as a possible option for their future career.

For the former objective, some forms of *sharing* of the available experts (through visits, short courses, research collaborations) appear the most viable option in the immediate future. The idea had already been envisaged since the first TCWA [36], but has not yet reached an implementation level. It clearly requires the design of innovative approaches in the relationships between the institutions where the experts are based and the others that would *share* their expertise.

The possibility of attracting students into this research field depends on a number of factors. The most important factor is exposure and, therefore, the presence of computational chemistry research activities in the institutions where they are studying. In order to develop interest in a certain area, students need to somehow come into contact with it, and this is possible only if research in that area is active in their institution. As already mentioned, the two aspects are interdependent: the presence of postgraduate students is essential to a *healthy* development of research in a certain area, but, on the other hand, students need to come into contact with that area prior to selecting it. This interdependence further underlines the relevance of establishing theoretical/computational chemistry research in African tertiary institutions.

It is particularly important that students are trained in their institution, or in other institutions in the continent. In the latter case, maintaining close links with their original institution would be essential to ensure that they anticipate a future role for themselves in that institution and prepare for it. The training within the continent, under conditions that are *normal* in the continent, is expected to ensure important benefits:

- Within the continent, the training can be better shaped to meet the requirements and the challenges that those students are more likely to encounter in their professional future in their institutions. For instance, the training should include the fostering of abilities to train other researchers or to start a new research activity *from scratch* and take a leading role in its conduction.
- By studying in the continent, students would automatically acquire the perception that research of this type can be carried out locally. This, in turn, would help reduce the brain drain. Some institutions have already experienced cases of

students being sent to other continents to train in computational chemistry, and not returning back after the training. Besides the frequent search for "greener pastures", a factor that contributes to their not coming back is the concern about anticipated difficulties (or even impossibility) to continue with a research in which they have developed interest, once they would be back in their original institution. This concern can be removed if students receive a training that prepares them to initialise research, to be able to work under conditions that may have limited resources, and to produce satisfactory results notwithstanding contextual difficulties [14]—a training that is likely possible only within the continent.

The design of training options needs to take into account the fact that computational chemistry is a highly specialised area integrating expertise from chemistry, physics and mathematics. Because of this, the training of new specialists is particularly demanding, as the student needs to develop comfortable familiarity with the approaches of all these disciplines, in order to develop creativity within theoretical/computational chemistry. The acquisition of such familiarity needs to be *integrated* (to integrate the perspectives of the three disciplines) since the very beginning and, therefore, the training needs to be done by a computational chemist (as someone who has already acquired this type of integrated familiarity). As a consequence, computational chemistry is an area in which the student–supervisor interactions are continuous and develop rapidly into collaboration patterns. This is obviously positive for the outcomes it produces, but poses limitations to the number of students that a supervisor can mentor, because of the extensive time that needs to be devoted to each student. In a situation in which potential supervisors are scarce, careful preliminary selection becomes a necessity, to ensure that only students with real potentialities and genuine interest are accepted for training projects.

Economic aspects do not need to be underestimated. The inducement associated with the availability of bursaries specifically for computational chemistry might play important roles (such bursaries could be made available, e.g. by the pharmaceutical industry, or within large-scale projects like those for the development of nanotechnology). Making such bursaries available may also be considered particularly significant because—given the nature of computational chemistry and the considerations expressed in the previous paragraph—only very good students can be accepted for postgraduate studies in computational chemistry and, therefore, the bursaries would contribute to the development of really promising future specialists.

The current scarcity of specialists also requires strong networking, for new trainees to be put in a position to initialize research in their institutions. A freshly graduated student (including a student freshly attaining his/her Ph.D. degree) is not yet a fully independent researcher. If left alone, he/she might not manage to initialise and lead new research activities. The phenomenon has already occurred, in some contexts, for other areas of chemistry and for other science disciplines, resulting in situations in which intermediate-age researchers are mostly absent, and the ensuing generation gap risks to become a drawback for a country's research capacity, as the senior researchers gradually reach retirement age. For computational chemistry, the impact is likely to be heavier, as a newly trained researcher, if left alone, might be totally lonely in his/her institution (if he/she is the first computational chemist coming back to it), and it would be extremely difficult (practically impossible) to conduct research alone. Maintaining collaboration contacts with the former supervisor and developing new contacts with other researchers in the continent becomes an essential factor. Therefore, it is important to establish patterns to facilitate such contacts, so as to provide all the necessary support for a newly trained researcher to be in a position to fulfil the role of research initiator and leader that is expected from him/her.

Other types of currently scarce expertise will also become essential as the research capacity grows, and will probably need to be *shared* between institutions, through innovative patterns. The experts that will be needed soon after the development takes off will be system managers—persons with specific training in the type of system management that is required for computational chemistry. In long-established research centres, the system manager is usually a theoretical/computational chemist who has acquired this additional expertise. The option proves optimal and, therefore, it can be envisaged that, once a sufficient number of computational chemists have been trained and research can expand to higher/broader sophistication levels, some of the new trainees will have to be offered the opportunity to acquire system-managing expertise.

3.5 Feasibility Assessment for the Initialization and Development of Computational Chemistry under Underprivileged Conditions

The development of computational chemistry research in recent years at UNIVEN is apt for a feasibility assessment [57], as its contextual situation is not much different from many other African institutions. UNIVEN is a *historically disadvantaged university* (or Historically Black University, HBU) located in a rural area in the Northeast of South Africa (Limpopo Province). The disadvantages include features such as inadequate facilities, chronic underresource, poor students and staff retention and the like. Despite all this, it has been possible to develop computational chemistry research in recent years. The obtained results include publications in reputable journals [37–43], conference presentations [44–55] and the training of a postgraduate student, who has completed his B.Sc. and M.Sc. [56] and is currently close to completing his Ph.D. studies. A quick overview of the development pathway and options can better highlight the feasibility assessment ensuing from them.

The development of computational chemistry research has followed a pattern aimed at maximising the matching between the steps that are necessary for research capacity building starting *ex novo* (or *from scratch*) and the standard patterns for the investigation of biologically active compounds. The focus on the study of biologically active molecules was chosen since the beginning, as the most apt for a university located in an area with rich biodiversity and rich traditional medicine knowledge. The first molecule investigated was the caespitate molecule that had been isolated from a plant utilised in traditional medicine in South Africa and exhibits antibacterial, antituberculosis and antifungal properties [58–60]. The research developed from the study of the caespitate molecule *in vacuo* [37] (that also involved the selection and study of model structures) to the study of the same molecule in solution [39] and the study of the parent compound (phloroglucinol, 1,3,5-trihydroxybenzene [38]); it has currently reached the study of a representative number of compounds of the same class that includes molecules with a variety of biological activities (antibacterial, antifungal, antimalarial, antiviral, antioxidant, antidepressant, etc.); this stage—highly demanding in terms of computational time, because of the size and characteristics of the molecules of interest—is close to completion, with some results already published [40, 42] and others still in progress of being analysed. The next envisaged stages include the investigation of structure–activity relationships—a typical component of the study of biologically active compounds in view of the understanding of their pharmacological potentialities.

The factors that have been essential to the initialisation and development of the research activity have been the presence of a specialist, the presence of a highly dedicated postgraduate student and the existence of links with a long-established group (the *Institute for Physico-Chemical Processes—Molecular Modelling Lab* in Pisa, Italy, and the Department of Chemistry of the University of Pisa) enabling interchanges that have resulted in transfer of expertise from specialists with long experience in the study of biologically active molecules, as well as technical support. The main difficulty was the minimal size of the research group (one professor and one student) that by itself would unavoidably limit the diversity range of the generation of ideas, normally born from interactions. This difficulty was to a substantial extent overcome through the above-mentioned links and to frequent participation in international conferences; such participation enabled the option of presenting results at a conference before preparing them for submission for publication, so that the preparation for submission could benefit from the comments and interactions at the conference, thus reducing the impact of the minimal sizedness of the research group.

As mentioned earlier, this development can be viewed as representative for feasibility assessment because of the context and manner in which it was realised. The institution is a non-privileged one, experiencing realities and constraints frequent in several other non-privileged contexts. The initialisation and development had to build *from scratch*, as there had never been prior research activity in this area. The development has been realised in a strictly economical manner, because of very little funding (in the first years, the only funds received were those enabling the purchase of the essential Gaussian [61] computational software package). The steps that have been followed have integrated the patterns for capacity building with the patterns for the study of biologically active compounds—an integration that has obvious advantages: active research since the first moment (particularly important for postgraduate students), hands-on development of expertise (important from an educational point of view) and production of research outputs (important in view of subsequent applications for funding, as well as to gain support within the university community). Analogous patterns could be easily utilised for initialisation and capacity building in other institutions. Utilising compounds from natural sources offers ideal options, since research on natural products is already active in many institutions in the continent and traditional medicine has a rich variety of remedies still to be investigated.

Financial aspects do not constitute an absolute deterrent. The initial capital investment is much lower than that for other (experiment-based) areas of chemistry research. In the take-off stage, one important software package (such as Gaussian in the development at UNIVEN) and some high RAM, high-speed personal computers are sufficient for research to develop up to interesting standards and to produce publishable results. After the initialisation stage, an increase of computational facilities (increase of total availability of computational time, and increase in individual computers' power) becomes desirable: it can be built gradually, thus avoiding high financial strains in a short period of time. Other options may be explored to reduce costs, e.g. the possibility of obtaining common licenses of software packages for groups of institutions of even groups of countries in the continent. Moreover, the availability of broadband Internet access might also enable the sharing of computational facilities based in other institutions. Similarly, possibilities for sharing access to computational chemistry journals can be explored, so as to reduce the costs of accessing literature.

3.6 Attracting Attention and Disseminating Information

Developing computational chemistry first of all requires interest in doing so. There is increasing interest in the area from young chemists, mostly prompted by the presence of computational results in articles in a variety of areas (study of organic compounds, study of biologically active molecules, material studies, nanotechnologies). However, for the interest to grow into the decision of undertaking the development of computational chemistry research in a given institution, it is important to disseminate information about the nature of computational chemistry, about the feasibility of the development within current situations and about options to overcome the drawback from the scarcity of experts. The TCWA have attracted attention and disseminated information among their participants. However, basic information needs to reach the entire community of African chemists and researchers in areas that can interface with computational chemistry. This is fundamental in many respects, from ensuring the needed encouragement and support by university communities to practical aspects, including funding. For instance, the author has recently experienced a rejection of a proposal for funding, with a motivation whose technical aspects clearly show that the persons who took the decision do not have adequate familiarity with computational work in chemistry. This can easily be ascribed to the scarcity of specialists (for which it may be difficult to ensure the presence of a specialist in an evaluating panel), but it also highlights another barrier that might arise and need to be overcome when initializing or developing computational chemistry research—the risk that the persons in charge of evaluating a proposal may not have enough familiarity with the specific features of computational chemistry research to be in a position to attain informed evaluations.

The dissemination of information requires the design of viable options, to be sufficiently effective. It is envisaged that the best vehicle could be a book specifically meant for the African context. The design of the features of such a book appears quite challenging. It needs to be easy enough to be accessible to chemistry students and to practising chemists, including those who prefer to refrain from materials with extensive mathematics presence and, therefore, it should nearly avoid mathematics. It should not be a simplified textbook of computational chemistry, but a book informing about what computational chemistry is and what it can do. It should highlight the African perspectives from as many points of view as possible, to provide a sufficiently informative picture of the relevance of developing computational chemistry research in African institutions. And it should give enough information on the way of proceeding of computational chemistry research to make potential evaluators sufficiently aware of the specificities of computational work on molecules. Explorations for a viable design are currently in progress.

4 Discussion and Conclusions

The information outlined in the previous sections, though limited to an overview of the most basic aspects, highlights the importance of computational chemistry in modern chemistry and in interfacing areas such as pharmacology or material science, thus highlighting the importance of its presence in universities and other research centres. It also considers the gap between Sub-Sahara African and the other continents: although computational chemistry research is now active in most tertiary institutions in the other continents, and chemistry students get exposure to computational chemistry approaches since their undergraduate level, both computational chemistry research and students' exposure to its foundations and approaches are still scarce in Sub-Sahara African, mainly because of dire scarcity of specialists. On the other hand, the need to develop it is increasingly acknowledged, above all by young chemists. Under such circumstances, it becomes important to utilise the available expertise to foster the training of new specialists and the initialisation of new research. The following aspects are considered particularly important for the training of new specialists:

- that the training is done in the continent, possibly in conditions not too different from those in which the students will work on coming back to their institutions of origin;

- that students are trained to be initiators, so that they feel ready for the challenges of initialising research in computational chemistry even if they are the first ones, or the only ones, to do so in their institution;
- to promote linkages and networks, so that young specialists undertaking the initialization of capacity building in their institutions can have adequate opportunities for extensive exchanges of views and information, can share intellectual and investigation challenges and can benefit from the support of more experienced colleagues (first of all their former supervisor).

The features typical of computational chemistry research enhance the feasibility of its initialization and development:

- The comparative low financial demands of this research area decrease the impact of one of the most frequent constrains in non-privileged institutions (the financial one).
- The research capacity building process can be designed and developed in close correspondence with the main stages of the investigation of compounds of the type of interest in the given institution (e.g. biologically active compounds), so that the capacity building process practically coincides with the realisation of a full research project; this is expected to increase both the confidence of the persons engaged in the capacity building and the support from the rest of the community in the institution.
- The dominant dependence of the capacity building process on human resources underlines the importance of networking and partnerships, to overcome drawbacks from continent-wide scarcity of experts. On the other hand, it ensures that the training of new experts becomes a guarantee of development, as human resources constitute the major capital for the development.

In summary, the development of computational chemistry research in Sub-Saharan Africa tertiary institutions is realistic and feasible within the current circumstances of the institutions. The most important requirement is the training of new specialists that can initialize and conduct research activities. The drawbacks from the current scarcity of specialists can be overcome through innovative ways of sharing the experts currently available. The initialization/development of computational chemistry research will bring significant contributions to chemical research in general, and to the roles of chemistry for sustainable development in particular.

References

1. Macquer P J (1766) Dictionnaire de Chymie. Paris
2. Dirac PAM (1929) Quantum mechanics of many-electron systems. Proc. Roy. Soc A123:714–733
3. Tomasi J (1996) Quantum chemistry: the new frontiers. In: Ellinger Y, Defranceschi M (eds) Strategies and applications in quantum chemistry. Kluwer Academic Publishers, Dordrecht, pp 1–30

4. Bushelyev SN, Stepanov NF (1989) Elektronnaya struktura y biologhicheskaya aktivnost molecul. Khimiya Snanye, Moscow
5. Schäfer A (2000) Industrial challenges for quantum chemistry. In: Grotendorst J (ed) Modern methods and algorithms of quantum chemistry, John von Neumann Institute for Computing, Jülich, NIC Series, vol. I, 1–5
6. Gillom RD (1989) Semiempirical and ab initio calculations of charged species used in the physical organic chemistry course. J Chem Ed 66(1):47–50
7. Canales C, Egan L, Zimmer M (1992) Molecular modelling as an inorganic chemistry exercise. J Chem Ed 69(1):21–22
8. Casanova J (1993) Computer based molecular modelling in the curriculum. J Chem Ed 70(11):904–909
9. Delaware DL, Fountain KR (1996) Computational chemistry in the first organic chemistry course. Applications in an active learning situation. J Chem Ed 73(2):116–119
10. Lipkowitz KB, Pearl GM, Robertson DH, Schults FA (1996) Computational chemistry for the inorganic curriculum. J Chem Ed 73(2):105–107
11. Lundell J, Aksela M, University of Helsinki, Finland. Private correspondence
12. Casavecchia G (2004) La didattica in 3D: come studiare le proteine al computer. IV Conferenza Nazionale sull'insegnamento della chimica, Assisi (Italy), 9–11 December
13. Mammino L (2003) Interfaces between theoretical chemistry and environmental chemistry. 5th TCWA and ESAECW, Dar es Salaam (Tanzania), 1–5 December
14. Mammino L (1993) Significance and perspectives of applied quantum chemistry. Int J BioChemiPhys 2(1&2):158–160
15. Mammino L (2007) Computational chemistry in chemistry and pharmacology research capacity building. An African perspective. In: Proceeding of the 6th TCWA and ESAECW, Windhoek (Namibia), 5–9 December 2005, pp 1–27
16. Mitscher LA, Gerhart MA, Rao GSR, Khanna I, Veysoglu T, Drake S (1983) A modern look at folkloric use of anti-infective agents. Phytochem 22:573–578
17. Mitscher LA, Park VH, Clark D, Beal JL (1980) Antimicrobial agents from higher plants. J Nat Prod 43:259–265
18. Mitscher LA, Gerhart MA, Rao GS (1984). In Krogsgaard-Larsen P, Brogger Christensen S, Kofod H (eds) Natural Products and Drug Development, Munksgaard, Copenhagen, 193–212
19. Mammino L (2005). Computational chemistry as part of capacity building in malaria research in the African continent. Fourth MIM Pan-African Malaria conference, Yaoundé (Cameroun), 13–18 November
20. Mammino L (2009) Teaching physical chemistry in disadvantaged contexts: challenges, strategies and responses. In Gupta-Bhowon M, Jhaumeer-Laulloo S, Li Kam Wah H, Ramasami P (eds.) Chemistry Education in the ICT Age, Springer, Netherlands, 197–223
21. Mammino L (1995) Teaching/learning theoretical chemistry at undergraduate level. Southern Africa J Math Sc Educ 2(1&2):69–88
22. Mammino L (2005) Method-related aspects in an introductory theoretical chemistry course. J Mol Struct (Theochem) 729:39–45
23. Mammino L (1995) The teaching of theoretical chemistry in African universities: problems and perspectives. First TCWA, Nairobi (Kenya) 20–24 February
24. Mammino L (1996) Computational chemistry in research and in education. Second TCWA, Nairobi (Kenya) 25–29 August
25. Mammino L (1998) Theoretical chemistry in the chemistry curriculum. Third TCWA and ESAECW, Nairobi (Kenya) 2–6 November
26. Mammino L (1998) Exploring new approaches to the teaching of theoretical chemistry. Third TCWA and ESAECW, Nairobi, Kenya, 2–6 November
27. Mammino L (1998) Theoretical chemistry courses and interactive teaching. 15th ICCE, Cairo, (Egypt), 9–14 August
28. Mammino L (1998) Enseñanza interactiva en cursos de química teorica. ALDEQ 11:275–278
29. Mammino L (2000) Making mathematics student-friendly in theoretical chemistry courses. 16th ICCE. Budapest (Hungary), 5–10 August

30. Mammino L (2003) Addressing the abstractness perception in theoretical chemistry courses. J Mol Struct (Theochem) 621:27–36
31. Mammino L (2002) Pedagogical significance of the presentation, to students, of the explorations of non-standard orbital basis sets. XXVIII QUITEL, Montevideo (Uruguay), 1–8 September
32. Mammino L (2004) Mentioning fuzzy logic in theoretical chemistry courses: motivations and extent. J Mol Struct (Theochem) 709:231–238
33. Mammino L (2006) The recent history of theoretical chemistry presented from a method-related perspective. J Mol Struct (Theochem) 769(1–3):61–68
34. Leach RA (2001) Molecular Modelling: Principles and Applications. Pearson Prentice Hall
35. Mammino L, Kabanda MM (2006) Semiempirical methods as introduction to computational chemistry. 18th ICCE. Seoul (Korea), 12–18 August
36. Mammino L, Khanra A (1995) Mobile teaching in theoretical chemistry: a proposal. First TCWA, Nairobi, Kenya, 20–24 February
37. Mammino L, Kabanda MM (2007) Model structures for the study of acylated phloroglucinols and computational study of the caespitate molecule. J Mol Struct (Theochem) 805:39–52
38. Mammino L, Kabanda MM (2008) A computational study of the interactions of the phloroglucinol molecule with water. J Mol Struct (Theochem) 852:36–45
39. Mammino L, Kabanda MM (2008) A computational study of the interactions of the caespitate molecule with water. Int J Quant Chem 108:1772–1791
40. Mammino L, Kabanda MM (2009) A study of the intramolecular hydrogen bond in acylphloroglucinols. J Mol Struct (Theochem) 901:210–219
41. Mammino L (2009) Could geometry considerations help take into account solute-solvent hydrogen bonding in continuum solvation models? Chem Phys Lett 473:354–357
42. Mammino L, Kabanda MM (2009) A computational study of the effects of different solvents on the characteristics of the intramolecular hydrogen bond in acylphloroglucinols. J Phys Chem A 113(52):15064–15077
43. Mammino L, Kabanda MM (2010) A computational study of the carboxylic acid of phloroglucinol *in vacuo* and in water solution. Int J Quant Chem 110(3):595–623
44. Mammino L, Kabanda MM (2005) A study of intramolecular hydrogen bonding in the caespitate molecule. WATOC 2005, Cape Town (South Africa), 16–21 January
45. Mammino L, Kabanda MM (2005) Semi-empirical methods for the study of medium-size molecules. Applications to the study of the caespitate molecule, 6th TCWA and ESAECW, Windhoek (Namibia), 5–9 December 2005, pp 28–54
46. Mammino L, Kabanda MM. (2006) Interactions of the phloroglucinol molecule with water molecules. XXXII QUITEL, Côtes de Carthage, Tunisie, 1–6 September
47. Mammino L, Kabanda MM (2007) A study of the interactions of the caespitate molecule with water. XXXIII QUITEL, La Habana (Cuba), 17–21 September
48. Mammino L, Kabanda MM (2007) Level of theory and basis set selection in the study of a medium-size molecule. The case of caespitate. *10 ICCA*, Benghazi (Libya), 18–21 November
49. Mammino L, Kabanda MM (2007) Studying a biologically active molecule computationally: the case of caespitate. 7th TCCA and ESAECC, Victoria Falls, (Zimbabwe), 3–7 December
50. Kabanda MM, Mammino L (2007) Criteria for the conformational analysis of a highly flexible molecule: the case of caespitate. 7th TCCA & ESAECC, Victoria Falls, (Zimbabwe), 3–7 December
51. Mammino L, Kabanda MM (2008) Computational study of the carboxylic acid of phloroglucinol in vacuo and in water solution. XXXIV QUITEL, Cetraro (Italy), 3–8 July
52. Mammino L, Kabanda MM (2008) Computational study of acylphloroglucinols—a promising class of biologically active compounds. WATOC 2008, Sydney (Australia), 14–19 September
53. Mammino L, Kabanda MM (2009) A study of the intramolecular hydrogen bond characterising acylphloroglucinols. 13th ICQC, Helsinki (Finland), 22–27 June
54. Mammino L, Kabanda MM (2009) Adducts of acylated phloroglucinols with explicit water molecules: similarities and differences over a sufficiently representative number of different structures. XXXV QUITEL, San Andres (Colombia), 18–22 September

55. Mammino L, Kabanda MM (2009) Computational study of nodifloridin-A and nodifloridin-B, with highlight of the peculiarities of acylated phloroglucinol derivatives. In: Bulucea CA, Mladenov V, Pop E, Leba M, Mastorakis N (eds) Recent Advances in Biology, Biophysics. WSEAS Press, Bioengineering and Computational Chemistry, pp 58–63
56. Kabanda MM (2007) A Computational Study of the Caespitate Molecule. University of Venda, South Africa, Thesis for the M.Sc Degree
57. Mammino L. (2009) Feasibility assessment for the realization and development of malaria-related computational chemistry research in Sub-Sahara African. 5th MIM Pan-African Malaria Conference, Nairobi (Kenya), 2–9 November
58. Mathekga ADM, Meyer JJM, Horn MM, Drewes SE (2000) An acylated phloroglucinol with antimicrobial properties from *Helychrisum caespititium*. Phytochem 53:93–96
59. Mathekga ADM. (2001) Antimicrobial Activity of *Helichrysum* Species and the Isolation of a New Phloroglucinol from *Helichrysum Caespititium*. Doctoral Thesis at the University of Pretoria
60. Meyer JJ, Lall N, Mathegka ADM (2002) In vitro inhibition of drug-resistant drug-sensitive strains of *Mycobacterium tuberculosis* by *Helicrysum caespititium*. South African J Bot 68:90–93
61. Frisch MJ, Trucks GW, Schlegel HB, Scuseria GE, Robb MA, Cheeseman JR, Montgomery JA, Vreven T, Kudin KN, Burant JC, Millam JM, Iyengar SS, Tomasi J, Barone V, Mennucci B, Cossi M, Scalmani G, Rega N, Petersson GA, Nakatsuji H, Hada M, Ehara M, Toyota K, Fukuda R, Hasegawa J, Ishida M, Nakajima T, Honda Y, Kitao O, Nakai H, Klene M, Li X, Knox JE, Hratchian HP, Cross JB, Adamo C, Jaramillo J, Gomperts R, Stratmann RE, Yazyev O, Austin AJ, Cammi R, Pomelli C, Ochterski JW, Ayala PY, Morokuma K, Voth GA, Salvador P, Dannenberg JJ, Zakrzewski VG, Dapprich S, Daniels AD, Strain MC, Farkas O, Malick DK, Rabuck AD, Raghavachari K, Foresman JB, Ortiz JV, Cui Q, Baboul AG, Clifford S, Cioslowski J, Stefanov BB, Liu G, Liashenko A, Piskorz P, Komaromi I, Martin RL, Fox DJ, Keith T, Al-Laham MA, Peng CY, Nanayakkara A, Challacombe M, Gill PMW, Johnson B, Chen W, Wong MW, Gonzalez C, Pople JA (2003) GAUSSIAN 03. Gaussian, Inc., Pittsburgh, PA

Geochemistry for Sustainable Development in Africa: Zimbabwe Case Study

M. L. Meck

Abstract Geochemistry is the geology and chemistry concerned with the chemical composition of, and chemical reactions taking place within, the Earth's crust. While the geology and chemistry of Africa is known the chemical reactions that are taking place are not fully documented yet if documented the geochemistry can be used as a tool for sustainable development in Africa. Most of Africa is in tropical and subtropical regions where a long history of chemical weathering takes place thus changing the surface chemistry and making it particularly fragile. A case study from Zimbabwe is presented here to illustrate how geochemistry can be used for sustainable development of Africa. The study assessed tailings dumps' potential to cause environmental problems related to their geochemistry. An overview of the general levels of potential toxic elements in different dump types is given by this study and the types of dumps and mines that are associated with certain risk elements are outlined. As Africa has the largest tropical area of any continent, it is likely to have many chemical weathering taking place thus a need to continuously study its geochemistry. A catalogue of information regarding Zimbabwean tailings dumps and their geochemistry as well as characteristic of immediate environment was constructed as part of a Masters study. This information was used to predict and model possible dispersion and pollution patterns that are likely to result from the tailings dumps found in the country. Possible environmental problems related to the geochemistry of the dumps are outlined. Different stakeholders who may need to redress problems associated with mine tailings dumps in Zimbabwe can use the information gathered during the course of this study. Short-and long-term impacts of the mines and their waste can also be deduced from the information. The results from this research indicate

M. L. Meck (✉)
Department of Geology, University of Zimbabwe, Mt. Pleasant,
PO Box MP167 Harare, Zimbabwe
e-mail: maideyimeck@yahoo.com; mabvira@science.uz.ac.zw

A. Gurib-Fakim and J. N. Eloff (eds.), *Chemistry for Sustainable Development in Africa*, DOI: 10.1007/978-3-642-29642-0_7,

significantly higher levels of potentially toxic elements in the base metals, minor metals, gold, sulphur and platinum group metal dumps compared to the soils around these dumps. The levels most of potentially toxic elements encountered within the dumps during the course of this study have significant implications to the mining industry and particularly to tailings disposal in terms of the potential to pollute the environment. The major output of the study is data that can be used for sustainable development in ways of managing the environment to ensure continual existence of the mining industry in a sustainable way.

1 Introduction

Zimbabwe has been a major mining country since the beginning of the twentieth century and mining activities can be traced several centuries before. Mining is necessary for both the development of the country and as a source of foreign currency. The mining industry has become pivotal to the Zimbabwean economy and can be expected to remain so into the future. However, if mining is to be guaranteed continual existence, it has to co-exist with other industries that share the same resources, such as agriculture and tourism. Thus, mining must be done in a manner that does not impact negatively upon the environment. This calls for minimisation of negative effects that might arise from mining and affect other sectors of the economy. For the purpose of this study negative effects are defined to be those effects that have recognisable detrimental impacts on the environment (living organisms and their habitat) and are synonymous with environmental pollution. Studying the geochemistry is therefore vital.

Studies by Engdahl and Hedenvind [4], Maponga [8], Mohiddin [13], Roberts [16], Thixton [22]. Mangwiro [7], Mandingaisa [6], Ngwenya [14], Ruzive [18], Ravengai [15], Lupankwa et al. [5] show that the Zimbabwean environment has had a fair share of mining-related pollution to warrant geochemical analyses. Since the Rio Summit in 1992, Zimbabwe, like all the other countries, has come under pressure to comply with sustainable environmental programmes [9]. Mining companies, in general, have responded to environmental challenges by developing charters and codes of conduct that minimise environmental contamination. Irrespective of the current efforts of the Zimbabwean government and mining fraternity, there is a considerable legacy of mining pollution as a result of past mining. Geochemical pollution occurring at non-working ("orphaned") mines may be very difficult and costly to redress. The most intractable and potentially costly environmental problems are predominantly those involving geochemical pollution. They include those involving changes of chemical forms of possibly harmful elements from inert forms to forms that are bioavailable. Mining, in its endeavour to win the desired elements, usually involves removal of the element from a relatively unreactive form in the ore to one that is more biologically accessible [23]. This increases the metal/element concentration that can be taken up by organisms. Most metals if taken up by living organisms in large quantities are poisonous. Hence, the change of unreactive forms to accessible forms has the potential for poisoning organisms. Mobility and bioavailability of metals are greatly increased by the development of acid mine drainage (AMD), which poses an acute threat of both acidification and metals release.

A major potential source of environmental problems in Zimbabwe is "mine dumps" comprising mill tailings and waste rock because they contain potentially toxic metals and sulphides. Zimbabwe has more than 10,000 mines, the majority of which are no longer working. Some degree of mineral processing has taken place at almost all these mines and, as a result, there are tailings and other waste materials stored at the mine sites. Thus, there is need to document the quantities of metals and sulphides in these "dumps". The Zimbabwean government metallurgical laboratory has recorded the locations of mines with stored mine waste but no information is available regarding the nature of the waste materials and its possible short- and long-term effects on the environment.

The present study was undertaken to provide baseline geochemical information regarding the nature of mine dumps in Zimbabwe wastes with the aim of gauging the prevalence of geochemical mining-related pollution in Zimbabwe. Possible dispersion patterns for contaminant metals and their likely dispersal were considered. The information gathered in the study includes the mine dumps' location, geochemical parameters and pertinent information on the immediate environment. The data gathered in this research are regarded as vital in characterising sources of geochemical problems in the Zimbabwean mining environment. The data can be used to for appropriate remediation, and as a basis for policies and legislature on the environment for sustainable development.

2 Materials and Methods

Environmental problems related to the geochemistry of potential toxic elements vary from mine to mine and depend on a number of factors that include the element in question, the amount of the element, the environment where the element exist (soil pH and redox potential), the co-existing elements, the bioavailable fraction and the nature of the immediate environment. Meaningful conclusions can only be made when a substantial amount of the above-mentioned baseline information is available. This project therefore endeavoured to collect as much baseline data as possible.

The data collected included geographic setting, geological setting, rock type, processing method, ore type and commodities mined. The geochemistry of the dumps and surrounding soil includes total and bioavailable concentrations for the major elements with particular emphasis on the elements that have a potential of polluting the environment.

Selection criteria for elements to include in this study was based on the priority list by McBride [10] who prioritised arsenic, beryllium, antimony, cadmium, chromium, copper, lead, mercury, nickel, selenium, silver and zinc as the elements that are likely to be problematic to human health. Due to the fact that Zimbabwe mines a variety of commodities in a range of different host rocks, it is host to all these problematic elements. This study however decided to analyse the effects of arsenic, antimony, cadmium, chromium, copper, lead, mercury, nickel, selenium and zinc. Beryllium and silver were left out as they occur in quantities that can be deemed to have negligible effects in the Zimbabwean mines. Molybdenum and cobalt which occur in a number of mines in Zimbabwe were added to the list as these two are known to be particularly deleterious to plants [3].

Taking into consideration the potential pollutant elements found in the Zimbabwean mines it was decided to include all the commodities mined in Zimbabwe, which occur in quantities that are likely to have a significant impact on the environment. The commodities gold, antimony, chromite, cobalt, copper, platinum, tungsten, lead, nickel, selenium and zinc were thus studied. Sulphur was included in the list of the commodities to be studied because of its potential to form acid that in turn produces AMD. The criteria used to select the mines varied from one group of commodity to another but included polluting type, production capacity, processing methods, geological setting, geographic setting and ore type. However, the main criterion used for selection was the lithology type. Mines were selected from different lithologies so as to ascertain the different effects the potential toxic elements have on different lithologies. Mines with polymetallic ores were also included with the aim of ascertaining the effect of interaction of the elements. Different methodologies were used to sample dumps soils, water and leachate samples. In each case, the most appropriate/suitable way for each sample type was employed to get statistically and scientifically sound data

3 Sampling of Dumps

For the dumps surficial material was sampled. This was deemed sufficient for this study, as surficial material is believed to have the greatest impact on geochemical-related problems. Among the methods evaluated were the methods by Meurig [12] that describes sampling strategies and when to use composite or individual samples, those by Runnels et al. [17], Bennett et al. [1], Smith et al. and Mend [11] who describe different sampling methods used in different situations. After the considerations, the sampling strategy by Smith et al. [19] was chosen as suitable for use in this project.

The following flow diagram shows the method as adopted from Smith et al.

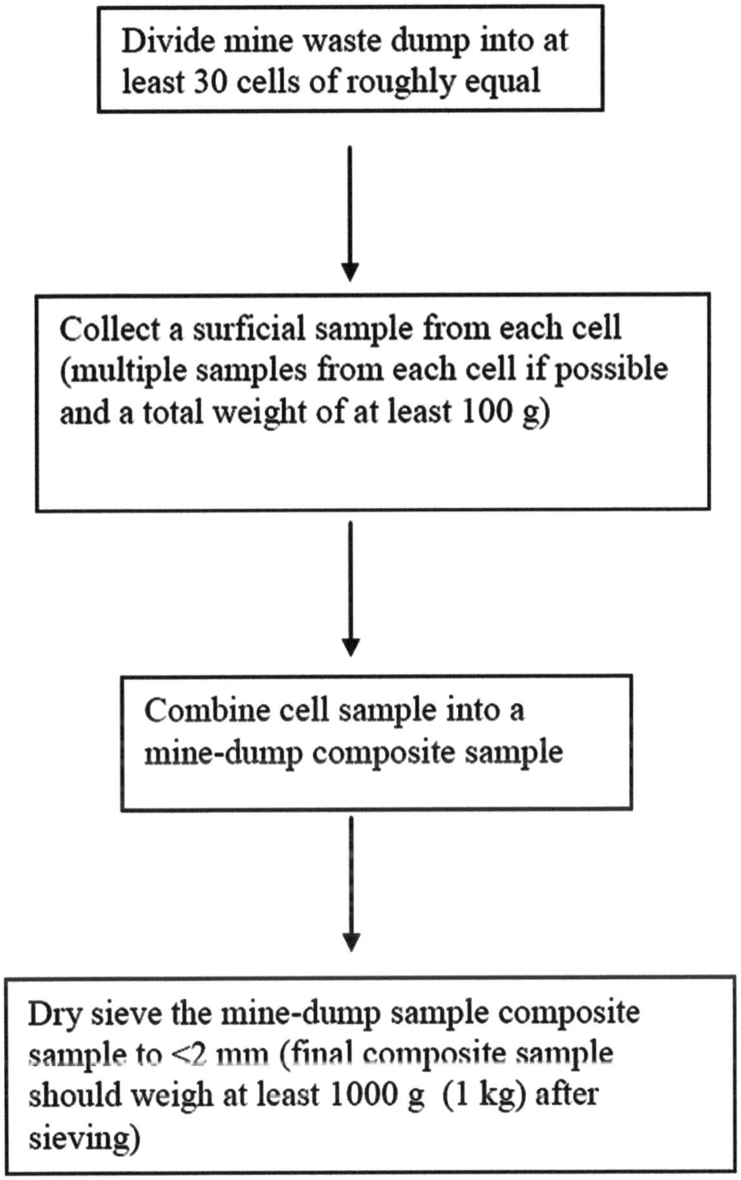

The control soil samples were taken from the same host rock as the host for the dump. These samples were collected at distances visually estimated to be away from pollution effects of the dumps. In most cases, these were upstream of the dump in the direction away from the main wind directions but in the mineralized zone that forms the mine. These samples were taken as one random sample in the area.

The samples were dried in an oven at 80 °C. The temperature was maintained at 80 °C to minimise possible loss of volatile elements such as mercury, selenium and sulphur. Higher temperatures are also known to result in an increase in the extractability of elements such as Cu [20] and would affect the bioavailability of the elements, which needed to be measured. All samples were then disaggregated by hand and organic material (leaves, twigs and roots) was manually removed. The samples were then ground and pulverised to pass through a 180 μm sieve. To minimise contamination of the samples with elements of interest, an agate mortar was used for pulverising the samples. The +80 mesh was discarded and the −80 mesh material was used for the analysis. The degree of grinding was standardised as it affects the extractability of elements such as Zn [20].

4 Sample Analysis

All the samples were then divided into two to make duplicates for analyses. The duplicates, known standards and blanks, were used to check the accuracy and errors of the machines used in analysing the samples. To avoid systematic errors the samples were analysed in a random order. Machine drift was checked using a solution of 2 ppb Ir, 2 ppb Rb and 1 % nitric acid. The Ir and Rb values are then determined together with the elements of interest. As Ir and Rb are usually very low in natural samples, their value should remain almost constant throughout the analysis if there is no drift in the machine. Any change (increase or decrease) in these values during analysis indicates a drift in the machine. The drift if present can be calculated and used to correct the results obtained by the particular machine.

5 Measurement of Element Concentration

For environmental geochemistry, a measure more useful for most purposes is an estimation of the availability or lability of the element, which is a partial element concentration. This measure is regarded as important because it is related to the mobility of the element, uptake by plants and extractability by chemical treatments [10]. This portion of the concentration, which is in most case equivalent to the bioavailable portion, occur adsorbed on surfaces, loosely bound in clays, trapped in oxide phase and associated with organic matter therefore can be extracted by partial extraction. The loose binding of this portion of concentration makes it susceptible to uptake by receptors. It is therefore not necessary to totally decompose soils, tailing and sediments samples to determine the concentration of interest. Total element concentration obtained by decomposition of the samples is still useful as it gives a measure of the maximum or upper limit to be expected in any case. This study adopted both partial decomposition and total extraction but paying particular attention to the bioavailable portion partial concentration.

Though the bioavailable partial concentration is ideal for environmental purposes, it is not easy to get an extraction method that gives a precise value for the bioavailable portion. The methods used in partial extraction results only in estimates of lability or bioavailable concentration. The chief reason is that the extractability of different elements depends on their properties, which include their tendency to complex with organic matter, chemisorb on minerals, precipitate as insoluble sulphides, carbonates, phosphates or oxides or co-precipitate in other minerals. These properties cannot be mimicked in the laboratory and hence more often than not the amount extracted is not the amount that would be extracted in the natural environment. Moreover, current soil testing procedures for most elements bathe the samples in solution of complexing, acidifying or reducing chemicals. This alters sample properties that control the solubility of the element. In the process redox potential, pH, mineral solubility and organic matter solubility may all be modified. Information about elemental speciation and availability is therefore lost in the process of attempting to extract part or all of the bioavailable pool. Success in predicting availability is therefore limited.

Nevertheless, Soltanpour [20] elucidate the ability of ammonium bicarbonate-dietylenetriaminepentaacetic acid—(AB-DTPA) soil test to give one of the best estimates of bioavailable portions. According to Soltanpour, AB-DTPA extraction in conjunction with inductively coupled plasma (ICP) spectrometry can be used to screen soils contaminated by mine spoils for their potential elemental toxicity to plants, animals and humans. The list of elements whose bioavailability can be determined with AB-DTPA- ICP system includes Pb, Cd, Se, As, B, Mo, Zn, Fe, Cu, Mn, NO_3, S, N, P and K. The test developed by Soltanpour and Schwab [21] to extract labile elements and later used for extraction of potentially toxic elements has further been tested and approved for use by USEPA and the Colorado Department of Public Health. The (AB-DTPA) method of determining the bioavailable fraction was hence adopted for this study.

For both total and partial metal content the samples were analysed by ICPS. The (AB-DTPA) extraction is expected to give the best estimate of the bioavailable portion whereas the total will give the maximum to be expected in each case.

6 Results

Eleven potential toxic elements (copper, zinc, lead, cobalt, nickel, molybdenum, chromium, arsenic, cadmium, antimony and selenium) were found present in both the dumps visited and the soils around them. The elements have varying ranges of element concentrations because dumps and soils investigated in this study are associated with mines that have different mineralogy, different commodities and different processing methods. This makes precise comparison of the data and trends difficult and casts doubt on the validity of the conclusion reached on any comparison made without further dividing this data. It was therefore decided that the data be divided and analysed according to the types of commodities being mined for the dumps. The data are divided as data for:

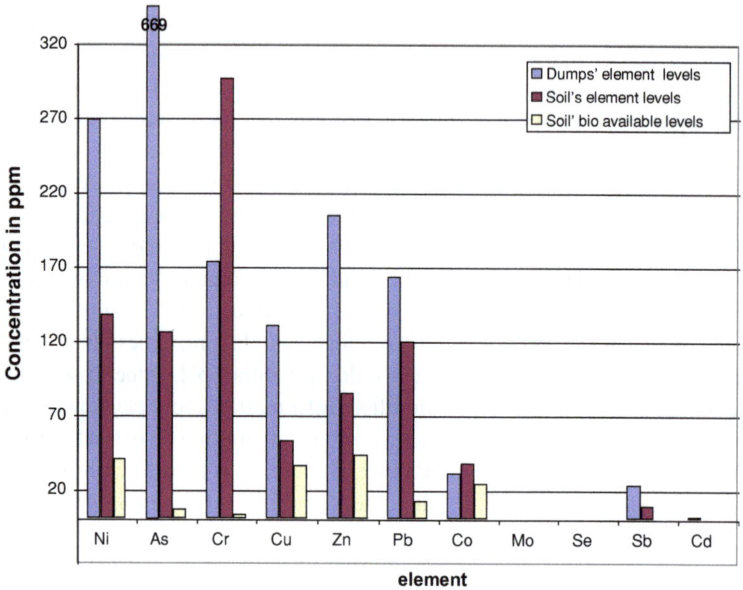

Fig. 1 Mean levels of potential toxic elements in minor metal tailings

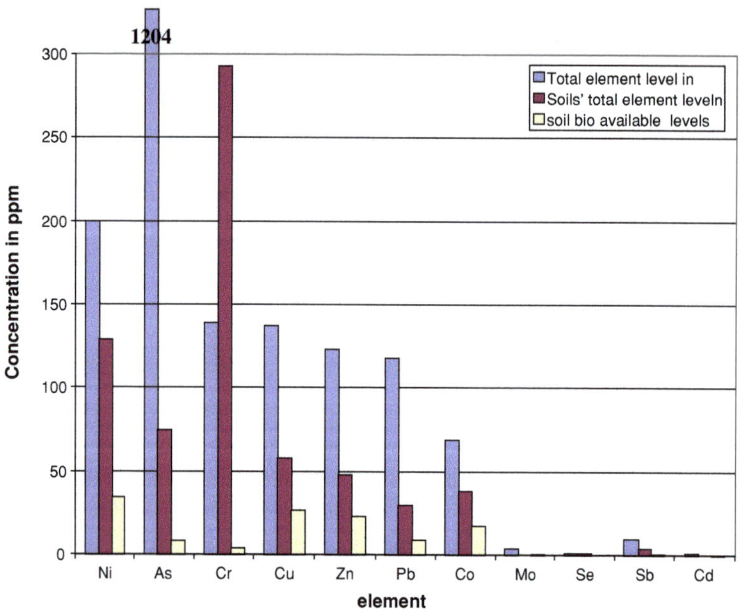

Fig. 2 Mean levels of potential toxic elements in gold tailings

(a) minor metals' tailings dumps
(b) gold tailings dumps
(c) base metals' tailings dumps
(d) chromite and asbestos tailings dumps
(e) platinum group elements tailings dumps
(f) sulphur tailings dumps.

The graphs below show the distribution of the potential toxic elements in the dumps and soils around them for each dump type. The soil bioavailable concentration is also shown on the graphs. The data used are the mean concentrations of the element for each dump type.

7 Summary of Observations Made on the Element Concentrations

The general trend observed in the six graphs in Figs. 1, 2, 3, 4, 5, 6 is that the dumps contain a higher concentration of the toxic elements compared to the soils around them. The data show that the elements' mean levels in the dumps are distinctively different from the mean levels in the soils around them. The data also show that in most cases the bioavailable proportion is lower than the total concentration in the soils.

However, several data (e.g. Cr in minor element dumps compared to Cr in soils and bioavailable Co in base metal soils compared to total) appear as if they do not follow the general trend. This is however a result of variable data that gives bigger means due to a small number of high values. If variance of the data is taken into consideration the apparent discrepancy disappears and the data adhere to the general trend.

The analysed control samples give an indication of the effective dump concentration which is obtained by taking away the natural backgrounds as given by the control samples from the total dump concentration. The data show that if the natural background is taken into consideration the chromite and asbestos mines actually have the least potential for polluting as they contain most elements is levels less than natural background and those that are higher are slightly above the natural background. Base metals and platinum group elements have almost all the elements generally higher than natural background levels

Chromium is lower than natural soil levels in most dumps so is antimony in platinum group elements. Arsenic is the element with the highest differential and it is more in gold, minor metals and sulphur tailings. Lead is visibly high in sulphur and base metals tailings

On the basis of the above facts on the effective levels of concentrations in the dumps, the effluent levels allowed in Zimbabwean terms and the safe drinking limits as set by USEPA the levels encountered can be summarised as given in Table 1: Since these elements occur in different proportions in the Earth the

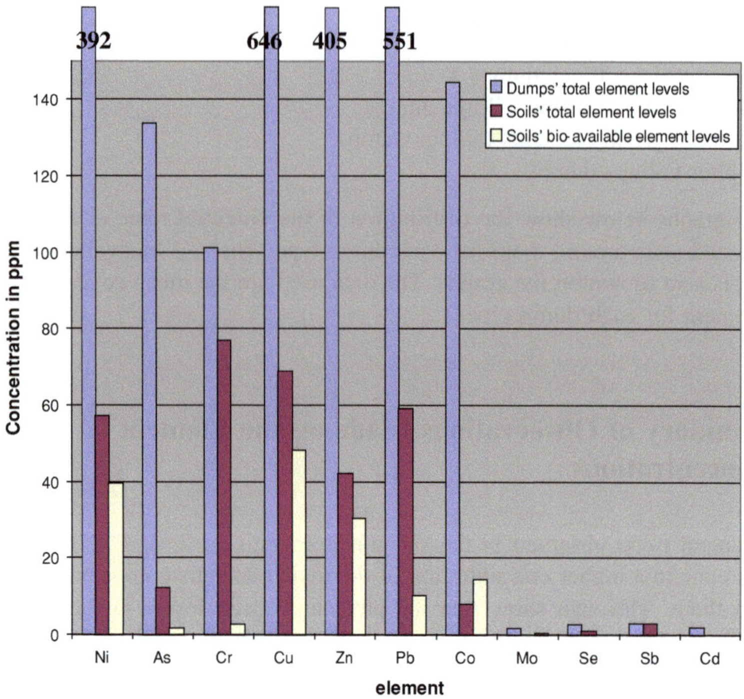

Fig. 3 Mean levels of potential toxic elements in base metal tailings

criteria for defining an element as high or low is element specific as a concentration that can be viewed as low for chromium will be definitely high for elements such as cadmium and selenium (Table 2, Table 3, Table 4).

8 Interpretation

These results were tested and validated by performing a student t-test on the data.

The above results show that the probability of the means being equal are very small and in some places almost nil which means that the levels of the elements of concern in the dumps can be considered distinctively different from the mean levels in the soils around them.

The observed trend implies that mining activities are either leaving significant levels of these metals in the dumps or are concentrating these elements. They are therefore possible sources of pollution to the environment and can pose a risk to the local population and ecosystems, who utilise the soils on and around these dumps. Since the soils were sampled in the mineralized zones, they can be treated as natural background (natural levels of the elements in the absence of mining

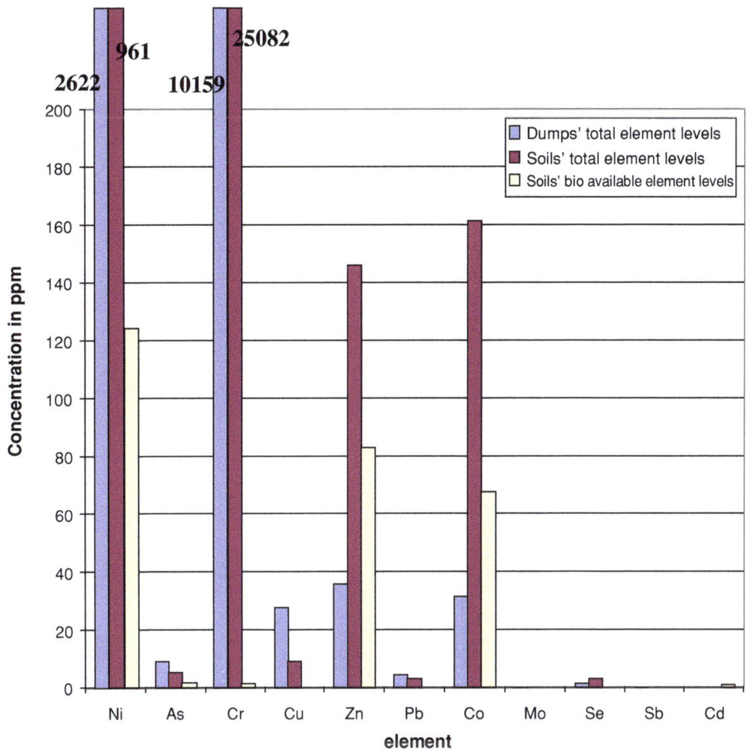

Fig. 4 Mean levels of potential toxic element in chromite and asbestos tailings

interference) thus can be used to measure the impacts of mining activities on the environment. The analysed control samples give an indication of the effective dump concentration which is obtained by taking away the natural backgrounds as given by the control samples from the total dump concentration. It should however be noted that the actual risk of the elements only exists when these elements become available to the environment and can disperse to the soils.

The trend discussed above has six exceptions where the levels in the dumps cannot be considered statistically different from the levels in the soils around the dumps. These are

(a) Cobalt in minor metals
(b) Selenium in minor metals dumps
(c) Antimony in base metals and chromite and asbestos mines
(d) Lead in minor metals tailings
(e) Chromium in chromite mines.

For these cases, the probability of the levels being equal is higher than the probability of being different suggesting that the levels of these elements in the

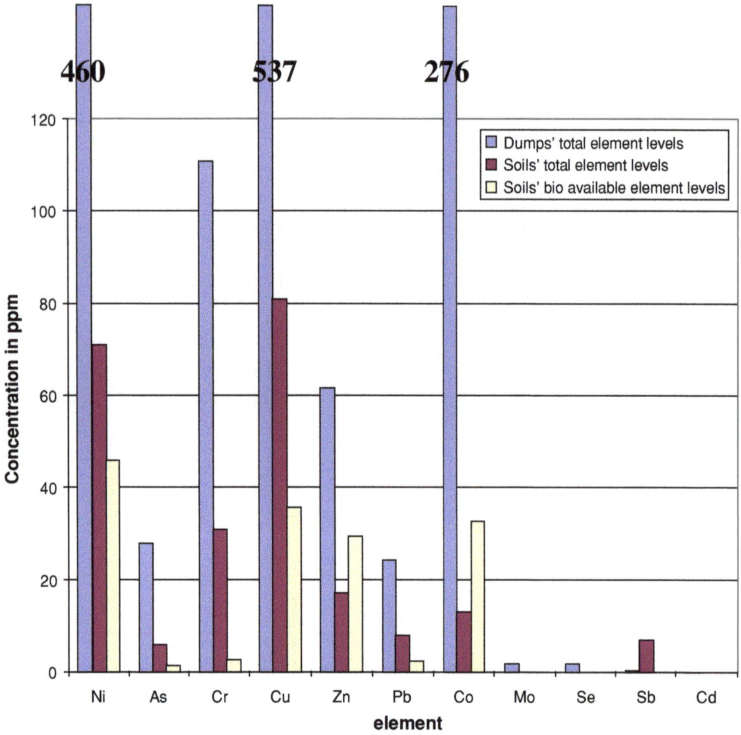

Fig. 5 Mean levels of potential toxic elements in platinum group metal tailings

dumps cannot be considered statistically different from the levels in the soils around them. The implications of these six cases are that mining activities have not affected the levels of these elements. If the bioavailability of the soils and dumps are the same, then the toxicity responsibility cannot be put on mining activities.

The t-tests above confirmed differences between levels in soils and dumps but it should be noted that though in the majority of cases the levels are higher in dumps, chromium and nickel in the chromite and asbestos dumps are an exception because these elements' mean concentrations are lower in dumps. This deviation may be due to the effect of mining, which takes out these minerals thereby rendering the dumps less concentrated in chromium and nickel. Ruzive [18] observed a similar trend in his work carried out in Mtorashanga.

9 Bioavailable Concentration

The differences in bioavailable levels and total concentrations were also tested with a student t-test to find out if cases of bioavailable concentration equal to total concentrations existed. The probabilities obtained from the t-tests are given in Table 5.

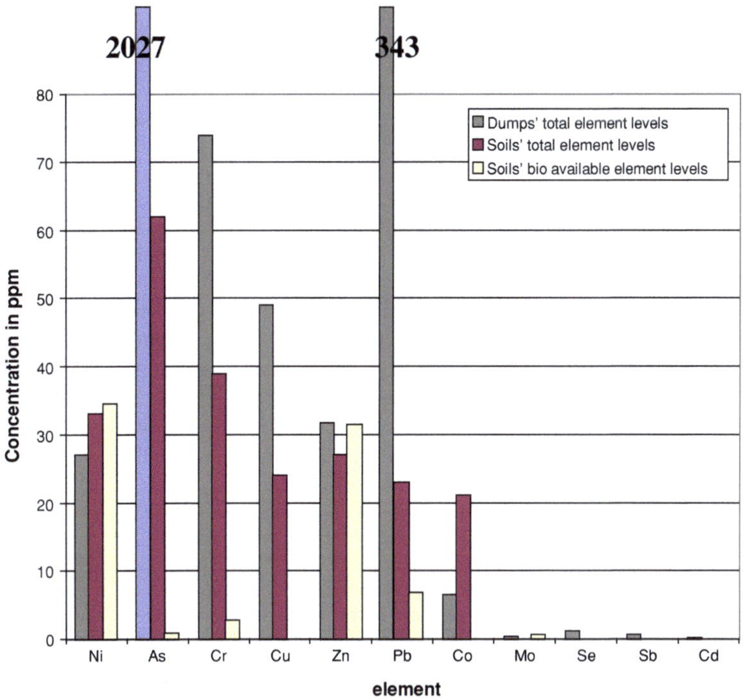

Fig. 6 Mean levels of potential toxic element in sulphur tailings

Table 1 Comparison of total element levels in different dump types

Mine-type Element	Minor metals	Gold	Base metals	Chromite and asbestos	Platinum group metals	Sulphur
Ni	Very high	High	Very high	Extremely high	Very high	Moderate
As	Extremely high	Extremely high	Extremely high	High	Very high	Extremely high
Cr	High	High	Moderate	Extremely high	Moderate	Low
Cu	Moderate	Moderate	Extremely high	Low	Extremely high	Low
Zn	High	High	Very high	Insignificant	Low	Insignificant
Pb	High	Moderate	Very high	Insignificant	Low	High
Co	High	High	High	High	Very high	Moderate
Mo	Insignificant	Moderate	Moderate	Insignificant	Moderate	Insignificant
Se	Moderate	Moderate	High	Moderate	High	Moderate
Sb	High	Moderate high	Moderate	Insignificant	Insignificant	Insignificant
Cd	Very high	Moderate	Very high	Insignificant	Insignificant	Insignificant

Table 2 Comparison of total element levels in soils around different dump types

Mine-type Element	Minor metals	Gold	Base metals	Chromite and asbestos	Platinum group metals	Sulphur
Ni	High	High	Moderate	Extremely high	Moderate	Moderate
As	Extremely high	Extremely high	High	High	High	Extremely high
Cr	Very high	Very high	Low	Extremely high	Insignificant	Low
Cu	Low	Moderate	Moderate	Insignificant	Moderate	Low
Zn	Moderate	Low	Insignificant	High	Insignificant	Insignificant
Pb	Low	Low	Low	Insignificant	Insignificant	Low
Co	High	High	Moderate	High	High	High
Mo	Insignificant	Insignificant	Insignificant	Insignificant	Insignificant	Insignificant
Se	Moderate	Moderate	Moderate	Very high	Insignificant	Insignificant
Sb	Moderate	Moderate	Moderate	Insignificant	Moderate	Insignificant
Cd	Moderate	Insignificant	Insignificant	Insignificant	Insignificant	Insignificant

Table 3 Comparison of bioavailable element levels in soils around different dump types

Mine-type Element	Minor metals	Gold	Base metals	Chromite and asbestos	Platinum group metals	Sulphur
Ni	Extremely high	Extremely high	Extremely high	Extremely high	Extremely high	Extremely high
As	High	High	High	High	High	High
Cr	Very high	Very high	Very high	High	Very high	Very high
Cu	Extremely high	Extremely high	Extremely high	Extremely high	Extremely high	Extremely high
Zn	Extremely high	Extremely high	Extremely high	Extremely high	Extremely high	Extremely high
Pb	Extremely high	Very high	Very high		High	Very high
Co	Extremely high	Extremely high	Extremely high	Extremely high	Extremely high	Extremely high
Mo	Insignificant	Insignificant	Insignificant	Insignificant	Insignificant	High
Se	Insignificant	Insignificant	Insignificant	Insignificant	Insignificant	Insignificant
Sb	Insignificant	Insignificant	Insignificant	Insignificant	Insignificant	Insignificant
Cd	Insignificant	Insignificant	Insignificant	Very high	Insignificant	

The results indicate that the bioavailable levels of the elements are in general distinctively different from the total concentration levels in the soils with only three exceptions. These exceptions are:

(a) Co in soils around base metals
(b) Ni in soils around platinum group metals and
(c) Se in soils around sulphur dumps.

Table 4 Probabilities of element levels being the same for dumps and soils around the dumps

	Ni	As	Cr	Cu	Zn	Pb	Co	Mo	Se	Sb	Cd
Minor metal	0.38	0.00	0.36	0.01	0.08	0.61	0.70	0.01	0.93	0.07	0.04
Gold	0.25	0.03	0.03	0.03	0.00	0.00	0.29	0.01	0.01	0.07	0.01
Base metal	0.01	0.24	0.52	0.02	0.29	0.29	0.09	0.01	0.14	0.98	0.20
Chromite and asbestos	0.00	0.22	0.06	0.29	0.01	0.37	0.07	0.03	0.06	0.80	0.14
Platinum group metals	0.04	0.06	0.01	0.00	0.03	0.05	0.09	0.00	0.02	0.37	0.27
Sulphur	0.15	0.00	0.15	0.06	0.44	0.05	0.00	0.38	0.31	0.11	0.12

Table 5 Probability of bioavailable element concentration being the same as the total element concentration for soils

Element Dump type	Ni	As	Cr	Cu	Zn	Pb	Co	Mo	Se	Sb	Cd
Minor metal	0.05	0.00	0.02	0.10	0.05	0.05	0.27	0.03	0.00	0.03	0.01
Gold	0.00	0.00	0.00	0.00	0.02	0.01	0.01	0.26	0.00	0.03	0.04
Base metal	0.33	0.01	0.00	0.28	0.38	0.16	0.61	0.30	0.00	0.23	0.04
Chromite and asbestos	0.00	0.21	0.00		0.09		0.16	0.00	0.00	0.03	0.48
Platinum group metals	0.55	0.02	0.00	0.18	0.40	0.02	0.39		0.08	0.36	0.03
Sulphur	0.26	0.00	0.00		0.37	0.01		0.02	0.5	0.07	

The probabilities that the bioavailable levels are equal to the total levels are greater than the probability that they are less implying that these levels cannot be considered statistically different. The 95 % confidence level was also used here. The cases where the bioavailable concentration may be equal to the total concentration indicate a high risk already associated with the soil (especially where the concentrations are high). This shows an already high bioavailability in the soils in the absence of mining activities. Tailings dumps of the mines are therefore going to add potential toxic elements to the soils that already have high bioavailable levels. The effect of this addition to the soils will depend on how much is being added, how much bioavailable toxic elements are already available in that soil, the combination of elements that are in the soils and whether the additions are in bioavailable forms.

Soils that have 100 % of a certain element in bioavailable form indicate conditions prone to bioavailability of that particular element. Addition of that particular element in any form (bioavailable or unavailable) will increase the already high concentrations existing in the soils especially if the element is being added in bioavailable form. The risk still exists even if the element is added in an unavailable form, as the likelihood of conversion to bioavailable form exists.

The lower bioavailable concentrations of the potential toxic elements compared to the total in the soils observed and confirmed by the above t-tests is an indication that in the absence of mining activities the natural levels of bioavailable potential toxic elements are low even in mineralised zones. Though the redox conditions created by mining activities will vary according to the methods employed and will determine the bioavailability of elements in each dump, most mining operations are however known to create conditions that result in the formation of bioavailable forms [2, 23]

The acidic dumps forms the bulk of the dumps analysed in this study and is associated with more elements that may cause problems. The acidic dumps are therefore of greater concern compared to the alkaline dumps. The results show that molybdenum, cadmium, antimony and selenium cannot be sourced from the dumps, as the effective dump concentration is very low. There is no real difference between natural background and dump concentration implying that they are part of the geochemical province where they are present. The study also showed that: nickel in minor metals and sulphur tailings, chromium in minor metals, gold, chromite and asbestos, zinc in chromite and asbestos, cobalt in minor metals, chromite and sulphur, and antimony in platinum group tailings are lower than averages in the soils around these dumps thus these elements cannot be considered problems arising from the dumps.

In the dump conditions where they are bioavailable arsenic chromium and lead are expected to have localised effects due to their immobility while copper, cobalt, zinc and nickel have a possibility of far-reaching effects due to the mobility.

The localised negative effects on the environment associated with elements of low mobility can be dealt with in the immediate areas where the mining is taking place. However, since these elements are immobile they lack dilution arising from dispersion thereby making their concentration higher in the area compared to the ones that disperse. The risk to the immediate areas may therefore be higher for the immobile elements. The mobile elements, though having far-reaching effects, their risk may be lessened by the dilution that arises from their dispersion.

The dumps can thus be characterised on the basis of concentrations and bioavailability for use in sustainable development. From this study it was noted that: minor metals dumps have the highest potential of polluting the environment followed by base metals, then gold dumps, then platinum group elements, then chromite and asbestos' dumps and finally the sulphur dumps. The base metals rank worse than gold mines though gold mines and base metals have similar redox conditions. This is because of the high levels of zinc, cobalt, copper and nickel in the dumps and soils around these dumps, which do not occur at gold mines. In terms of effective dump concentration the base metals have more elements that are higher than natural background. Sulphur tailings come lowest on this scale but this should not be taken to imply that the sulphur tailings are the least polluting mines in Zimbabwe as only one mine was considered and variations are not included.

10 Conclusion

In summary, the results show that molybdenum, selenium, cadmium, chromium are unlikely to cause problems in most dumps while antimony, copper, cobalt, zinc, nickel, arsenic and lead may be problematic in the majority of dumps in Zimbabwe. Effects of antimony, nickel, copper, cobalt, zinc and lead are likely to be significant in acidic dumps such as those for the base metals, minor metals, gold, sulphur and platinum group metals. The effects of arsenic are likely to be present in all dump types alkaline and acidic. Antimony, arsenic and lead effects are likely to be localised to the immediate environments as these elements have low mobility. Copper, cobalt, zinc and nickel are likely to have far-reaching effects due to the high mobility.

The levels of potentially toxic elements observed in dumps and soils during this study have significant implications to the mining industry and particularly to tailings disposal, therefore top sustainable development. Copper, cobalt, zinc lead and nickel are particularly deleterious in the Zimbabwean dump environment. Which of the above species will exist at one time depends on the redox conditions. The species present will also determine how, where and to what extent the element will be dispersed. The soluble species (free ions, simple radicals and inorganic complexes) are however likely to be the greatest dispersing/distributing form. Based on observations made during this study the environmental impacts that can be and are being experienced by the local communities and eco-systems include a decrease in pH, an increase in suspended and dissolved solids, effects on water quality organisms and plants and clogging of water bodies.

This study has demonstrated that the chemical reactions taking place are site and commodity specific. It should be also noted that surface chemistry changes due to chemical weathering taking place in this subtropical country will change the nature and degree and of environmental problems. Thus for sustainable development the geochemistry should be continual assessed and environmental problems re modelled in line with prevailing geochemistry. As a recommendation, geochemical analyses should always be done on all scales. Regional surveys have already been carried out in some countries, and with increased national and international funding they can be extended to cover the rest of the land surface of the globe.

References

1. Bennett MW, Kempton PJ, Maley A (1997) Application of geological block models to environmental management. In proceedings of the fourth International Conference on Acid Rock Drainage, Vancouver, Canada, May 31 June 1997, pp 293–303
2. Chihota JS (1995) Distribution of Arsenic and related heavy metals around a gold mine in Zimbabwe. Environmental engineering, Royal Institute of Technology, Stockholm, Sweden (Unpublished M.Sc thesis)
3. DWAF (Department of Water Affairs and Forestry) (1996) South Africa water quality guidelines 1: domestic use 2nd edition. Government Printer, Pretoria
4. Engdahl D, Hedenvind H (1998) Environmental impacts caused by small-scale alluvial gold mining. Environmental Engineering Department Royal Institute of Technology. Stockholm (Unpublished M.Sc Thesis)
5. Lupankwa K, Love D, Mapani B, Mseka Sd Meck M (2006) Influence of the trojan Nickel Mine on Surface water quality, Mazowe Valley, Zimbabwe: runoff chemistry and acid generation potential of waste rock. Phys Chem Earth 31(2006):789–796
6. Mandingaisa O (1998) Effects of evaporation ponds on groundwater: AMD disposal at Iron Duke Mine Glendale Zimbabwe. geology department university of Zimbabwe (Unpublished B.Sc Honours Thesis)

7. Mangwiro B (1998) Characterizing the source, type of placer accumulation and environmentally sustainable production of gold In: The Angwa River NW Zimbabwe geology department university of Zimbabwe (Unpublished B.Sc Honours Thesis)
8. Maponga OP (1995) Gold Panning along mazowe river and its tributaries. University of Zimbabwe. (Unpublished Institute of Mining Research. Confidential Report C669)
9. Maponga OP (2000) Self-regulation and environmental management in the mining industry in Zimbabwe—a survey of issues. Abstract in environmental issues and management of waste in energy and mineral production. In: Singhal RK, Mehrotra AN, Balkema AA (Eds) Rotterdam. Brookfield
10. McBride MB (1994) Environmental chemistry of soils. Oxford University Press, New York
11. Mend A (1989) Field sampling manual for reactive sulphide tailings. mine environment neutral drainage (MEND) Program Report 4.1.1 Prepared by Canect Environmental Control Technologies Limited
12. Meurig PJ (1987) Applied mineralogy: a quantitative approach mineral resources. Engineering Department. Imperial College, London
13. Mohiddin HL (1997) Small scale mining and gold panning in Zimbabwe Institute of Ecology and Resource Management University of Edinburgh. (Unpubl M.Sc Thesis)
14. Ngwenya G (1997) Environmental effects of mining and mine waste disposal at a mining complex in the midlands greenstone belt. Geology Department University of Zimbabwe. (Unpublished B.Sc Honours Thesis)
15. Ravengai S (2001) Evaluation of seepage and acid generation potential from evaporation ponds: implications and management options for water quality and aquatic life. Iron Duke Mine, Mazowe district Zimbabwe. Geology Department University of Zimbabwe. (Unpublished B.Sc Honours Thesis)
16. Roberts AE (1996) Environmental impact of chromite mining. institute of mining research. University of Zimbabwe Report 159:38–39
17. Runnells DD, Smelds MJ Jones RL (1997) Methodology for adequacy of sampling of mill tailings and mine waste rock. In proceedings of Tailings and Mine Waste 97. Rotterdam: Balkema, pp 561–563
18. Ruzive B (2000) An assessment of the contribution of mine dumps to siltation and Environmental pollution by Heavy metals in Mtorashanga Northern Great Dyke Geology Department University Of Zimbabwe. (Unpublished B.Sc Honours Thesis)
19. Smith KS, Ramsey CA, Hageman P (2000) Sampling strategy for the rapid screening of mine waste dumps on abandoned mine lands. www.crustal.usgs.gov/minewaste/pdfs/leinz1.pdf
20. Soltanpour PN (1991) Determination of nutrient availability and elemental toxicity by AB-DTPA Soil test and ICPS in advances in soil science vol 16 Springer, New York
21. Soltanpour PN, Schwab AP (1977) A new soil test for simultaneous extraction of macro and micronutrients in alkaline soils. Commu Soil Sci Plant Anal 8(3):195–207
22. Thixton DH (1999) Managing mercury in mining. Chamb mines J 1999
23. Watkins R (2000) Environmental geochemistry of mining pollution: University of Zimbabwe. (Unpublished Geology Department course notes for Environmental Geochemistry)

Relevance of Nanotechnology to Africa: Synthesis, Applications, and Safety

Ndeke Musee, Lucky Sikhwivhilu and Mary Gulumian

Abstract In this chapter, two nanotechnology-based applications relevant to Africa in promoting sustainability and achievement of the Millennium development goals (MDGs) are presented. The applications comprise the provision of therapeutic treatment of diseases (HIV/AIDS and malaria) and the treatment of contaminated water through purification, remediation, and disinfection process to promote access to clean water to millions of African inhabitants without clean drinking water. Extensive examination of the available scientific literature suggests that nanotechnology can potentially improve the provision of health and water services in the African continent. While the authors agree these benefits are of great relevance to the continent, the chapter gives insights into the concerns related to the potential risks posed by nanotechnology-based products both to humans and other ecological systems. In addition, the chapter seeks to outline the chemistry underpinning the development of nanotechnology and its relevance in achieving sustainable

N. Musee (✉)
CSIR, P. O. Box 395, Pretoria 0001, South Africa
e-mail: nmusee@csir.co.za

N. Musee
Department of Chemical Engineering,
University of Johannesburg, P.O. Box 524,
Auckland Park, Johannesburg 2006, South Africa

L. Sikhwivhilu
Advanced Materials Division,
DST/Mintek Nanotechnology Innovation Centre,
Private Bag X3015, Randburg, Johannesburg 2125, South Africa

M. Gulumian
NIOH, P. O. Box 4788, Johannesburg 2000, South Africa

M. Gulumian
Haematology and Molecular Medicine,
University of the Witwatersrand, Johannesburg 2000, South Africa

A. Gurib-Fakim and J. N. Eloff (eds.), *Chemistry for Sustainable Development in Africa*, DOI: 10.1007/978-3-642-29642-0_8,

development within the context of developmental challenges in Africa. Finally, as the future socioeconomic status will be mostly defined by nanotechnology capabilities, Africa should be alert to these changes and take advantage, particularly, at this early development phase of nanotechnology development.

1 Introduction

1.1 What is Nanotechnology?

Nanotechnology is the science of manipulating materials at very small scales, i.e., at the atomic and molecular levels. Nanotechnology allows for the design, synthesis, and control at the length scale range of 1–100 nm. This is about 10,000 times smaller than the width of human hair and is in the range of the size of a large protein structure. Nanotechnology is a broad interdisciplinary area of research, development, and industrial activity in which the unifying characteristic is one of size.

The prospects of research in nanotechnology also known as the science of the ultra small, were first made known by the Nobel Prize winning physicist, Richard Feynman in 1959. Since then the concept of nanotechnology surfaced with the idea that atoms could be used to build structures, with absolute control over properties and functions. While materials in the micrometer scale often exhibit physical properties the same as that of the bulk form, it is interesting to note that materials in the nanometer scale may exhibit physicochemical properties uniquely different from their counterpart bulk materials of the same chemical composition. For example, bulk semiconductors become insulators when the characteristic dimension is sufficiently small (within the nanometer region). The underlying difference is size dependency with the effect becoming more prominent when the particles are 100 nm or less in diameter.

As the particle size of materials decreases to nano level a change in physical phenomena is inevitable. Quantum size effect takes precedence as the electronic properties of solids are altered with this reduction in particle size to the nanoscale—accompanied by dramatic increase in the number of atoms at the surface of the material. This results in change in numerous physical properties such as mechanical, electrical, optical, surface properties. Nanoparticles, due to their small size, have a high surface-to-volume ratio and this leads to improved mechanical, thermal, and catalytic properties.

1.2 Nanotechnology: A Chemistry Perspective

Synthesis of materials at nanometre scale is now a routine practice and an active field of research. Materials with one or more dimensions below 100 nm, the normal rules of physics and chemistry no longer apply and many materials start to exhibit novel and unique properties. They may have large surface area, become stronger, more conductive, with improved optical properties amongst others. As such nanomaterials are also defined in terms of novelty and uniqueness of properties they possess as a result of small dimensions.

In Africa, nanotechnology has already been recognized and utilized as an important tool for industrial development and a means to improve the lives of ordinary people through more efficient health care services, safe drinking water, and low cost clean energy.

Although the advanced concept of nanotechnology is still novel, research on the nanometer scale is not new at all. For example, biological systems and the engineering of many materials such as colloidal dispersions and metallic quantum dots have been extensively studied at a nanometer level. Catalysis is also known to be the milestone field when it comes to the use of nanoparticles. This occurred long before the inception of the field of nanotechnology. The high surface area of catalyst particles means that the surface effects are far more important than in the bulk substance. Nanotechnology comprises molecules, atoms, and quantum dots, and is dominated by surface effects such as Van der Waals forces of attraction, hydrogen bonding, electronic charge, ionic bonding, covalent bonding, etc. The vastly increased ratio of surface area to volume gives new possibilities to surface-based sciences such as catalysis. To illustrate the contribution of chemistry to the development of Africa; in the following sections, a few examples of nanomaterials chemistry are presented.

1.3 Carbon Nanotubes

The discovery of carbon nanotubes has led to the dawn of discoveries of nanostructured materials and technologies [1]. The main drive has been their unique structure and properties [2]. Carbon nanotubes have versatile applications in areas such as electronic devices, tools in nanotechnology, hydrogen storage, water purification, etc. [3]. There are two forms of manufactured carbon nanotubes, single-walled carbon nanotubes (SWCNT) and multi-walled carbon nanotubes (MWCNT). The SWCNT is made up of a single layer of graphene sheet rolled up as cylindrical shapes. The rolling up of these graphene layer(s) results in a unique electronic structure. SWCNT has an internal diameter of approximately 1 nm and a length of several microns. However, MWCNT is made up of two or more concentric layers with various lengths and diameters [4].

Various methods have been utilized to generate carbon nanotubes and include arc evaporation, laser ablation, chemical vapor deposition (CVD), etc. [5]. The synthetic yields of SWCNT are often very low but can be improved by adding cobalt or other transition metals such as nickel, iron, and molybdenum. In all the synthetic routes, generation of carbon species (atoms) involves heating the precursor at 600–1200°C under inert ambience and the use of catalytic metal [6]. Both SWCNT and MWCNT contain a proportion of amorphous carbon (nontubular) and metal residual as impurities. Post treatment of these materials is, thus, important to remove all contaminants. Post treatment procedures include gas phase, liquid phase, and intercalation technique. It is important to preserve the aspect ratio of tubes during the purification process (post treatment) in order to retain the inherent unique properties of the materials. Some of these properties include mechanical, thermal, photochemical, and electrical properties which are of industrial use [7–11].

1.4 Gold Nanoparticles

Colloidal gold has increasingly received considerable attention since they hold promise in technological advancement and development. For example, due to their nanosized dimensions gold colloids exhibit interesting optical properties [12]. Depending on the particle size, shape, and agglomeration gold colloids can be red, violet, or blue. Gold colloids also find applications in areas of histochemistry and cytochemistry where they are used as electron–dense labeling agents [13]. Moreover, due to high thermal and electrical conductivity they also find application in electronics [14].

Gold is also widely used in biocatalysis and catalysis. Nano gold has been widely used to construct biosensors due to its excellent ability to immobilize biomolecules such as proteins, enzymes, and antibodies while maintaining the biocatalytic activities of those biomolecules [15–17]. Supported gold catalysts give a high catalytic activity in numerous reactions such as the reduction of nitrogen oxide (1), low temperature of oxidation of CO (2), and the epoxidation of propene (3) [18]. The high catalytic activity of gold is not uniquely strange as bulk gold is quite inert. Reactive molecules such as CO and H_2 do not adsorb on the surface of gold. However, this behavior dramatically changes when gold is highly dispersed as nanosized particles on certain metal oxides such as TiO_2 [19–21].

The success to achieve highly dispersed gold nanoparticles is underpinned by methods and conditions of synthesis. Typically, colloidal gold nanoparticles are prepared by treatment of $HAuCl_4$ with $NaBH_4$ or with Ag (4). Furthermore, for catalysis purposes incipient wetness impregnation is unsuitable to produce well-dispersed gold catalysts. In order to obtain high activity, the catalyst has to be prepared via co-precipitation or deposition precipitation [22].

$$4NO + 4NH_3 + O_2 \xrightarrow{Cat} 4N_2 + 6H_2O \qquad (1)$$

$$CO + 1/2O_2 \xrightarrow{Cat.} CO_2 \qquad (2)$$

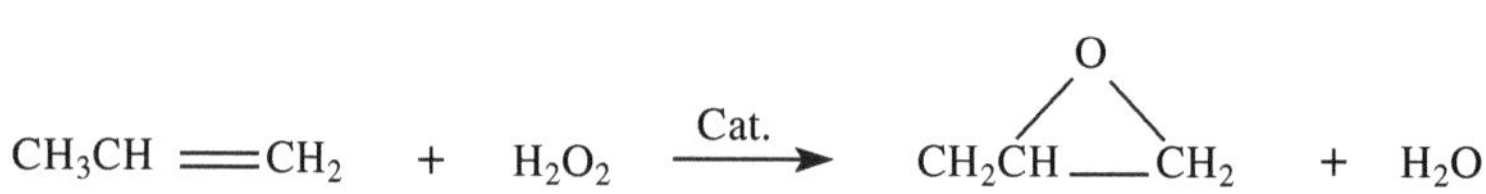

(3)

$$3Ag\ (s) + HAuCl_4(aq) \rightarrow Au\ (s) + 3AgCl\ (aq) + HCl\ (aq) \qquad (4)$$

1.5 Titanium Dioxide Nanoparticles

Because of its availability and relatively low costs; titanium dioxide has been widely used in catalysis as a catalyst or support. Moreover, its low surface area is an important limitation for catalytic applications [23, 24]. In order to alleviate this problem, scientists have developed a variety of new synthesis procedures to prepare titanium dioxide with a higher specific surface area. Nevertheless, for many other applications, titanium dioxide can also be modified by insertion of metal ions to enhance both performance and activity [25, 26].

Titanium dioxide is known to exist in three main crystallographic structures, namely; anatase, rutile, and brookite. Among the three polymorphs of titanium dioxide, anatase is known to have a higher photoactivity than both rutile and brookite [27, 28]. It is therefore, advantageous to produce the anatase powder with a high degree of crystallinity and high surface area (or small grain size) to enhance the photocatalytic activity [29].

There are various methods available for the production of small TiO_2 particles, but the sol–gel method has been widely used since it is deemed relatively inexpensive. Unfortunately, the sol–gel derived material is amorphous in nature and often requires a further treatment to induce crystallization [30]. Elevated treatment temperatures (higher than 350° C) are often necessary to expedite the transition from the amorphous material to the crystalline anatase phase but such high temperatures result in increased size of the nanoparticles and subsequently a decrease in the surface area [31, 32].

1.6 Titanium Dioxide Nanotubes

Nanotechnology has made the synthesis of nanotubular materials possible, with carbon nanotubes at the forefront. Nanotubes of many semi-conducting metal oxides such as TiO_2 have also attracted much attention [33]. Due to the outstanding physical and chemical properties of the novel TiO_2 nanotubes many applications have been anticipated. The morphological feature and high surface area of the tubes have offered TiO_2 materials a variety of applications in catalysis. Nanotubular TiO_2 materials are currently used as a support in areas such as catalysis [34]. They have also been used as a photocatalyst in the treatment of water and due to the optical properties of the material, nanotubes of TiO_2 have been incorporated into flat panels for improved display [35].

The ever-growing interest in titania-derived nanotubes is driven mainly by the large surface area, chemical stability, nontoxicity, and the modest production costs of the tubular materials [36]. Conventionally, the sol–gel method has been widely used to synthesize TiO_2-derived nanotubes using templates such as organogels or porous anodic alumina [37, 38]. These methods are relatively inexpensive and offer numerous advantages since the final products are pure and homogeneous [32, 34, 39, 40]. However, the removal of the homogeneously mixed template is invariably a challenging task [41].

Hydrothermal synthesis has increasingly been used to generate TiO_2 nanotubes. The advantages of using hydrothermal synthesis include the utilization of relatively fewer chemical reactants. The products obtained have high levels of purity. The use of water as solvent renders the process environmentally friendly.

Recently, there has been a growing effort to utilize microwave heating in search for methods that allow the production of nanomaterials with well-controlled properties. Microwave heating is particularly useful for reactions carried out under elevated pressures, because of contactless delivery of energy to the reading fluids, high energy density possible, and short heating times leading to nanoparticles weakly agglomerated, with high crystallinity and narrow grain size distribution.

1.7 Zinc Oxide

Zinc oxide is a very useful material and finds versatile applications in technologies such as gas sensors, UV light emitters, varistors, pigments, and catalysts to mention a few. It has also shown potential applications in UV blocking in cosmetics and suncreens, optoelectronics, and is an excellent candidate for blue to UV lasing [42–44]. Zinc oxide is a semiconductor material with wide direct band gap energy of 3.37 eV and a large exciton binding energy (60 meV) which emits blue light in bulk form because of the bombardment with electron beam at cryogenic temperature [45, 46] However, when particles are reduced to nanoscale level the material starts to emits even at low temperature with low threshold.

The ever-growing interest in nZnO is largely driven by chemical stability and modest production costs. Although the nanoparticles may provide improved specific microstructural, surface, and chemical properties further modification by introducing metal ions may alter both the electrical and optical properties of the material [47, 48]. On the other hand, several methods of synthesizing ZnO nanocrystals have been developed including thermal transport and condensation, template and surfactant assisted liquid methods, among others [49, 50].

Although these methods are relatively inexpensive and generate products that are pure and homogeneous, they are limited in a number of ways. For example, the removal of the homogeneously mixed template is always a challenging task [49]. However, hydrothermal synthesis has proven to be a facile method for the synthesis of nZnO. The use of microwave assisted hydrothermal is also an alternative option that offers many advantages such as low cost, short reaction time, high purity, and large-scale production. Studies have shown that the morphology of nanoparticles of nZnO may be influenced by factors such as synthesis method and temperature [51]. The hydrothermal synthesis of nZnO using hydrothermal synthesis is carried out in the presence of a strong alkaline solution such as NaOH or KOH. The overall reaction is summarized in Eq. 5.

$$[Zn(OH)_4]^{-2} \xrightarrow{heat} ZnO + H_2O + 2OH^- \tag{5}$$

1.8 *Catalysis and Nanoparticles in Water Purification and Other Applications*

The science of catalysis is driven by technology, as it has been from the beginning of studies on catalysts. The primary objective in catalysis research is to design catalysts that can achieve perfect selectivity and desirable activity. It is commonly acceptable that selectivity is much more difficult to achieve and control. Thus, a reaction that yields a perfect selectivity would not generate spin-off products thereby reducing energy and process requirements for separation and purification [52].

Many of the products that support our everyday lives—fuels, fertilizers, construction materials, medicines and artificial fibres, to name but a few—involve heterogeneous catalytic processes at one or several stages in their manufacture. By heterogeneous catalysis, we mean that the catalyst has a distinct and separate phase from that of the reactants so that the catalytic reaction takes place at a boundary, or interface, between phases [53]. In most cases of practical interest the catalyst is a solid while the reactants are in the fluid phase (gas, liquid or solution), so that under these conditions the catalytic reaction takes place at the surface of the solid. The subject of heterogeneous catalysis is very broad and embraces a wide field of physical, chemical, and engineering sciences [54].

A good catalyst should have active sites where chemical transformation occurs to enhance bond formation and breaking. The interaction between the surface of the catalyst and the reactant molecules is crucial as it affects the adsorption and desorption of the reacting species. In the past two decades, the use of homogeneous catalysts has dramatically declined due to the difficulties associated with the separation of reaction products and costs thereof. Solid catalysts, on the other hand, are environmentally benign and more compatible with increasingly demanding environmental requirements [54, 55].

2 Nanotechnology Applications: The African Context

For nanotechnology applications to be relevant and sustainable in the Africa context, the technology should offer social and economic utility values. In this section, we argue that the nanotechnology can enhance sustainability within the context of solving the African continent challenges given it can provide novel solutions to societal needs that accelerate the achievement of the millennium development goals (MDGs), and secondly, proactively support socioeconomic developmental challenges. Salamanca-Buentello et al. [56] ranked the top ten applications of nanotechnology that can support sustainable development issues in the developing world including Africa. Similarly, the United Nations Environment Development [57] report suggested nanotechnology has the potential to contribute to the target set for achieving the UN MDGs specifically in the areas of water, energy, health, and the environment. Given the MDGs were set in an attempt to lift the livelihoods of millions of people in the developing countries—including

Africa, these areas of need appear as the most appropriate areas for the nanotechnology applications in a resources-limited region as Africa.

Conversely, this case is strengthened by the well known environmental, economic, and social impacts of poor water supply and sanitation [58] as well as the health and welfare of people—especially the vulnerable groups like the elderly, children, poor, and those with comprised health systems [59]. It is in this context that, several African countries have initiated activities on nanotechnology research and development (e.g. South Africa, Egypt, and Botswana), or have expressed interest in the technology (e.g. Tanzania, Ghana, Senegal, Kenya, Swaziland, and Zimbabwe) [60]. Over the last 5 years, the authors contend that the situation is likely to have changed; however, little has been published to expound on the account of the intended uses of nanotechnology in most Sub-Saharan countries except South Africa.

Therefore, using the South Africa National nanotechnology strategy (NNS) [61] and the Ten-Year Research Plan [62] as guiding pillars of addressing socially related development needs using nanotechnology—the application fields include health, water, and energy. Conversely, the industrial-based application includes the beneficiation of minerals as well as supporting advanced materials manufacturing, and chemical and bioprocessing [61, 62]. As the focus of the chapter is to examine the contribution of nanotechnology to support sustainable development in the broader Africa context—only the social-related applications are summarized.

2.1 Nanotechnology: Water-Related Applications

Nanotechnology is a suitable platform to offer sustainable solutions to address challenges related to; water scarcity, increasing quantities of portable water, and improving clean water supply for domestic and industrial uses in the developing countries—including Africa [63]. For example, Hillie et al. [64] report identified specific opportunities associated with nanotechnology applications, especially those with potential utility value in the developing world. The authors identified the use of nanofiltration as a suitable technology for the treatment of water in developing countries. To contextualize the role nanotechnology can play in improving the supply of clean water in the developing countries (with focus on Africa), it is essential to first examine the extent of the problem.

In this section, several examples are presented to illustrate the contribution of nanotechnology in enhancing water quality and quantity particularly in the rural dwelling settings. Broadly, nanotechnology finds potential application in water purification through water treatment, bioremediation, purification, and disinfection processes. Detailed and excellent review of nanotechnology applications and potential long-term opportunities in water treatment has been presented by Cloete and co-workers [65].

2.2 Water Purification

Conventionally, numerous and diverse water treatment technologies have been developed over the years to purify water including sedimentation, activated carbon, coagulation–flocculation, adsorptive filtration using ion exchange resins, and biological processes. However, many of these processes have shown serious limitations in removing dissolved salts as well as soluble inorganic (e.g. heavy metal ions) and organic pollutants from contaminated water. Baker [66] summarized the fundamental principles of membrane technology and their applications in water and wastewater treatment, desalination as well as water reclamation. Examples of membranes comprise of microfiltration (MF), ultrafiltration (UF) and nanofiltration (NF), and reverse osmosis (RO).

Several nanotechnologies have been investigated for potential applications in water purification systems. For example, Srivastava et al. [67] presented results on the production of nanostructured membranes—carbon nanotubes filters—characterized by ease of manufacturing them and controlling their cylindrical geometry—comprising of radially aligned carbon nanotubes. These carbon nanotubes-based filters were shown to be effective in removing 25 nm-sized polio virues (*Poliovirus sabin 1*) and bacterial pathogens (*Escherichia coli* and *Staphylococus aureus*) from contaminated water. The filters performance was found higher in comparison to that of the conventional membranes due to their high thermal and mechanical stability, large surface area, reusability (achievable through ultrasonication or autoclaving), high influx, and relatively low production cost. On the other hand, branched dendrimers supported in polymers comprising of a central core, repeating units, and terminal functional groups were also used [68]. There are mainly two types of dendrimers; hyperbranched polymers [69] and cyclodextrins [70]. The diversity of these nanomaterials' in water purification is achieved through functionalization; for example, cyclodextrin polyurethanes copolymerized with functionalized multi-walled carbon nanotubes as adsorbents exhibited capability to remove organic pollutants present at nanograms per liter (ng $\ell^{-}1$) or lower in water [71, 72].

Other studies have shown potential applications of cyclodedrins for the removal of pesticides [73] and organic pollutants from contaminated water [74]. Removal of organic pollutants is enhanced using $nTiO_2$ porous ceramic filters impregnated with an alkylated poly(propylene imine) dendrimer, poly(ethylene imine) hyperbranched polymer or β-cyclodextrin, thus resulting in hybrid organic/inorganic filter modules. The modules are characterized by high mechanical strength and high surface area. The results of testing the filters showed the polycyclic aromatic hydrocarbons (PAHs) pollutants removal exceeded 95 %, whereas those of trihalogen methanes (THMs), monoaromatic hydrocarbons (benzene, toluene, xylene), and pesticides (simazine)) were also removed efficiently above 80 %. Finally, other studies have demonstrated the effectiveness of using metal-based nanoparticles such as silver and gold to remove pesticides (e.g. endosulfan, malathion, and clorpyrifos) [75]. The findings suggested the reaction products were environmentally benign, no by-products were generated, and exhibited high selectivity of nanoparticles in treating the targeted pollutants in comparison to the counterpart bulk form of noble metals, and the capability to target a large range of pesticides such as halocarbons (e.g., benzyl chloride, chloroform, and bromoform), organochlorine pesticides (e.g., endosulfan), and organophosphorus pesticides (e.g., chlorpyrifos, malathion). Using sensitive instrumentation, it was established that all such organics underwent complete mineralization at low concentrations [76, 77]

2.3 Water Remediation

In Africa, there are expansive contaminated water and soil environments due to large industrial, mining, agricultural, and poor waste disposal activities as well as leakages from petrochemical storage tanks. The most dominant pollutants include the toxic heavy metals (e.g. arsenic, mercury, cadmium, lead, etc.), toxic effluents such as the acid mine drainage (ADM), pesticides, and organic pollutants. Most pollutants exhibit high biopersistence, bioaccumulation, and long half-lives (not easily biodegradable) in the environment. As a result, they cause diverse adverse effects to humans such as cancer, liver damage, and deformations in babies. In addition, they disrupt the environmental ecosystems through distortion of community populations to plants, invertebrates, and vertebrates. However, majority of the traditional technologies like solvent extraction, activated carbon adsorption, and common chemical oxidation, whilst effective, has serious limitations as they are costly, laborious, and time-consuming [78]. In addition, whereas the biological degradation techniques are environmentally friendly and cost-effective; but are time-consuming [79].

It is within the constraints of the current technologies such as in ability to remove the toxic contaminants from the environment to safe levels, long lead time requirements, and high costs [80] that motivated research efforts in search of alternative efficient techniques of cleaning contaminated regions. The use of nanomaterials has become one active area of research efforts of remediating contaminated soils and underground waters due to their strong reduction effect in transforming toxic into less or nontoxic pollutants. This is enhanced by their large surface area, high reactivity, and sequestration properties [81]. At present, a large number of nanoparticles and nanomaterials have been researched as potential candidates for remediating industrial effluents, groundwater, surface water and drinking water, and recently have been comprehensively reviewed [65, 82].

For illustrative purposes, the use of zero-valent iron nanoparticles (nZVI)—which is the most studied metallic nanoparticle for the environmental remediation [83], is briefly discussed. Findings show that nZVI are very effective for the transformation and detoxification of a wide variety of common environmental contaminants like chlorinated organic solvents, organochlorine pesticides, and PCBs. In addition, nZVI are effective in reducing other contaminants such as nitrate [84, 85], perchlorate [86], chromate [87], arsenite [88] to benign by-products. And finally, nZVI has shown great potential in removing heavy metals (e.g. lead, copper, arsenic, nickel, etc.) from solution or dramatically reducing their reactivity [87–89]. Keum and Li [90] showed the effectiveness of the nZVI in remediating pesticides through the reduction of debromination of polybrominated diphenyl ethers (PBDEs) where over a 40 days period 92 % of BDE congener 209 was transformed into lower bromo congeners. In addition, under 5 days, hexa- to heptabromo BDEs were found the most abundant products, but the tetra- to pentabromo congeners were dominant after 2 weeks. More importantly, no oxidation products were detected in all these experiments. The results suggest that a stepwise debromination from n-bromo to (n-1)-bromodiphenyl ethers was the dominant reaction in all congeners. Debromination of PBDEs by zero-valent iron has high potential values for remediation of PBDEs in the environment [91]. The potential of nZVI for remediating underground water has reported [92]. Elliot and Zhang [92] reported that trichloroethene (TCE) was reduced up to 96 % after injecting 1.7 kg of nZVI into the underground water. Despite the remarkable results reported to date on the use of nZVI for remediation in groundwater, findings in soil environments are lacking or limited. Therefore, in order to improve the large-scale application of the nZVI for remediation applications, several challenges merit redress. These include avoiding or eliminating the aggregation of nZVI using several stabilizers such as water-soluble starch [93], hydrophilic carbon or polyacrylic acids [94], sodium carboxymethyl cellulose (CMC) [95], and polymers [96].

2.4 Water Disinfection

The current conventional water disinfection techniques exhibit numerous limitations. This has formed the basis for active research efforts towards the development of alternative technologies including the application of nanotechnology in water disinfection have been initiated over for the last few years. The proposed technologies are likely to generate great appeal to the developing countries because of their low cost, do not require complex technological support (simple to apply), produces no harmful by-products, and can support decentralized or point-of-use water treatment and reuse systems. In this chapter, few examples are presented to illustrate the utility of nanotechnology for water disinfection (for detailed review see [65, 82].

Nanoparticles have been shown to possess strong antimicrobial properties (e.g. nTiO2, nZnO, nAg and fullerol) through diverse mechanisms including photocatalytic production of reactive oxygen species that damage cell components and viruses ability to compromise the bacterial cell envelope (e.g. peptides, chitosan, carboxyfullerene, CNTs, nZnO, nAg), interruption of energy transduction (e.g. nAg and aqueous nC60), and the inhibition of enzymatic activity and DNA synthesis (e.g. chitosan) [82]. The nanoscale-based disinfectants have proven to possess similar or superior antibacterial activities in comparison to the conventional chemical disinfectants. For instance, $nTiO_2$ on a thin film in a UV reactor have showed higher bactericidal effect (3600–8500 ml/cm^2) [97, 98] in comparison to chloramine (95–180 ml/cm^2) [99] or ozone (0.0007–0.02 ml/cm^2) [100].

The antibacterial properties of nAg have been proposed to be due to their size and shape [101, 102]. Results of the effect of nAg in the size range 1–100 nm on the growth of four Gram-negative bacteria (*Escherichia coli*, *Vibrio cholerae*, *Pseudomonas aeruginosa*, and *Salmonella typhus*) suggested that growth of the bacterial cultures (5 $\times$ 10^7 colony forming units (cfu)/ml) was inhibited at concentrations exceeding 75 μg/ml. On the other hand, previous characterization of the interactions of the nAg with the bacteria using high angle annular dark field (HAADF) scanning transmission electron microscopy (STEM) showed that nanoparticles attached to the surface of the cell membrane or located inside the bacteria were mainly in the size range of 1–10 nm [101]. For the first time, shape-dependent antibacterial properties were reported by Pal et al. [102] suggesting that truncated triangular silver nanoplates exhibited stronger biocidal action when compared with spherical and rod-shaped nanoparticles and with Ag^+ (in the form of $AgNO_3$).

The antibacterial activity of silver ion (Ag^+) and related silver species, and to a lesser extent nAg have been studied extensively [101–107]; however, the exact mode of action for the nAg on bacteria is presently partially unknown. Several propositions have been advanced to explain this phenomenon. First, the catalytic oxidation by metallic silver and reaction with dissolved monovalent silver ion was viewed as likely contributing factor to the bactericidal effect [108]. Alternatively, this was explained based on the electronspin resonance spectroscopy studies of nAg showing the damage of the bacterial membranes may have been induced owing to the formation of free radicals [104, 109].

Another form of nanoparticles used in water disinfectant is the $nTiO_2$ through the photodegradation process due to UV light irradiation, and the complete mineralization of toxic organic pollutants. This phenomenon, successfully, was applied in environmental technology for wastewater and groundwater treatment, removal of benzothiophene from diesel fuel, degradation of air pollutants (e.g. nitrogen oxide, sulforoxides, and volatile organic compounds), and in the killing of bacteria, fungi, algae, and viruses [110–115]. For example, using *E. coli* as the model organism, the bactericidal activity of TiO_2/UV photocatalysts properties have been demonstrated with enhanced effect due to use of TiO_2 nanorods [116]. Alternatively, this is achieved through doping using either iron (Fe(III)) [117] or silver (Ag) [118]. However, despite the large specific surface area of $nTiO_2$ the development of commercial applications is slow chiefly because of their tendency to aggregate and coalesce very easily forming larger particles. Such undesirable effect lowers the catalyst efficiency, and requires complex procedures to separate and recover the $nTiO_2$ particles from the reactant mixture [119].

The forgoing few examples illustrate the nanotechnology-based solutions potential in addressing water-related challenges in the developing world including Africa. Therefore, the potential for nanoparticles to offer alternative technologies to support the provision of clean water, increase water access, and clean up the environment in the context of the developing countries is beyond debate. However, there are both technical and environmental risk aspects that are outstanding—which in the view of the authors should adequately be addressed before full scale commercialization of these technologies.

2.5 Nanotechnology: Health-Related Applications

Nanotechnology offers numerous advantages, in comparison, to the traditional drug design, delivery, and medical diagnostics approaches primarily due to the possibility of controlling and manipulating structure at the molecular level. For example, in pharmaceuticals—nanotechnology has made the dissolution rates higher and faster, have increased the bioavailability of many drugs, and improved the stability of sensitive agents. This is because it has been shown that nanoparticles can be engineered to (1) recognize diseases at the cellular level (2) provide visual imaging of affected organs (through imaging studies), and (3) deliver the therapeutic compounds. Currently, there are enormous scientific efforts in improving the treatment of cancer using nanotechnology-enabled therapeutic drugs; however, the high prevalence of epidemic infectious diseases particularly the human immunodeficiency virus (HIV) responsible for causing acquired immunodeficiency syndrome (AIDS) (HIV/AIDS), tuberculosis (TB), and malaria in the developing countries—merits special attention in order to reduce, manage, or eliminate the present unacceptable disease burden these diseases have impacted in the society particularly in Africa as shall be presented in the latter section. HIV/AIDS and malaria will be discussed in the context of nanotechnology application to illustrate this possibility.

2.6 Diagnostics for the HIV/AIDS

Presently, the HIV/AIDS is a global challenge and though the conventional drug delivery approaches including the highly active antiretroviral therapy (HAART) have shown the capability to increase the life span of the patients, however, serious limitations to achieve total eradication of the HIV remains. One way of bringing the pandemic under control is to exploit the capability of nanotechnology platforms currently at preclinical, and research and development phase. Comprehensive reviews on the potential of nanomedicines for the treatment of HIV/AIDS have been presented recently [120–125] with special emphasis on their application in the developing world. In this chapter, only a few salient aspects are highlighted. The application of nanotechnology in the treatment of HIV/AIDS is partly premised in addressing the limitations of the current antiretroviral therapies [122], and secondly to enhance the treatment effectiveness, efficiency, reduction of costs [124].

For example, it has been reported that HAART is generally less ineffective for the treatment of central nervous system (CNS) complications than other AIDS-related illnesses [126]. As a result, even though HAART remains fairly effective against the CNS illnesses; however, it is ineffective to more severe cases such as HIV-associated dementia (HAD). Consequently, this leads to high numbers of HAD patients living with HIV/AIDS as their lives are prolonged. In addition, the inability of these drugs from completely or reducing the viral loads in the CNS considerably potentially allows latent infection [127] which can promote the development of drug resistance—multi-drug resistance (MDR) and as result limit the effectiveness of the antiretroviral therapy [128]. Because of the inefficiencies of the present HAART regimens, it is estimated that up to 7.7 years may be required for uninterrupted treatment to eliminate the HIV viral load during the early phases of infection [129]. Clearly, such long periods of treatment are undesirable due to potential increase in risks like liver dysfunctions, peripheral neuropathy, metabolic complications, as well as high drug costs, patients' noncompliance, and possible drug–drug interactions [130]. And finally, the current HAART is limited or fail to deliver the drugs into the brain [124]. Also, the current conventional methods are limited in providing sustained release and delivery of active molecules to the target sites [120].

To address the above challenges in pursuit of improving the management and treatment of the HIV/AIDS, several nanotechnology-based treatment therapies have been proposed and are under research and development or preclinical phase as mentioned earlier. To enhance the delivery of the ARVs into the CNS, a number of nanocarriers have been studied in in vitro or in animal models [131, 132]. Three nanocarriers broadly classified as polymer/dendrimer based, lipid based and micelle based have been developed. However, only a few are suitable candidates for drug delivery into the brain have been identified. The criteria for the selection of a nanocarrier include; being nontoxic, fully biodegradable as well as producing well-characterized, and harmless degradation products.

One of the innovative and feasible pathways of using nanocarrier-based delivery for the HIV/AIDS drugs is the multi-functionalization of the nanocarrier system [122]. Multi-functionalization of the nanocarrier allows the incorporation of several therapeutic agents (e.g. HAART) into one formulation for maximum clinical effect. Secondly, it allows the drugs to be effectively targeted to the desired site using the multi-functional nanocarriers. It appears from the extensive review of studies reported to date (see for example reviews by Gupta and Jain and Wong et al. [122, 124]), that the solid lipid nanoparticles (SLN) and liposomes are among the most suitable candidate nanocarriers for the HIV/AIDS drug therapeutic delivery. This is because they are biocompatible, safe, versatile and flexible, stable as well as technically feasible for fabrication at industrial scale—and at reasonable costs. That means the SLN and liposomes offer tremendous possibilities in the clinical application of nanotechnology for the treatment of HIV/AIDS infections in the coming years. Therefore, it is premised that such a drug delivery nanocarrier-based system would find great appeal in the developing world—including Africa—not only because of superior clinical efficacy, lower costs, but also due to ease of administration particularly in very remote regions.

2.7 Diagnostics for Malaria

Malaria is the most prevalent parasitic infectious disease to humans caused by the *plasmodium* genus. The *plasmodium* genus are broadly classified as; *Plasmodium falciparum, Plasmodium vivax, Plasmodium malariae* and *Plasmodium ovale*—and transmitted to humans by female mosquito vector of the *Anopheles* genus. Malaria is the leading parasitic infectious disease imposing the highest mortality and morbidity globally, and particularly in the developing countries [133, 134].

One of the challenges is the intractable difficulties in the disease control because of the parasite complex lifecycle [135, 136] and the increasing number of immune-compromised patients that suffer from malaria and HIV/AIDS co-infections [137]. Another factor is the complexity of the recommended regimens—which are usually a combination of two or more drugs. The latter factor contributes to rapid increase in the drug costs and a reduction in patient compliance mainly associated with severe side effects [138]. Furthermore, extrinsic factors like technical and operational failures in implementing campaigns to fight malaria, poor quality of medicines distributed in different countries, drug–drug interactions, the unavailability of less toxic drugs, resistance of the vector to insecticides, and socioeconomic conditions of affected populations aggravate the difficulties for the eradication of malaria in the world [139]. Up to now, a suite of tools and methods have been developed to reduce or eliminate the widespread dissemination of malaria through measures such as the prevention of infection, preventive treatment using antimalarial drugs, and treatment with artemisinin-based combination therapy (ACT) [138].

It is in the context of the above setting compounded by the small number of new drugs or innovative antimalarial medicines approved since 1990, the search for more efficient and less toxic antimalarials, the development of a successful vaccine, and the design of nanotechnology-based delivery systems applied to drugs and antigens are likely to be the main strategies in combating this disease in future. The research efforts towards the development of nanocarriers are currently at the research and development stage where several nanosized delivery systems have shown positive effectiveness in animal models and prophylaxis of malaria. Nanocarriers are useful tools to improve the pharmacokinetic profile of effective drugs characterized by poor water solubility, low bioavailability, and high intolerable toxicity which posed serious limitations to the pharmacotherapy [140, 141].

A number of nanocarriers namely; colloidal liposomes, polymeric nanoparticles, lipid nanoparticles including lipid drug conjugate (LDC) nanoparticles, solid lipid nanoparticles (SLN), and polymer chitosan have been proposed for the optimization of anti-malarial drugs delivery to treat malaria infections. An example of promising anti-malarial agents is the phosphorothioate antisense oligodeoxynucleotides (ODNs) in silencing malarial topoisomerase II gene especially to address the increasing drug resistance of *P. falciparum* [142]. However, the drug's effectiveness is hampered by poor stability and limited intracellular penetration. To overcome these limitations, Foger et al. [142] showed that ODNs complexed with the biodegradable polymer chitosan to form solid nanoparticles increases the susceptibility of *P. falciparum*. In that study, nanoparticles were found to protect ODNs from nuclease degradation. *P. falciparum* K1 strain exposed to the chitosan/ODN-nanoparticles for 48 h in order to examine the effects of chitosan/antisense (AS) and chitosan/sense (S) oligodeoxynucleotide nanoparticles on malaria parasite growth.

Both negatively and positively charged antisense nanoparticles as well as free antisense ODNs (in a final concentration of 0.5 μM) showed sequence specific inhibition compared with sense sequence controls. Notably, nanoparticles were much more sequence specific in their antisense effect than free ODNs. For instance, nanoparticles with negative surface charge exhibited a significantly stronger inhibitory effect (~ 87 % inhibition) on the parasite growth in comparison to the positive ones (~ 74 % inhibition) or free ODNs (~ 68 % inhibition) [142]. Furthermore, the use of nanoparticles demonstrated more pronounced sequence specific antisense effects as compared to free ODNs likely due to sustained release of ODNs from chitosan nanoparticles hence lowering the initial concentration—results which are in agreement with earlier findings [143, 144]. This confirms ODNs inhibit cellular gene expression in a specific sequential matter at low concentrations.

Another example is the encapsulation of the quinine (CQ) in long-circulating liposomes. The drug release kinetics were evaluated mimicking physiological fluids and endosome environments [145]. CQ was highly encapsulated in neutral conventional and PEGylated liposomes using a transmembrane pH gradient method to improve the encapsulation efficiency. Initially, blank conventional liposomes were prepared with SPC and CHOL, or PEGylated liposomes using 5 mol % PEG 2000 grafted onto DSPE-PEG2000 and CHOL. Thereafter the entrapment of the drug was performed in neutral liposomes by an imposed pH gradient. Finally, the CQ-loaded liposomes were lyophilized using trehalose as cryoprotectant. The highest CQ efficiency of encapsulation was 99 % before lyophilization and 87 % after hydration of the lyophilized form for an internal pH of 3.6 (0.2 M). The in vitro release profile of CQ was different, considering the pH of the internal phase of liposomes and the pH of the release medium (7.4 and 5.5). For conventional liposomes, the cumulative release of CQ at physiological pH 7.4 was 30 % within 6 h, while more than 90 % of the drug was released from liposomes at pH 5.5. In contrast, PEGylated liposomes significantly reduced the release of CQ from liposomes at pH 5.5 [145].

The above two examples have shown the possibilities provided by the pharmaceutical nanotechnology in improving the efficacy of the antimalaria drugs to overcome the challenges of the conventional delivery systems such as poor solubility, chemical instability, inadequate bioavailability profiles, and intolerable toxicity. This is due to the possibility of manipulating physicochemical properties of nanocarriers (e.g. surface reactivity, surface charge, etc.) aimed at improving antimalaria drugs selectivity; however, these drugs development are at research and development phase, and only after some years before their full potential to treat infectious malaria disease will be realized. Recent detailed review of the current nanotechnology-based antimalaria drugs advances have been presented by Santos-Magalhães and Mosqueira [138].

3 Health Effects of Nanomaterials

Concerns have been raised that the very desirable properties of nanoparticles exploited for their beneficial applications might also be the reason for their toxicity. With the knowledge that many applications of nanotechnologies may pose new risks, international agencies have established working groups to consider their potential risks to health and environment during manufacture, use, and disposal in order to ensure their beneficial applications without sacrificing their safety.

The hazardous properties and their health effects of newly synthesized nanoparticles and nanomaterials have been investigated using in vitro and in vivo animal systems. Using experimental animals, the behavior of nanoparticles and nanomaterials and their distribution to target organs following exposure via inhalation, ingestion or through the skin, were studied. Using cells that represented these target organs, their toxicity, and their partition within different cellular compartments were investigated. These studies were essential for proper risk assessment which requires understanding of the toxicological hazards associated with these materials and of the levels of exposure, which may produce these toxicities.

It is important to emphasize, at this stage, that many of the nanoproducts may pose no risk to health or to the environment if they are fixed to or are incorporated within a material. It is the exposure to free manufactured nanoparticles and nanotubes in the production stage or during their release from the fixed or embedded composites that will be of major concern. A short summary of the identified adverse effects of free nanoparticles and nanotubes will therefore be presented in this section that are presently being synthesized to create awareness on their toxicity and urge the researchers as well as industry to take the necessary precautions to make their products safer in their applications.

3.1 Carbon Nanotubes

Potential health effects of carbon nanotubes were of concern due to their similarity to asbestos fibers [146]. Great number of in vitro and in vivo studies has therefore been conducted to assess their toxicity and pathogenicity. For example, in vivo experiments with SWCNT have shown their ability to translocate [147] and distribute to different organs, as well as accumulate in the cell nucleus [148]. Intravenous injection of acid-oxidized MWCNTs and Tween-80-dispersed MWCNTs has produced severe liver damage with the latter producing more severe effects [149]. Ocular instillation could produce no adverse effects [150] but exposure through the skin could result in dermal penetration and toxicity due to generation of free radicals, oxidative stress, and inflammation [151].

Inhalation studies have found that MWCNTs could produce pathological changes in the lung [152] as well as alteration in systemic immune function [153].

Intraperitoneal instillation to MWCNT of mice could result in asbestos like, length dependent, pathogenic behavior including inflammation and formation of granulomas [154] and mesotheliomas in p53 heterozygous mice [155] and in Fischer 344 rats [156]. Furthermore, an investigation of molecular signaling on normal mesothelial and malignant mesothelial cells in vitro has revealed that SWCNTs were able to produce similar molecular events as asbestos. Using the same cell types, these authors could also show that MWCNT containing low iron content generated only negligible amounts of reactive oxygen species, but has caused a parallel activation of two important transcription factors. It was, therefore, concluded as in SWCNT that MWCNT are biologically potent activators of molecular events in normal cells associated with mesothelioma development [157, 158].

Their incubation with human lung fibroblasts, could cause direct fibrogenic effects [159] and with human bronchial epithelial BEAS 2B cells could produce genotoxic effects [160]. These cytotoxic responses of cells in culture were found to be dependent on the presence of functional groups where an increase of sidewall functionalization decreased SWCNT toxicity to human dermal fibroblasts [161]. Multi-walled carbon nanotubes could also produce similar effects as to those observed with SWCNT. The incubation of these nanotubes with human epidermal keratinocytes in vitro could decrease cell viability and a significant increase in an inflammation marker (interleukin-8) [162]. A comparative study between the cytotoxicty of single walled and multi-walled carbon nanotubes on alveolar macrophages in Guinea pigs [163] and genotoxicity on murine macrophage cell line RAW 264.7 [164] have shown that SWCNT has higher toxicity than MWCNT.

If released in the environment, nanomaterials might be inhaled by populations and cause damage to the deepest regions of the respiratory tract, i.e., the alveolar compartment. To model this situation, we studied the response of A549 human pneumocytes after exposure to aluminium oxide or titanium oxide nanoparticles, and to multi-walled carbon nanotubes. The influence of size, crystalline structure, and chemical composition was investigated. After a detailed identification of nanomaterial physico-chemical characteristics, cells were exposed in vitro and viability and intracellular accumulation were assessed. In our conditions, carbon nanotubes were more toxic than metal oxide nanoparticles. Our results confirmed that both nanotubes and nanoparticles are able to rapidly enter into cells, and distribute in the cytoplasm and intracellular vesicles. Among nanoparticles, we demonstrate significant difference in biological response as a function of size, crystalline phase, and chemical composition. Their toxicity was globally lower than nanotubes toxicity. Among nanotubes, the length did not influence cytotoxicity, neither the presence of metal catalyst impurities [165].

3.2 Quantum Dots

Under certain conditions, QDs may also pose risks to human health and the environment [166]. For example, release of QD precursor core materials such as cadmium and selenium was found to be one of the mechanisms involved in their toxicity where this toxicity was found to be directly related to the accessibility of the core cadmium atoms to the surrounding medium and to the permeability to oxygen and protons of the different extra layers of materials that are added to the core (shell and ligands) [167, 168]. Subsequently, it was proposed that a shell will reduce QD toxicity by delaying the oxidation of the core [169]. Moreover, the ability of QDs to generate free radical species during excitation including the hydroxyl ($^{\bullet}OH$) and superoxide ($^{\bullet}O_2{}^{-}$) free radicals as well as the production of singlet oxygen ($^1O_2$) were also examined [170] that could affect DNA [171] was proposed to be due to the degradation of the ZnS shell upon irradiation, illustrating once more the importance of a well-protected core [172]. Mechanisms of toxicity by QDs could therefore be summarized by their ability to generate free radicals and also through the shell/core degradation [170] resulting in the potential risks in their biologic and clinical applications [173].

Additional physicochemical properties that were found to determine these toxicities included size, charge, concentration, outer coating bioactivity (capping material, functional groups), and oxidative, photolytic, and mechanical stability. In addition, it has been found that factors such as concentration, species, and exposure time may also influence their toxicity. The importance of the functional groups of surface coating on the toxicity of QD was also investigated. For example, the genotoxicity of the surface modified coating functional groups ZnS/CdSe—OH/COOH and ZnS/CdSe—NH_2/OH groups using the Comet assay could show that ZnS/CdSe —COOH exhibited a distinct toxicity in WTK1 cells [174]. On the other hand, SiO_2-coated QDs were more water soluble and noncytotoxic [175]. In HepG2 cells, a surface modified with Fluronic® 68 (F-68) and sodium dodecyl sulfate (SDS) could reduce the cytotoxicity of CdSe QDs, while surface modification with cetyltrimethylammonium bromide (CTAB) showed significant cell damage [176]. Thus, both the intrinsic natures of QDs and the external environmental conditions should be considered when evaluating QD toxicity [166].

3.3 Silver Nanoparticles

In vitro skin penetration of silver nanoparticles, coated with polyvinylpyrrolidone, with intact and damaged human skin was studied and was found that silver nanoparticle absorption through intact and damaged skin was very low but detectable [177]. Silver nanoparticles were also found to be cytotoxic to cultured human fibroblasts and keratinocytes [178] and to rat liver derived cell line (BRL 3A) [179] in vitro. On the other hand, silver nanoparticles were also found to interact size-dependently with HIV-1 via preferential binding to the gp120 glycoprotein knobs and thus inhibiting the virus from binding to host cells, as demonstrated in vitro [180].

The in vivo toxicity and pathogenicity of silver nanoparticles was also investigated. A 28-day inhalation exposure study in Sprague–Dawley rats found that both male and female rats did not show any significant changes in the hematology and blood biochemical values in either male or female rats with no significant changes in body weight relative to all the concentrations of silver nanoparticles during entire experimental period except hyperplasia in bile-duct [181]. A 28-day oral exposure, again in Sprague–Dawley rats, could not find any hyperplasia in bile-duct, indicating that a higher dose with prolonged exposure is needed to induce these responses [182]. A subchronic 90-day inhalation study could, however, find that lungs and liver were the major target tissues for prolonged silver nanoparticle accumulation. Bile-duct hyperplasia could also be noted in both the male and female animals with a higher accumulation of silver nanoparticles in the female kidneys [183].

3.4 Gold Nanoparticles

Size and functional groups on the surfaces of gold nanoparticles were found to determine their toxicity. Initial reports suggested that gold nanoparticles with functional cationic side chains are likely to be toxic to mammalian and bacterial cells [184], whereas the toxicity of anionic surface modifier was found to be dependent on the property of functional group present. On the other hand, gold nanoparticles with citrate and biotin surface modifier did not reveal obvious toxic, nonetheless, glucose and cysteine were found to be [185]. On other hand, layer-by-layer polyelectrolyte (PE) coated gold nanorods were found to be of low toxicity and therefore recommended for applications such as thermal cancer therapy [186].

In addition to the functional groups, the size of the gold nanoparticles did also influence their toxicity. For example, it was found that in spite of efficient uptake into human cells by endocytosis, AuNPs with 18 nm showed little cytotoxicity [187]. Consistent with these observations, others have found that particles of 1–2 nm in size were highly toxic and AuNPs larger than 15 nm were not toxic [188]. Cytotoxicity also depended on the type of cells used. For example, 33 nm citrate-capped gold nanospheres were found to be noncytotoxic to baby hamster kidney and human hepatocellular liver carcinoma cells, but cytotoxic to a human carcinoma lung cell line at certain concentrations [189]. Moreover, 10-, 20-, and 40-nm gold particles conjugated with anti-protein A antibodies for targeting the bacterial surface were shown to be selectively toxic to Gram-positive bacterium *Staphylococcus aureus* [190]. Recent excellent review has summarized data on cytotoxicity of many types of nanomaterials including gold nanoparticles [191].

Several groups have also examined the translocation and tissue distribution of gold nanoparticles. Rats exposed to gold particles with diameters of 5–8 nm have shown that a large portion of the deposited gold particles was retained in the lung tissue and also alveolar macrophages. In addition, a low but significant increase of gold was found in the blood indicating systemic particle translocation [192]. A similar systemic translocation was also found following ingestion of colloidal gold particles in mice. In addition, absorption in the animals' brain, lungs, heart, kidneys, intestines, stomach, liver and spleen, were more pronounced for 4 and

10 nm nanoparticles, in comparison with 28 and 58 nm particles [193]. On the other hand, Paciotti et al. [194] studied colloidal gold nanoparticles injected intravenously in mice in which they had implanted colon tumour cells. Nanoparticle distribution occurred preferentially at the tumour site, without significant accumulation in the liver, the spleen, or the animals' other organs. Similarly, Hainfeld et al. [195] have shown that gold nanoparticles in solution, injected intravenously into mice with induced breast tumors, were found in the kidneys 5 min after injection, and then, were located preferentially at the tumor site and, to a lesser degree, in the liver. It was concluded that this type of formulation may be of low toxicity. However, the acute toxicity of 13 nm PEG-coated gold nanoparticles has recently been confirmed [196]. The intracellular distribution of gold nanoparticles was also found to be size dependent for their localization into vesicles [197] with [198] or without [199] nuclear penetration.

3.5 Titanium Dioxide Nanoparticles

Short-term inhalation studies of male Wistar rats were exposed to different concentrations of TiO_2 nanoparticles for 6 h/day for 5 days resulted in morphological changes in the lung, with the highest tested concentrations of 50 mg/m^3 producing an increase in lung weight and in inflammation associated with dose-dependent increases in bronchoalveolar lavage fluid (BALF) total cell and neutrophil counts, total protein content, enzyme activities, and levels of a number of cell mediators [200]. Translocation and distribution of TiO_2, subsequent to pulmonary exposure has also been investigated, with a particular focus on the transfer of particles to the brain. In vivo studies have indicated that following nasal instillation, both forms of TiO_2 nanoparticles, rutile and anatase have translocated to the brain, with their accumulation within the cerebral cortex, thalamus, and hippocampus. It was postulated that this transport may have occurred via the olfactory bulb neuronal transport as it was unlikely to be mediated via penetration into the cardiovascular system and via the blood [201]. It was later confirmed that hippocampus was the main target of accumulation and toxicity of TiO_2 [202]. On the other hand, systemically available TiO_2 nanoparticles, as simulated by the i.v. injection, were trapped mainly in the liver and spleen [203].

Parameters that affected the toxicity of TiO_2 were also investigated. For example, size and the crystalline nature of the TiO_2 nanoparticles were found to determine their toxicity in vivo. For example, results have shown that inert particles can become biologically active when nanoscaled [204] and that nanoscale titanium dioxide could increase pulmonary inflammation parameters 10 times more than administration of fine particles of the same products [205]. With in vivo inhalation studies with rats, TiO_2 nanoparticles could also be shown that the nanoscale TiO_2 could be retained by the lung and their clearance was found to be much slower than their larger counterparts [206, 207–208]. Similar inhalation studies with mice, the highly crystalline anatase of TiO_2 were found to be more toxic than a mixture of rutile and anatase [209]. The pathological response following pulmonary exposure to TiO_2 within the lung was found to be emphysema-like in nature with more pronounced lesions in areas where particles preferentially accumulated, with an inflammatory response [210]. Inhalation studies using rat, mice, and hamster found that rat was the most sensitive and hamster the least sensitive species to the effects of TiO_2 indicating that the choice of species may influence the pulmonary response to TiO_2 nanoparticles [211]. Maternal exposure to anatase TiO_2 nanoparticles was also investigated. It was found that such an exposure has caused in the offspring and the changes in the expression of genes associated with brain development, cell death, response to oxidative stress, and mitochondria in the brain during the perinatal period, and those associated with inflammation and neurotransmitters in the later stage [212].

In vitro, investigations using a variety of lung cell models were conducted to assess TiO_2 toxicity. Using the human lung bronchial epithelial cells BEAS-2B, a dose and time dependent decrease in cell viability was observed by 21 nm TiO_2 nanoparticles with a concomitant increase in reactive oxygen species production and depletion in glutathione, implying that an oxidative mechanism was involved in this cytotoxic response [213]. Using alveolar epithelial type 2 cell line A549, crystal structure-dependent cytotoxicity of TiO_2 was observed where it was found that the greater the anatase content of the sample, the greater the ability of the TiO_2 nanoparticles to induce cell death [165]. Using rat tracheal explants, it could be shown that 21 nm TiO_2 nanoparticles could stimulate growth factor expression (such as platelet-derived growth factor (PDGF)) and could enhance procollagen expression [214]. Finally, using rat alveolar macrophages obtained from bronchoalveolar lavage, a dose-dependent decrease in cell viability could be observed [215]. These, in vitro, investigations have therefore confirmed the observation with the in vivo experiments that size and the crystalline nature of TiO_2 nanoparticles will determine their toxicity and that oxidative stress is the main mechanism for this toxicity. Finally, In vitro studies have also shown that coating of TiO_2 may influence their toxicity. For example, it was shown that uncoated nanosized anatase TiO_2 and fine rutile TiO_2 are more efficient than SiO_2-coated nanosized rutile TiO_2 in inducing DNA damage in human bronchial epithelial BEAS 2B cells, whereas only nanosized anatase is able to slightly induce micronuclei [216].

The use of TiO_2 in sunscreens and cosmetics has prompted number of investigations to assess the dermal penetration and toxicity of TiO_2 nanoparticles. In vivo, and in vitro, have shown that the majority of the TiO_2 was contained within the stratum corneum, with minimal distribution within the epidermis [217]. Using HaCaT keratinocytes, human dermal fibroblast cells, SZ95 sebaceous gland cells and primary human melanocytes, it could be seen that the TiO_2 nanoparticles were internalized and were evident in the cytoplasm and perinuclear region of fibroblasts and melanocyte and that this internalization was associated with an increase in intracellular calcium. No particle uptake or alterations in calcium signaling were observed within keratinocytes or sebocytes. On the other hand, a dose and time dependent decrease in cell proliferation was evident within all cell types, and an increase in cell death (via apoptosis) within fibroblasts was apparent [218]. Similarly, using mouse L929 fibroblasts, it was also possible to show time and dose-dependent decrease in cell viability induced by TiO_2 nanoparticles. In addition, it was possible to show the internalization of these nanoparticles by phagocytosis, and their containment within lysosomes with an increase in ROS production and decreases in GSH and SOD activities [219].

Finally, a number of ecotoxicological studies were carried out using short-term toxicity of TiO_2 NPs to algae, daphnids, juvenile and adult zebrafish, and were found to be of low toxicity [220]. However, independent of direct toxic effects, the presence of TiO_2 nanoparticles was found to cause indirect effects, for example, by influencing toxicity and bioaccumulation of other pollutants present in the aquatic environment. Indirect effects can also be a result of environmental pollutants adsorbing onto the particles. For example, the enhanced accumulation of cadmium (Cd) and arsenic (As) found in carp (*Cyprinus carpio*) in the presence of TiO_2 NPs [221] was attributed to facilitated transport into different organs. On the other hand, subacute toxicity studies of TiO_2 NPs on this same aquatic organism, have shown depletion in antioxidant enzyme activities and the elevation of LPO with a concomitant necrosis and apoptosis hepatocytes with gill pathologies including edema and thickening of gill lamellae as well as gill filaments, indicating a potential risk from TiO_2-NPs released into the aqueous environment [222]. As only a few studies describe TiO_2 NP toxicity to algae [223], the potential carrier effect of TiO_2 for other pollutants through a series of experiments, designed to reveal the influence of TiO_2 particles on the algal toxicity of cadmium was most recently investigated. It was found that in addition to particle toxicity, potential interactions with existing environmental contaminants are also of crucial importance in assessing the potential environmental risks of nanoparticles [224].

3.6 Zinc Oxide Nanoparticles

Given the intensive application of nanoscale zinc oxide (nZnO), growing concerns have been expressed about its health and environmental impacts. The effect of ZnO nanoparticles on numerous prokaryotic and eukaryotic as well as on plant cell systems was, therefore, investigated. For example, Gojova et al. have shown that in vitro incubation of human aortic endothelial [225] and mouse neuroblastoma cells [226] with nZnO have caused 50 % reduction in their viability; Incubation of mesothelioma MSTO-211H or rodent 3T3 fibroblast with these nanoparticles have caused their complete death [227]. On the other hand, they were only slightly toxic to human T cells [228] but very toxic to human pulmonary cell lines: A549 and THP-1 cells [229]. In vivo, exposure to nZnO has produced potent but reversible inflammation which was resolved by 1 month postinstillation exposure [230].

The toxicity of nZnO against number of pathogenic bacteria was also investigated. Once again, a comparative toxicity study among different metal oxides has shown that ZnO was the most toxic nanoparticles against *Bacillus subtilis*, *Escherichia coli* and *Pseudomonas fluorescens*, [231] and *Streptococcus agalactiae* and *Staphylococcus aureus* [232] and thus confirming the bactericidal action of nZnO on both gram-negative and gram-positive bacteria. Their use as aerosols to control airborne disease (e.g., influenza, tuberculosis, and pneumonia) and prevent bacterial fouling was, therefore, suggested [233].

Comparative ecotoxicity studies on metal oxide nanoparticles have indicated that ZnO nanoparticles were always most toxic to *Pseudokirchneriella subcapitata* algal growth [223, 234] and to zebrafish embryos and larvae [235] and that this toxicity could be attributed to the solubilized Zn^{2+} [223, 234]. A similar comparative phytotoxicity study on metal oxide nanoparticles has also confirmed that nZnO was most toxic on the development of *Arabidopsis thaliana* (Mouse-ear cress) assessed by their effect on seed germination, root elongation, and number of leaves. In this instance, however, zinc dissolution could not solely account for the observed toxicity as direct exposure to nanoparticles significantly could contribute to this phytotoxicity [236]. A similar effect of ZnO nanoparticles on seed germination and root growth of six higher plant species (radish, rape, ryegrass, lettuce, corn, and cucumber) was also reported [237].

Several of these studies have indicated that the solubilized Zn^{2+} is the cause of these toxic effects. For example, this was suggested to be the cause of toxicity of nZnO to crustaceans *Daphnia magna* and *Thamnocephalus platyurus* and protozoan *Tetrahymena thermophila* [238]. Other studies have shown that metal NPs may be more toxic than either their ionic forms or their parent compounds [239, 240]. For example, it was shown that nZnO killed zebrafish embryos (50 and 100 mg/L), retarded the embryo hatching (1–25 mg/L), reduced the body length of larvae, and caused tail malformation after the 96 hpf exposure. $Zn_{(dis)}$ only partially contributed to the toxicity of nZnO [241]. The same was true for nematode *Caenorhabditis elegans* where the oxide solubility influenced the toxicity of ZnO, but nanoparticle-dependent toxicity was also observed [242].

Similar contradictory observations were also made for the toxicity of nZnO to mammalian cells. For example, in mouse neuronal stem cells, it was concluded that the nZnO toxicity comes from the dissolved Zn^{2+} in the culture medium or inside cells [243] confirming earlier studies that the dissolved Zn^{2+} from zinc-containing compounds would more easily affect the function of neuro cells, causing neurotoxicity [244, 245]. The same was proposed for the toxicity of ZnO to RAW 264.7 and BEAS-2B cell lines [246].

3.7 Biodegradable Polymeric Nanoparticles

When using nanoparticles as vehicle for drug delivery, the health effects of the residual material after drug delivery should be considered. It is, therefore, desirable to use nanoparticles that are biodegradable. Poly(ester) nanoparticles are said to be nontoxic and biodegradable [247]. On the other hand, surface modified PLGA such as chitosin-PLGA nanoparticles were found to be toxic to certain cell types [248].

The cytotoxicity of both PAMAM and poly(propylene imine) (PPI) dendrimers is very much dependent on size and also on the nature and density of the surface charged groups where in general, cationic charges are found to be more toxic [249]. For example, a concentration-dependent tendency to cause hemolysis and changes in erythrocyte morphology has been linked to the presence of $-NH_2$ groups. In contrast to PAMAM dendrimers, PPI dendrimers with DAB and DAE cores did not show generation dependence for the hemolytic effect [250]. On the other hand, dendrimers with carboxylic acid terminal functional groups, are not toxic to zebrafish embryos [251]. It is, therefore, recommended to modify the surface amine groups of PAMAM and PPI dendrimers with neutral or anionic moieties to avoid the toxicity and liver accumulation associated with their polycationic surfaces [250, 252]. Subsequently, when PEGylated dendrimer was evaluated for acute toxicity in vivo, no toxicity, mortality, or abnormal blood chemistry was observed in doses up to 2.56 g/kg ip and 1.28 g/kg iv [253].

4 Conclusions

In this chapter, through the use of basic building blocks of nanostructures entailing the development of nanoscale materials hinged on advances in chemistry have demonstrated their potential in addressing part of the MDGs—with particular emphasis on the African continent. This is through the application of nanotechnology-based products and services aimed at improving; the provision of clean water, the quality of the environment, and the treatment of pandemic diseases particularly HIV/AIDS and malaria—which are between the chief causes of mortality and morbidity in Africa annually. This is because the materials fabricated based on nanotechnologies are expected to be of low cost, high efficiency, and able to provide effective applications particularly for treating diseases and cleaning of drinking water in very remote regions. Such aspects are contrary to the current technologies both in water and treatment of diseases.

On the other hand, while the benefits of nanotechnology-based applications in drug therapeutics and water treatment are beyond debate, the available scientific literature suggests that evidence exists for some manufactured nanoparticles and nanotubes being more toxic per unit mass than particles of the same macrochemicals. This implies that nanoparticles potentially present a greater risk to both humans and the ecosystems. The fundamental mechanisms of toxicity of nanoparticles may not be very different: the capacity to induce inflammation through the release of free radicals in response to a dose that is adequate to overcome the body's natural defenses. However, the difference is expected largely due to two size-dependent factors; namely, the relatively large surface area of nanoparticles, given equal mass, and their probable ability to penetrate cells more easily and in a different way. To pose a risk, these nanoparticles must come into contact with humans or the environment in a form and quantity that can cause harm. Currently, the main risk of human exposure to manufactured nanoparticles and nanotubes is in a few workplaces (including academic research laboratories) and through the use of a small number of skin preparations that contain free nanoparticles. However, the current lack of available research means that the scale of this risk cannot be fully determined.

An additional concern is that increasing the use of engineered nanoparticles in industrial, for example, treatment of water and nanomedicines will, very likely, lead to the release of such materials into different environmental compartments (i.e. soil, air, water, and sediments). Assessment of the potential risks of such nanoparticles in the environment requires an understanding of their mobility, reactivity, ecotoxicity and persistency, however, currently it is very limited or nonexistent. From our review, several nanoparticles have shown their capability to cause adverse effects to organisms at different trophic levels. In spite of these challenges, it is in the view of the authors Africa stands tremendously if it gets actively involved in engaging applications enhanced by nanotechnology-driven capabilities. Therefore, it is recommended that the focus by the African authorities and researchers in the continent should not only be limited to the societal benefits of nanotechnology but also on the potential downside aspects such as toxic effects to humans and other ecosystems. Benefits of this approach include the optimal exploitation of nanotechnology unique capabilities that promotes safe, responsible, and sustainable development in the target communities.

References

1. de Heer WA, Bonard J-M, Fauth K et al (1997) Electron field emitters based on carbon nanotube films. Adv Mater 9:87–89
2. Tans SJ, Devoret MH, Dai H et al (1997) Individual single-wall carbon nanotubes as quantum wires. Nature 386:474–477
3. Dai H, Hafner JH, Rinzler AG et al (1996) Nanotubes as nanoprobes in scanning probe microscopy. Nature 384:147–150
4. Gao G (2004) Synthesis, properties and applications. Imperial College Press, London
5. Endo M, Takeuchi K, Igarashi S et al (1993) The production and structure of pyrolytic carbon nanotubes (PCNTs). J Phys Chem Solids 54:1841–1848
6. Rinzler AG, Liu J, Dai H et al (1998) Large scale purification of single-wall carbon nanotubes: process, product, and characterization. Appl Phys A 67:29–37
7. Amelinckx S, Zhang XB, Bernaerts D et al (1994) A formation mechanism for catalytically grown helix-shaped graphite nanotubes. Science 265:635–639
8. Setlur AA, Dai JY, Lauerhaas JM, Chang RPH (1998) Formation of filled carbon nanotubes and nanoparticles using polycyclic aromatics hydrocarbon molecules. Carbon 36:721–723
9. Bladh K, Falk LKL, Rohmund F (2000) On the iron-catalysed growth of single-walled carbon nanotubes and encapsulated metal particles in the gas phase. Appl Phys A Mater Sci Process 70:317–322
10. Diener MD, Nichelson N, Alford JM (2000) Synthesis of single-walled carbon nanotubes in flames. J Phys Chem B 2000 104:9615–9620
11. Loiseau A, Willaime F (2000) Filled and mixed nanotubes: from TEM studies to the growth mechanism within a phase-diagram approach. Appl Surf Sci 164:227–240
12. Puddephatt RJ (1978) The chemistry of gold. Elsevier Science, Amsterdam
13. Hyatt AD, Eaton BT (eds) (1993) Immune-gold electron microscopy in virus diagnosis and research, CRC Press. Boca Raton, pp 178–454

14. Hayat MA (1989) (ed) Colloidal gold: principles, methods and applications. Academic Press, San Diego
15. Alivisatos AP (1996) Perspectives on physical chemistry of semiconductor nanocrystals. J Phys Chem 100:13226–13239
16. Chan WCW, Nie S (1998) Quantum dot bioconjugates for ultra sensitive non isotopic detection. Sci 281:2016–2018
17. Mckenzie KJ, Marken F (2003) Accumulation and reactivity of the redox protein cytochrome in mesoporous films of TiO2 phytate. Langmuir 19:4327–4331
18. Haruta M (1997) Size- and Support-dependency in catalysis of gold. Catal Today 36:153–166
19. Haruta M, Yamada N, Kobayashi T et al (1989) Gold catalysts prepared by coprecipitation for low-temperature oxidation of hydrogen and of carbon monoxide. J Catal 115:301–309
20. Haruta M, Tsubota S, Kobayashi T et al (1993) Low-temperature oxidation of CO over gold supported on TiO_2, α-Fe_2O_3, and Co_3O_4. J Catal 144:175–192
21. Wang J, Koel BE (1998) IRAS studies of NO_2, N_2O_3, and N_2O_4 adsorbed on Au(111) surfaces and reactions with co-adsorbed H_2O. J Phys Chem A 102:8573–8579
22. Okumura M, Tanaka K, Ueda A et al (1997) The reactivities of Dimethylgold(III)Beta-Diketone on the Surface of TiO_2–a novel preparation method for Au catalysts. Solid State Ion 95:143–149
23. Wei Z, Xin Q, Guo X et al (1990) Titanaia-modified hydrodesulphurization catalysts I : effect of preparation techniques on morphology and properties of TiO_2-Al_2O_3 Carrier. Appl Catal A 63:305–309
24. Weibel A, Bouchet R, Knauth P (2006) Electrical properties and defect chemistry of anatase (TiO_2). Solid State Ion 177:229–236
25. Pajonk GM (1991) Aerogel catalysts. Appl Catal 72:217–266
26. Reluert J, Quijada R, Arias V (1998) Porous titania obtained through polymer incorporated composites. Chem Mater 10:3923–3927
27. Fox MA, Dulay MT (1993) Heterogeneous photocatalysis. Chem Rev 93:341–357
28. Sopyan I, Murasawa S, Hashimoto S et al (1998) Highly efficient TiO_2 film photocatalyst: degradation of gaseous acetaldehyde. Chem Lett 10:723–726
29. Zhang Z, Wang C–C, Zakaria R et al (1998) Role of particle size in nanocrystalline TiO_2-based photocatalysts. J Phys Chem B 102:10871–10878
30. Wang C–C, Ying JY (1999) Sol-Gel Synthesis and hydrothermal processing of anatase and rutile titania nanocrystals. Chem Mater 11:3113–3120
31. Huang W, Tang X, Wang Y et al (2000) Selective synthesis of anatase and rutile via ultrasound irradiation. Chem Commun 15:1415–1416
32. Liu H, Yang W, Ma Y et al (2002) Promoted phase transition of titania nanoparticles prepared by a photo-assisted sol-gel method. New J Chem 26:975
33. Hoyer P (1996) Formation of a titanium dioxide nanotu be array. Langmuir 12:1411–1413
34. Liu SM, Gan LM, Liu LH et al (2002) Synthesis of single-crystalline TiO_2 nanotubes. Chem Mater 14:1391–1397
35. Wang YQ, Hu GQ, Duan XF et al. (2002) Microstructure and formation mechanism of titanium dioxide nanotubes. Chem Phys Lett 365:427–431
36. Ivanovskaya VV, Enyashin AN, Ivanovskii AL (2003) Electronic structure of single-walled TiO_2 and VO_2 nanotubes. Mendeleev Commun 13:5–7
37. Kasuga T, Hiramatu M, Hirano M et al (1997) Synthesis and functionalization titania nanotube. Mater Res 12:607–609
38. Jung JH, Kobayashi H, van Bommel KJC et al (2002) Creation of novel helical ribbon and double-layered nanotube TiO_2 structures using an organogel template. Chem Mater 14:1445–1447
39. Imai H, Takei Y, Shimizu K et al (1999) Direct preparation of anatase TiO_2 nanotubes in porous alumina membranes. J Mater Chem 9:2971–2972
40. Dignam MJ, Moskovits M (1973) Influence of surface roughness on the transmission and reflectance spectra of adsorbed species. J Chem Soc, Faraday Trans 2(69):95–78
41. Mahata N, Raghavan KV, Vishwanathan V et al (2001) Phenol hydrogenation over palladium supported on magnesia: relationship between catalyst structure and performance. Phys Chem Chem Phys 3:2712–2719

42. Mahmood FS, Gould RD, Salih MH (1995) D.C. properties of ZnO thin films prepared by r.f. magnetron sputtering. Thin Solid Films 270:376–379
43. Gorla CR, Emanetoglu NW, Liang S et al (1999) Structural, optical, and surface acoustic wave properties of epitaxial ZnO films grown on (0112) sapphire by metalorganic chemical vapor deposition. J Appl Phys 85:2595–2602
44. Look DC (2001) Recent advances in ZnO materials and devices. Mater Sci Eng, B 80:383–387
45. Hvam JM (1973) Exciton-exciton interaction and laser emission in high-purity ZnO. Solid State Commun 12:95–97
46. Klingshirn C (1975) Room-temperature stimulated emission of ZnO: alternatives to excitonic lasing. Phys Status Solidi B 71:547–556
47. Caillaud FA, Smith A, Baumard JM (1992) Additives content in ZnO films prepared by spray pyrolysis. J Eur Ceram Soc 9:447–452
48. Musić S, Dragčević D, Malijković M et al (2003) Influence of chemical synthesis on the crystallization and properties of zinc oxide. Mater Chem Phys 77:521–530
49. Wang X, Li Y (2002) Selected-control hydrothermal synthesis of α- and β-MnO_2 single crystal nanowires. J Am Chem Soc 124:2880–2881
50. Chu D, Zeng Y, Jiang D (2007) Controlled growth and properties of Pb^{2+} doped ZnO nanodisks. Mater Res Bull 42:814–819
51. Sun Y, Xia Y (2002) Shape-controlled synthesis of gold and silver nanoparticles. Science 298:2176–2179
52. Kirk-Othmer (1993) Encyclopedia of chemical technology***, vol 5. Wiley, Canada
53. Christoffel EG (1989) Laboratory studies of heterogeneous catalytic processes. Elsevier, Oxford
54. Corma A (1995) Inorganic solid acids and their use in acid-catalyzed hydrocarbon reactions. Chem Rev 95:559–614
55. Ardizzone S, Bianchi CL, Ragaini V et al (1999) SO_4-ZrO_2 catalysts for the esterification of benzoic to methylbenzoate. Catal Lett 62:59–65
56. Salamanca-Buentello F, Persad DL, Court EB et al (2005) Nanotechnology and the developing world. PLoS Med 2:383–386
57. UNEP (2007) UNEP Emerging Challenges—Nanotechnology and the Environment, Geo Yearbook 2007. Nairobi, Kenya, UNEP
58. Montgomery MA, Elimelech M (2007) Water and sanitation in developing countries: including health in the equation. Environ Sci Technol 41:17–24
59. Eshelby K (2007) Dying for a drink. BMJ 334:610–612
60. Maclurcan DC (2005) Nanotechnology and Developing Countries, Part 2: what realities? AzoNano J Nanotechnol. doi:10.2240/azojono0105
61. DST (2005) National nanotechnology strategy. Department of Science and Technology, Pretoria
62. DST (2010) Nanoscience and nanotechnology 10-year research plan. Department of Science and Technology, Pretoria
63. Hillie T, Hlophe M (2007) Nanotechnology and the challenge of clean water. Nat Nanotechnol 2:663–664
64. Hillie T, Munasinghe M, Hlope M et al. (2006) Nanotechnology, water and development, Paper commissioned as part of the Global Dialogue on Nanotechnology and the poor: opportunities and risks, Meridian Institute's, Web link: http://www.merid.org/nano/waterpaper
65. Theron J, Walker JA, Cloete TE (2008) Nanotechnology and water treatment: applications and emerging opportunities. Crit Rev Microbiol 34:43–69
66. Baker RW (2004) Membrane technology and applications. Wiley, England
67. Srivastava AK, Srivastava ON, Talapatra S et al (2004) Carbon nanotube filters. Nat Mat 3:610–614
68. Tully DC, Frechet JMJ (2001) Dendrimers at surfaces and interfaces: chemistry and applications. J Chem Commun 14:1229–1239
69. Gao C, Yan D (2004) Hyperbranched polymers: from synthesis to applications. Prog Polym Sci 29:183–275

70. Del Valle EMM (2004) Cyclodextrins and their uses, A review. Process Biochem 39: 1033–1046
71. Mamba BB, Krause RW, Malefetse TJ et al (2007) Monofunctionalized cyclodextrin polymers for the removal of organic pollutants from water. Environ Chem Lett 5:79–84
72. Salipira KL, Mamba BB, Krause RW et al (2008) Cyclodextrin polyurethanes polymerised with carbon nanotubes for the removal of organic pollutants in water. Water SA 34: 113–118
73. Sawicki R, Mercier L (2006) Evaluation of mesoporous cyclodextrin-silica nanocomposites for the removal of pesticides from aqueous media. Environ Sci Technol 40:1978–1983
74. Arkas M, Allabashi R, Tsiourvas D et al (2006) Organic/inorganic hybrid filters based on dendritic and cyclodextrin "nanosponges" for the removal of organic pollutants from water. Environ Sci Technol 40:2771–2777
75. Nair AS, Pradeep T (2003) Halocarbon mineralization and catalytic destruction by metal nanoparticles. Curr Sci 84:1560–1564
76. Predeep TA, Anshup (2009) Nobel metal nanoparticles for water purification: a critical review. Thin Solid Films 517:6441–6478
77. Pradeep TA (2009) In: Savage N, Diallo M, Duncan J, Street A, Sustich R (eds) Nanotechnology applications for clean water. William Andrew Publication, USA, p 2009
78. Schwarzenbach RP, Escher BI, Fenner K et al (2006) The challenge of micropollutants in aquatic systems. Sci 313:1072–1077
79. Ahluwalia SS, Goyal D (2007) Microbial and plant derived biomass for removal of heavy metals from wastewater. Bioresour Technol 98:2243–2257
80. Savage N, Diallo MS (2005) Nanomaterials and water purification: opportunities and challenges. J Nanopart Res 7:331–342
81. Tratnyek PG, Johnson RL (2006) Nanotechnologies for environmental cleanup. Nanotoday 1:44–48
82. Li Q, Mahendra S, Lyon DY et al (2008) Antimicrobial nanomaterials for water disinfection and microbial control: potential applications and implications. Water Res 42:4591–4602
83. Wang CB, Zhang WX (1997) Synthesizing nanoscale iron particles for rapid and complete dechlorination of TCE and PCBs. Environ Sci Technol 31:2154–2156
84. Liou YH, Lo SL, Kuan WH et al (2006) Effect of precursor concentration on the characterics of nanoscale zerovalent iron and its reactivity with nitrate. Wat Res 40:2485–2492
85. Sohn K, Kang SW, Ahn S et al (2006) Fe(0) nanoparticles for nitrate reduction: stability, reactivity, and transformation. Environ Sci Technol 40:5514–5519
86. Cao J, Elliott D, Zhang WX (2005) Perchlorate reduction by nanoscale iron particles. J Nanopart Res 7:499–506
87. Ponder SM, Darab JG, Mallouk TE (2000) Remediation of Cr(VI) and Pb(II) aqueous solutions using supported, nanoscale zero-valent iron. Environ Sci Technol 34:2564–2569
88. Kanel SR, Manning B, Charlet L et al (2005) Removal of arsenic (III) from groundwater by nanoscale zero-valent iron. Environ Sci Technol 39:1291–1298
89. Li XQ, Zhang WX (2006) Iron nanoparticles, the core–shell structure and unique properties for Ni (II) sequestration. Langmuir 22:4638–4642
90. Keum YS, Li QX (2004) Reduction of nitroaromatic pesticides with zerovalent iron. Chemosphere 54:255–263
91. Zhang L, Fang M (2010) Nanomaterials in pollution trace detection and environmental improvement. Nano Today 5:128–142
92. Elliot DW, Zhang WX (2001) Field assessment of nanoscale biometallic particles for ground water treatment. Environ Sci Technol 35:4922–4926
93. He P, Zhao DY (2005) Preparation and characterization of a new class of starch–stabilized bimetallic nanoparticles for degradation of chlorinated hydrocarbons in water. Environ Sci Technol 39:3314–3320
94. Schrick B, Hydutsky BW, Blough JL, Mallouk TE (2004) Delivery vehicles for zerovalent metal nanoparticles in soil and groundwater. Chem Mater 16:2187–2193

95. He F, Zhao DY, Liu J et al (2007) Stabilization of Fe-Pd nanoparticles with sodium carboxymethyl cellulose for enhanced transport and dechlorination of trichloroethylene in soil and groundwater. Ind Eng Chem Res 46:29–34
96. Saleh N, Sirk K, Liu YQ et al (2007) Surface modifications enhance nano iron transport and NAPL targeting in satulated porous media. Envrion Eng Sci 24:45–57
97. Kikuchi Y, Sunada K, Iyoda T et al (1997) Photocatalytic bactericidal effect of TiO_2 thin films: dynamic view of the active oxygen species responsible for the effect. J Photochem Photobiol A Chem 106:51–56
98. Choi H, Stathatos E, Dionysiou D (2007) Photocatalytic TiO_2 films and membranes for the development of efficient wastewater treatment and reuse systems. Desalination 202: 199–206
99. Hoff JC (1986) Inactivation of microbial agents by chemical disinfectants. EPA/600/S602-686/067. U.S. Environmental Protection Agency, Cincinnati, OH
100. Hunt NK, Marinas BJ (1997) Kinetics of *Escherichia Coli* inactivation with ozone. Wat Res 31(6):1355–1362
101. Morones JR, Elechiguerra JL, Camacho A et al (2005) The bactericidal effect of silver nanoparticles. Nanotechnol 16:2346–2353
102. Pal S, Tak YK, Song JM (2007) Does the antibacterial activity of silver nanoparticles depend on the shape of the nanoparticle? A study of the Gram—negative bacterium *Escherichia coli*. Appl Environ Microbiol 73:1712–1720
103. Sondi I, Salopek-Sondi B (2004) Silver nanoparticles as antimicrobial agent: case study on E. coli as a model for Gram-negative bacteria. J Colloid Interface Sci 275:177–182
104. Kim JS, Kuk E, Yu KM et al (2007) Antimicrobial effects of silver nanoparticles. Nanomedicine 3:95–101
105. Shahverdi AR, Kakhimi A, Shahverdi HD, Minaian S (2007) Synthesis and effect of silver nanoparticles on the antibacterial activity of different antibiotics against Staphylococcus aureus and Escherichia coli. Nanomed Nanotechnol Biol Med 3:168–171
106. Shrivastava S, Bera T, Roy A et al. (2007) Characterization of enhanced antibacterial effects of novel silver nanoparticles. Nanotechnol 18(22). doi:10.1088/0957-4484/18/22/22510_3
107. Yoon KY, Byeon JH, Park JH et al (2007) Susceptibility constants of *Escherichia coli* and *Bacillus subtilis* to silver and copper nanoparticles. Sci Total Environ 373:572–575
108. Ju-Nam Y, Lead JR (2008) Manufactured nanoparticles: an overview of their chemistry, interactions and potential environmental implications. Sci Total Environ 400:396–414
109. Danilczuk M, Lund A, Saldo J et al (2006) Conduction electron spin resonance of small silver particles. Spectrochim Acta A 63:189–191
110. Mills A, LeHunte S (1997) An overview of semiconductor photocatalysis. J Photochem Photobiol A Chem 108:1–35
111. Lee S, Nakamura M, Ohgaki S (1998) Inactivation of phage Qβ by 254 nm UV light and titanium dioxide photocatalyst. J Environ Sci Health A 33:1643–1655
112. Inaba R, Fukahori T, Hamamoto M et al (2006) Synthesis of nanosized TiO2 particles in reverse micelle systems and their photocalytic activity for degradation of toluene in gas phase. J Mol Cat A: Chem 200:247–254
113. Shephard GS, Stockenstrom S, de Villiers D et al (2002) Degradation of microcystin toxins in a falling film photocatalytic reactor with immobilized titanium dioxide catalyst. Water Res 36:140–146
114. Kominami H, Murakami S, Kato J et al (2002) Correlation between some physical properties of titanium dioxide particles and their photocatalytic activity for some probe reactions in aqueous systems. J Phys Chem B 106:10501–10507
115. Toma D, Bertrand G, Chwa SO et al. (2006) Comparative study on the photocatalytic decomposition of nitrogen oxides using TiO_2 coatings prepared by conventional plasma spraying and suspension plasma spraying. Surf Coat Technol 200:5855–5862).
116. Joo J, Kwon SG, Yu T et al (2005) Large– scale synthesis of TiO_2 nanorods via nonhydrolytic sol-gel ester elimination reaction and their application to photocatalytic inactivation of *E. coli*. J Phys Chem B 109:15297–15302

117. Egerton TA, Kosa SA, Christensen PA (2006) Photoelectrocatalytic disinfection of *E. coli* suspensions by iron doped TiO2. Phys Chem Chem Phys 8:398–406
118. Page K, Palgrave RG, Parkin IP et al (2007) Titania and silver-titania composite films on glass-potent antimicrobial coatings. J Mat Chem 17(1):95–104
119. Yu JC, Yu J, Zhao J (2002) Enhanced photocatalytic activity of mesoporous and ordinary TiO2 thin films by sulfuric acid treatment. Appl Catal B Environ 36:31–43
120. das Neves J, Amiji MM, Bahia MF et al (2010) Nanotechnology-based systems for the treatment and prevention of HIV/AIDS. Adv Drug Deliv Rev 62:458–477
121. Gunaseelan S, Gunaseelan K, Deshmukh M et al (2010) Surface modifications of nanocarriers for effective intracellular delivery of anti-HIV drugs. Adv Drug Deliv Rev 62:518–531
122. Gupta U, Jain NK (2010) Non-polymeric nano-carriers in HIV/AIDS drug delivery and targeting. Adv Drug Deliv Rev 62:478–490
123. Sharma P, Garg S (2010) Pure drug and polymer based nanotechnologies for the improved solubility, stability, bioavailability and targeting of anti-HIV drugs. Adv Drug Del Rev 62:491–502
124. Wong HL, Chattopadhyay N, Wu XY et al (2010) Nanotechnology applications for improved delivery of antiretroviral drugs to the brain. Adv Drug Deliv Rev 62:503–517
125. du Toit LC, Pillay V, Choonara YE (2010) Nano-microbicides: challenges in drug delivery, patient ethics and intellectual property in the war against HIV/AIDS. Adv Drug Del Rev 62:532–546
126. Hammer SM, Eron JJ Jr, Reiss P et al (2008) International AIDS Society-USA, antiretroviral treatment of adult HIV infection: 2008 recommendations of the International AIDS Society-USA panel. JAMA 300:555–570
127. Blankson JN, Persaud D, Siliciano RF (2002) The challenge of viral reservoirs in HIV-1 infection. Ann Rev Med 53:557–593
128. Amiji MM, Vyas TK, Shah LK (2006) Role of nanotechnology in HIV/AIDS treatment: potential to overcome the viral reservoir challenge. Discov Med 6:157–162
129. Chun TW, Justement JS, Moir S et al (2007) Decay of the HIV reservoir in patients receiving antiretroviral therapy for extended periods: implications for eradication of virus. J Infect Dis 195:1762–1764
130. Temesgen Z, Warnke D, Kasten MJ (2006) Current status of antiretroviral therapy. Expert Opin Pharmacother 7:1541–1554
131. Spitzenberger TJ, Heilman D, Diekmann C et al (2007) Novel delivery system enhances efficacy of antiretroviral therapy in animal model for HIV-1 encephalitis. J Cereb Blood Flow Metab 27:1033–1042
132. Govender T, Ojewole E, Naidoo P et al (2008) Polymeric nanoparticles for enhancing antiretroviral drug therapy. Drug Deliv 15:493–501
133. Urdea M, Penny LA, Olmsted SS et al (2006) Requirements for high impact diagnostics in the developing world. Nature 444(Suppl 1):73–79
134. Yager P, Edwards T, Fu E, Helton K et al (2006) Microfluidic diagnostic technologies for global public health. Nature 442:412–418
135. Gardella F, Assi S, Simon F et al (2008) Antimalarial drug use in general populations of tropical Africa. Malar J 7:124
136. Greenwood BM, Fidock DA, Kyle DE et al (2008) Malaria: progress, perils, and prospects for eradication. J Clin Invest 118:1266–1276
137. Skinner-Adams TS, McCarthy JS, Gardiner DL et al (2008) HIV and malaria co-infection: interactions and consequences of chemotherapy. Trends Parasitol 24:264–271
138. Santos-Magalhães NS, Mosqueira VCF (2010) Nanotechnology applied to the treatment of malaria. Adv Drug Del Rev 62:560–575
139. Winstanley P, Ward S (2006) Malaria chemotherapy. Adv Parasitol 61:47–76
140. Vauthier C and P. Couvreur P (2007) Nanomedicines: a new approach for the treatment of serious diseases J Biomed Nanotechnol 3:223–234
141. Forrest ML, Kwon GS (2008) Clinical developments in drug delivery nanotechnology. Adv Drug Deliv Rev 60:861–862

142. Foger F, Noonpakdee W, Loretz B et al (2006) Inhibition of malarial topoisomerase II in Plasmodium falciparum by antisense nanoparticles. Int J Pharm 319:139–146
143. Barker RH Jr, Metelev V, Coakley A et al (1998) *Plasmodium falciparum*: effect of chemical structure on efficacy and specificity of antisense oligonucleotides against malaria in vitro. Exp Parasitol 88:51–59
144. Noonpakdee W, Pothikasikorn J, Nimitsantiwong W et al (2003) Inhibition of Plasmodium falciparum proliferation in vitro by antisense oligodeoxynucleotides against malarial topoisomerase II. Biochem Biophys Res Commun 302:659–664
145. Wong HL, Bendayan R, Rauth AM et al (2006) A mechanistic study of enhanced doxorubicin uptake and retention in multidrug resistant breast cancer cells using a polymer-lipid hybrid nanoparticle system. J Pharmacol Exp Ther 317:1372–1381
146. Pacurari M, Castranova V, Vallyathan V (2010) Single- and multi-wall carbon nanotubes versus asbestos: are the carbon nanotubes a new health risk to humans? J Toxicol Environ Health A 73:378–395
147. Wang H, Wang J, Deng X et al (2004) Biodistribution of carbon single-wall carbon nanotubes in mice. J Nanosci Nanotechnol 4:1019–1024
148. Pantarotto D, Singh R, McCarthy D et al (2004) Functionalized carbon nanotubes for plasmid DNA gene delivery. Angew Chem Int Ed Engl 43:5242–5246
149. Ji Z, Zhang D, Li L et al (2009) The hepatotoxicity of multi-walled carbon nanotubes in mice. Nanotechnol 20:445101
150. Huczko A, Lange H (2001) Carbon nanotubes: experimental evidence for a null risk of skin irritation and allergy. Fullerene Sci Technol 9:247–250
151. Murray AR, Kisin E, Leonard SS et al (2009) Oxidative stress and inflammatory response in dermal toxicity of single-walled carbon nanotubes. Toxicol 257:161–171
152. Ma-Hock L, Treumann S, Strauss V et al (2009) Inhalation toxicity of multiwall carbon nanotubes in rats exposed for 3 months. Toxicol Sci 112:468–481
153. Mitchell LA, Gao J, Wal RV et al (2007) Pulmonary and systemic immune response to inhaled multiwalled carbon nanotubes. Toxicol Sci 100:203–214
154. Poland CA, Duffin R, Kinloch I et al (2008) Carbon nanotubes introduced into the abdominal cavity of mice show asbestos-like pathogenicity in a pilot study. Nat Nanotechnol 3:423–428
155. Takagi A, Hirose A, Nishimura T et al (2008) Induction of mesothelioma in p53 $\pm$ mouse by intraperitoneal application of multi-wall carbon nanotube. J Toxicol Sci 33:105–116
156. Sakamoto Y, Nakae D, Fukumori N et al (2009) Induction of mesothelioma by a single intrascrotal administration of multi-wall carbon nanotube in intact male Fischer 344 rats. J Toxicol Sci 34:65–76
157. Pacurari M, Yin XJ, Ding M et al (2008) Oxidative and molecular interactions of multi-wall carbon nanotubes (MWCNT) in normal and malignant human mesothelial cells. Nanotoxicol 2:155–170
158. Pacurari M, Yin XJ, Zhao J et al (2008) Raw single-wall carbon nanotubes induce oxidative stress and activate MAPKs, AP-1, NF-kappaB, and Akt in normal and malignant human mesothelial cells. Environ Health Perspect 116:1211–1217
159. Wang L, Mercer RR, Rojanasakul Y et al (2010) Direct fibrogenic effects of dispersed single-walled carbon nanotubes on human lung fibroblasts. J Toxicol Environ Health A 73:410–422
160. Lindberg HK, Falck GC, Suhonen S et al (2009) Genotoxicity of nanomaterials: DNA damage and micronuclei induced by carbon nanotubes and graphite nanofibres in human bronchial epithelial cells in vitro. Toxicol Lett 186:166–173
161. Sayes CM, Liang F, Hudson JL et al (2006) Functionalization density dependence of single-walled carbon nanotubes cytotoxicity in vitro. Toxicol Lett 161:135–142
162. Monteiro-Riviere NA, Nemanich RJ, Inman AO et al (2005) Multi-walled carbon nanotube interactions with human epidermal keratinocytes. Toxicol Lett 155:377–384
163. Jia G, Wang H, Yan L, Wang X et al (2005) Cytotoxicity of carbon nanomaterials: single-wall nanotube, multi-wall nanotube, and fullerene. Environ Sci Technol 39:1378–1383

164. Kagan VE, Tyurina YY, Tyurin VA et al (2006) Direct and indirect effects of single walled carbon nanotubes on RAW 264.7 macrophages: role of iron. Toxicol Lett 165:88–100
165. Simon-Deckers A, Gouget B, Mayne-L'hermite M et al (2008) In vitro investigation of oxide nanoparticle and carbon nanotube toxicity and intracellular accumulation in A549 human pneumocytes. Toxicol 253:137–146
166. Hardman R (2006) A toxicologic review of quantum dots: toxicity depends on physicochemical and environmental factors. Environ Health Perspect 114:165–172
167. Aldana J, Wang YA, Peng X (2001) Photochemical instability of CdSe nanocrystals coated by hydrophilic thiols. J Am Chem Soc 123:8844–8850
168. Aldana J, Lavelle N, Wang Y et al (2005) Size-dependent dissociation pH of thiolate ligands from cadmium chalcogenide nanocrystals. J Am Chem Soc 127:2496–2504
169. Kim S, Bawendi MG (2003) Oligomeric ligands for luminescent and stable nanocrystal quantum dots. J Am Chem Soc 125:14652–14653
170. Tsay JM, Michalet X (2005) New light on quantum dot cytotoxicity. Chem Biol 12:1159–1161
171. Green M, Howman E (2005) Semiconductor quantum dots and free radical induced DNA nicking. Chem Commun (Camb)7:121–123
172. Ipe BI, Lehnig M, Niemeyer CM (2005) On the generation of free radical species from quantum dots. Small 1:706–709
173. Choi AO, Brown SE, Szyf M, Maysinger D (2008) Quantum dot-induced epigenetic and genotoxic changes in human breast cancer cells. J Mol Med 86:291–302
174. Hoshino A, Fujioka K, Oku T et al (2004) Physicochemical properties and cellular ttoxicity of nanocrystal quantum dots depend on their surface modification. Nano Lett 4:2163–2169
175. Selvan ST, Tan TT, Ying JY (2005) Robust, non-cytotoxic, silica-coated CdSe quantum dots with efficient photoluminescence. Adv Mater 17:1620–1625
176. Guo G, Liu W, Liang J et al (2007) Probing the cytotoxicity of CdSe quantum dots with surface modification. Material Lett 61:1641–1644
177. Larese FF, D'Agostin F, Crosera M et al (2009) Human skin penetration of silver nanoparticles through intact and damaged skin. Toxicol 255:33–37
178. Chen X, Schluesener HJ (2008) Nanosilver: a nanoproduct in medical application. Toxicol Lett 176:1–12
179. Hussain SM, Hess KL, Gearhart JM et al (2005) In vitro toxicity of nanoparticles in BRL 3A rat liver cells. Toxicol In Vitro 19:975–983
180. Elechiguerra JL, Burt JL, Morones JR et al (2005) Interaction of silver nanoparticles with HIV-1. J Nanobiotechnol 3:6
181. Ji JH, Jung JH, Kim SS et al (2007) Twenty-eight-day inhalation toxicity study of silver nanoparticles in Sprague-Dawley rats. Inhal Toxicol 19:857–871
182. Sung JH, Ji JH, Yoon JU et al (2008) Lung function changes in Sprague-Dawley rats after prolonged inhalation exposure to silver nanoparticles. Inhal Toxicol 20:567–574
183. Sung JH, Ji JH, Park JD et al (2009) Subchronic inhalation toxicity of silver nanoparticles. Toxicol Sci 108:452–461
184. Goodman CM, McCusker CD, Yilmaz T et al (2004) Toxicity of gold nanoparticles functionalized with cationic and anionic side chains. Bioconjug Chem 15:897–900
185. Connor EE, Mwamuka J, Gole A et al (2005) Gold nanoparticles are taken up by human cells but do not cause acute cytotoxicity. Small 1:325–327
186. Hauck TS, Ghazani AA, Chan WC (2008) Assessing the effect of surface chemistry on gold nanorod uptake, toxicity, and gene expression in mammalian cells. Small 4:153–159
187. Shukla R, Bansal V, Chaudhary M et al (2005) Biocompatibility of gold nanoparticles and their endocytotic fate inside the cellular compartment: a microscopic overview. Langmuir 21:10644–10654
188. Pan Y, Neuss S, Leifert A et al (2007) Size-dependent cytotoxicity of gold nanoparticles. Small 3:1941–1949
189. Patra HK, Banerjee S, Chaudhuri U, Lahiri P et al (2007) Cell selective response to gold nanoparticles. Nanomedicine 3:111–119

190. Zharov VP, Mercer KE, Galitovskaya EN et al (2006) Photothermal nanotherapeutics and nanodiagnostics for selective killing of bacteria targeted with gold nanoparticles. Biophys J 90:619–627
191. Lewinski N, Colvin V, Drezek R (2008) Cytotoxicity of nanoparticles. Small 4:26–49
192. Takenaka S, Karg E, Kreyling WG et al (2006) Distribution pattern of inhaled ultrafine gold particles in the rat lung. Inhal Toxicol 18:733–740
193. Hillyer JF, Albrecht RM (2001) Gastrointestinal persorption and tissue distribution of differently sized colloidal gold nanoparticles. J Pharm Sci 90:1927–1936
194. Paciotti GF, Myer L, Weinreich D, Goia D et al (2004) Colloidal gold: a novel nanoparticle vector for tumor directed drug delivery. Drug Deliv 11:169–183
195. Hainfeld JF, Slatkin DN, Smilowitz HM (2004) The use of gold nanoparticles to enhance radiotherapy in mice. Phys Med Biol 49:N309–N315
196. Cho WS, Cho M, Jeong J et al (2009) Acute toxicity and pharmacokinetics of 13 nm-sized PEG-coated gold nanoparticles. Toxicol Appl Pharmacol 236:16–24
197. Chithrani BD, Ghazani AA, Chan WC (2006) Determining the size and shape dependence of gold nanoparticle uptake into mammalian cells. Nano Lett 6:662–668
198. Pernodet N, Fang X, Sun Y, Bakhtina A et al (2006) Adverse effects of citrate/gold nanoparticles on human dermal fibroblasts. Small 2:766–773
199. Gu YJ, Cheng J, Lin CC et al (2009) Nuclear penetration of surface functionalized gold nanoparticles. Toxicol Appl Pharmacol 237:196–204
200. Ma-Hock L, Burkhardt S, Strauss V et al (2009) Development of a short-term inhalation test in the rat using nano-titanium dioxide as a model substance. Inhal Toxicol 21:102–118
201. Wang J, Chen C, Liu Y et al (2008) Potential neurological lesion after nasal instillation of TiO_2 nanoparticles in the anatase and rutile crystal phases. Toxicol Lett 183:72–80
202. Wang J, Liu Y, Jiao F et al (2008) Time-dependent translocation and potential impairment on central nervous system by intranasally instilled TiO_2 nanoparticles. Toxicology 254:82–90
203. van Ravenzwaay B, Landsiedel R, Fabian E et al (2009) Comparing fate and effects of three particles of different surface properties: nano-TiO(2), pigmentary TiO_2 and quartz. Toxicol Lett 186:152–159
204. Zhang Z, Kleinstreuer C, Donohue JF et al (2005) Comparison of micro- and nano-size particle depositions in a human upper airway model. J Aerosol Sci 36:211–233
205. Donaldson K, Stone V, Clouter A et al (2001) Ultrafine particles. Occup Environ Med 58:211–216
206. Ferin J, Oberdorster G, Penney DP (1992) Pulmonary retention of ultrafine and fine particles in rats. Am J Respir Cell Mol Biol 6:535–542
207. Oberdorster G, Ferin J, Lehnert BE (1994) Correlation between particle size, in vivo particle persistence, and lung injury. Environ Health Perspect 102(Suppl 5):173–179
208. Renwick LC, Brown D, Clouter A, Donaldson K (2004) Increased inflammation and altered macrophage chemotactic responses caused by two ultrafine particle types. Occup Environ Med 61:442–447
209. Grassian VH, O'Shaughnessy PT, Adamcakova-Dodd A et al (2007) Inhalation exposure study of titanium dioxide nanoparticles with a primary particle size of 2 to 5 nm. Environ Health Perspect 115:397–402
210. Chen HW, Su SF, Chien CT et al (2006) Titanium dioxide nanoparticles induce emphysema-like lung injury in mice. Faseb J 20:2393–2395
211. Bermudez E, Mangum JB, Wong BA et al (2004) Pulmonary responses of mice, rats, and hamsters to subchronic inhalation of ultrafine titanium dioxide particles. Toxicol Sci 77:347–357
212. Shimizu M, Tainaka H, Oba T et al (2009) Maternal exposure to nanoparticulate titanium dioxide during the prenatal period alters gene expression related to brain development in the mouse. Part Fibre Toxicol 6:20
213. Gurr JR, Wang AS, Chen CH et al (2005) Ultrafine titanium dioxide particles in the absence of photoactivation can induce oxidative damage to human bronchial epithelial cells. Toxicology 213:66–73

214. Churg A, Gilks B, Dai J (1999) Induction of fibrogenic mediators by fine and ultrafine titanium dioxide in rat tracheal explants. Am J Physiol 277:L975–L982
215. Kim JK, Lee WK, Lee EJ et al (1999) Mechanism of silica- and titanium dioxide-induced cytotoxicity in alveolar macrophages. J Toxicol Environ Health A 58:437–450
216. Falck GC, Lindberg HK, Suhonen S et al (2009) Genotoxic effects of nanosized and fine TiO_2. Hum Exp Toxicol 28:339–352
217. Schulz J, Hohenberg H, Pflucker F et al (2002) Distribution of sunscreens on skin. Adv Drug Deliv Rev 54(Suppl 1):S157–S163
218. Kiss B, Biro T, Czifra G, Toth BI et al (2008) Investigation of micronized titanium dioxide penetration in human skin xenografts and its effect on cellular functions of human skin-derived cells. Exp Dermatol 17:659–667
219. Jin CY, Zhu BS, Wang XF et al (2008) Cytotoxicity of titanium dioxide nanoparticles in mouse fibroblast cells. Chem Res Toxicol 21:1871–1877
220. Griffitt RJ, Luo J, Gao J et al (2008) Effects of particle composition and species on toxicity of metallic nanomaterials in aquatic organisms. Environ Toxicol Chem 27:1972–1978
221. Zhang X, Sun H, Zhang ZN et al (2007) Enhanced bioaccumulation of cadmium in carp in the presence of titanium dioxide nanoparticles. Chemosphere 67:160–166
222. Hao L, Wang Z, Xing B (2009) Effect of sub-acute exposure to TiO2 nanoparticles on oxidative stress and histopathological changes in Juvenile Carp (Cyprinus carpio). J Environ Sci (China) 21:1459–1466
223. Aruoja V, Dubourguier HC, Kasemets K et al (2009) Toxicity of nanoparticles of CuO, ZnO and TiO_2 to microalgae Pseudokirchneriella subcapitata. Sci Total Environ 407:1461–1468
224. Hartmann NB, Von der Kammer F, Hofmann T et al (2009) Algal testing of titanium dioxide nanoparticles-testing considerations, inhibitory effects and modification of cadmium bioavailability. Toxicol 269:190–197
225. Gojova A, Guo B, Kota RS et al (2007) Induction of inflammation in vascular endothelial cells by metal oxide nanoparticles: effect of particle composition. Environ Health Perspect 115:403–409
226. Jeng HA, Swanson J (2006) Toxicity of metal oxide nanoparticles in mammalian cells. J Environ Sci Health A Tox Hazard Subst Environ Eng 41:2699–2711
227. Brunner TJ, Wick P, Manser P et al (2006) In vitro cytotoxicity of oxide nanoparticles: comparison to asbestos, silica, and the effect of particle solubility. Environ Sci Technol 40:4374–4381
228. Reddy KM, Feris K, Bell J et al (2007) Selective toxicity of zinc oxide nanoparticles to prokaryotic and eukaryotic systems. Appl Phys Lett 90:2139021–2139023
229. Lanone S, Rogerieux F, Geys J et al (2009) Comparative toxicity of 24 manufactured nanoparticles in human alveolar epithelial and macrophage cell lines. Part Fibre Toxicol 6:14
230. Sayes CM, Reed KL, Warheit DB (2007) Assessing toxicity of fine and nanoparticles: comparing in vitro measurements to in vivo pulmonary toxicity profiles. Toxicol Sci 97:163–180
231. Adams LK, Lyon DY, Alvarez PJ (2006) Comparative eco-toxicity of nanoscale TiO_2, SiO_2, and ZnO water suspensions. Water Res 40:3527–3532
232. Huang Z, Zheng X, Yan D et al (2008) Toxicological effect of ZnO nanoparticles based on bacteria. Langmuir 24:4140–4144
233. Wu B, Wang Y, Lee YH, Horst A et al (2010) Comparative eco-toxicities of nano-ZnO particles under aquatic and aerosol exposure modes. Environ Sci Technol 44:1484–1489
234. Franklin NM, Rogers NJ, Apte SC et al (2007) Comparative toxicity of nanoparticulate ZnO, bulk ZnO, and $ZnCl_2$ to a freshwater microalga (*Pseudokirchneriella subcapitata*): the importance of particle solubility. Environ Sci Technol 41:8484–8490
235. Zhu X, Zhu L, Duan Z et al (2008) Comparative toxicity of several metal oxide nanoparticle aqueous suspensions to Zebrafish (*Danio rerio*) early developmental stage. J Environ Sci Health A Tox Hazard Subst Environ Eng 43:278–284
236. Lee CW, Mahendra S, Zodrow K (2010) Developmental phytotoxicity of metal oxide nanoparticles to *Arabidopsis thaliana*. Environ Toxicol Chem 29:669–675

237. Lin D, Xing B (2007) Phytotoxicity of nanoparticles: inhibition of seed germination and root growth. Environ Pollut 150:243–250
238. Blinova I, Ivask A, Heinlaan M et al (2010) Ecotoxicity of nanoparticles of CuO and ZnO in natural water. Environ Pollut 158:41–47
239. Navarro E, Piccapietra F, Wagner B et al (2008) Toxicity of silver nanoparticles to Chlamydomonas reinhardtii. Environ Sci Technol 42:8959–8964
240. Farre M, Gajda-Schrantz K, Kantiani L et al (2009) Ecotoxicity and analysis of nanomaterials in the aquatic environment. Anal Bioanal Chem 393:81–95
241. Bai W, Zhang Z, Tian W et al (2009) Toxicity of zinc oxide nanoparticles to zebrafish embryo: a physicochemical study of toxicity mechanism. J Nanopart Res. doi:10.1007/s11051-009-9740-9
242. Wang H, Wick RL, Xing B (2009) Toxicity of nanoparticulate and bulk ZnO, Al_2O_3 and TiO_2 to the nematode Caenorhabditis elegans. Environ Pollut 157:1171–1177
243. Deng X, Luan Q, Chen W et al (2009) Nanosized zinc oxide particles induce neural stem cell apoptosis. Nanotechnol 20:115101
244. Takeda A, Ohnuma M, Sawashita J et al (1997) Zinc transport in the rat olfactory system. Neurosci Lett 225:69–71
245. Persson E, Henriksson J, Tallkvist J et al (2003) Transport and subcellular distribution of intranasally administered zinc in the olfactory system of rats and pikes. Toxicol 191:97–108
246. Xia T, Kovochich M, Liong M et al (2008) Comparison of the mechanism of toxicity of zinc oxide and cerium oxide nanoparticles based on dissolution and oxidative stress properties. ACS Nano 2:2121–2134
247. Tracy MA, Ward KL, Firouzabadian L et al (1999) Factors affecting the degradation rate of poly(lactide-co-glycolide) microspheres in vivo and in vitro. Biomaterials 20:1057–1062
248. Nafee N, Schneider M, Schaefer UF et al (2009) Relevance of the colloidal stability of chitosan/PLGA nanoparticles on their cytotoxicity profile. Int J Pharm 381:130–139
249. Mishra V, Gupta U, Jain NK (2009) Surface-engineered dendrimers: a solution for toxicity issues. J Biomater Sci Polym Ed 20:141–166
250. Malik N, Wiwattanapatapee R, Klopsch R et al (2000) Dendrimers: relationship between structure and biocompatibility in vitro, and preliminary studies on the biodistribution of 125I-labelled polyamidoamine dendrimers in vivo. J Control Release 65:133–148
251. Heiden TC, Dengler E, Kao WJ et al (2007) Developmental toxicity of low generation PAMAM dendrimers in zebrafish. Toxicol Appl Pharmacol 225:70–79
252. Jevprasesphant R, Penny J, Jalal R et al (2003) The influence of surface modification on the cytotoxicity of PAMAM dendrimers. Int J Pharm 252:263–266
253. Chen HT, Neerman MF, Parrish AR et al (2004) Cytotoxicity, hemolysis, and acute in vivo toxicity of dendrimers based on melamine, candidate vehicles for drug delivery. J Am Chem Soc 126:10044–10048

Biotechnology and Nanotechnology: A Means for Sustainable Development in Africa

Geoffrey S. Simate, Sehliselo Ndlovu, Sunny E. Iyuke and Lubinda F. Walubita

Abstract In this twenty-first Century, Africa is still lagging behind both in the development and utilization of new cost-effective and high-productive technologies. This is expected to remain so for a long time to come. The lack of technological innovations and monetary investments are some of the key factors that are viewed as contributing to Africa's perpetual underdevelopment and economic instability. In this context, sound technology transfer and acquisition can play an active role in African development and economic sustainability. This chapter looks at how appropriate technology transfer can enhance economic development in Africa, with particular focus on two specific areas, namely Biotechnology and Nanotechnology. Concurrently, this chapter also reviews and discusses some of the key factors that generally impede technology transfer in Africa. Equally, this chapter also addresses some of the merits and demerits of technology transfer, in particular as related to the African continent and its quest for sustainable development and economic growth. Both Biotechnology and Nanotechnology are relatively new, but these are hot technologies that have great potential in many industrial sectors where they can serve to cost-effectively optimize operational processes, increase efficiency, and maximize productivity.

G. S. Simate · S. Ndlovu (✉) · S. E. Iyuke
School of Chemical and Metallurgical Engineering, University of the Witwatersrand, Johannesburg, South Africa
e-mail: Sehliselo.Ndlovu@wits.ac.za; simateg@yahoo.com

L. F. Walubita
TTI—Texas A&M University System, College Station, Texas, TX, USA

A. Gurib-Fakim and J. N. Eloff (eds.), *Chemistry for Sustainable Development in Africa*, DOI: 10.1007/978-3-642-29642-0_9,

1 Introduction

Although Africa is vastly endowed with a wealth of abundant natural resources and greater human resource potential it, however, remains as one of the poorest and technologically backward continents in the world. The key challenges for most African countries is the ability to productively use this wealth and proactively tap into the modern technology so as to build sustainable economies and compete favorably in the Global arena. In this chapter, we examine the importance and ramifications of technology transfer in the fields of Biotechnology and Nanotechnology; as related to Africa's sustainable development and economic growth.

Technology is defined as the practical knowledge, know-how, skill, and artifacts that can be beneficially utilized to develop a product or service and/or a new production/delivery system [1]. Technology is also considered as the knowledge, tools, equipment, and work techniques used by an organization in delivering its products or services to its consumers and/or clients [2]. Technology transfer itself involves the moving of technical knowledge, ideas, services, scientific inventions, and products from the origin of their development (or other locations) to where they can be productively and beneficially put into use [3]. Technology transfer and adaptation, if utilized and applied appropriately, can play a significant role in promoting innovation, entrepreneurship, as well as strengthening development and local economic ownership [4].

However, embracing and mastering technology presents considerable challenges for most African countries. This is partly because the acceptability and successful application of new technologies require, among others, people who have sound scientific backgrounds, technically literate, innovative, and entrepreneurial. Additionally, having a strong national base of scientific and engineering infrastructure such as accredited laboratories and Global connectivity is also a necessary requirement. Irrefutably, however, it still remains critically imperative that Africa should understand and take due cognisance of the urgency to find technologically based solutions to some of the problems it faces, in particular, by its technology dependent industries such as the mines, agriculture, communication, etc.

Presently, some of the innovative technologies that are enjoying growing demand in the developed world are Biotechnology and Nanotechnology. Similarily, these technologies can also add developmental and economic value to Africa. Biotechnology and Nanotechnology have the potential, in some of the technologically dependent industries, to create new innovative products, services, and value adding activities that can cost-effectively improve product performance, functionality, features, and quality; ultimately contributing to Africa's sustainable development and economic growth. These two technologies are also more attuned to technological and market trends, giving them the ability to respond more flexibly to changing conditions.

In a nutshell, the first century of the new millennium will belong not only to communications, or information technologies, but also to Biotechnology and

Nanotechnology, which will bring unprecedented advances in human and animal health, agriculture and food production, mining industries, manufacturing industries, and sustainable environmental management. In other words, these two technologies will yield significant benefits and predict not only the elimination of hunger, but also, almost as a collateral effect, the technical solution of most environmental problems.

With the above background, this chapter looks at how appropriate technology transfer, namely Biotechnology and Nanotechnology, can enhance economic development in Africa. The general merits and demerits associated with technology transfer are also discussed in this chapter. Accordingly, the chapter contents are organized as follows: the definitions and applications of Biotechnology and Nanotechnology are presented first, followed by a discussion of the challenges of technology transfer. A summary of the key points is then presented to conclude this chapter.

2 Biotechnology

According to the National Center for Biotechnology Information (NCBI), Biotechnology is the body of knowledge related to the use of organisms, cells, or cell-derived constituents for the purpose of developing products that are technically, scientifically, and clinically useful [5]. Biotechnology is characterized by a number of unique conditions. More importantly, it is a *cross-cutting* technology that is subject to wide applications, across sectors and biological boundaries. For example, a technique developed for and applied in human health can be used in agriculture and vice versa. Secondly, the development and application of Biotechnology requires a convergence of skills from a variety of disciplines. It requires appropriate combinations of biochemistry, genetics, information technology, engineering, and several other expertises. Thus, it qualifies as a *multi-disciplinary* field.

Finally, the industrial application of Biotechnology requires the acquisition of strong scientific and engineering capabilities, and the deployment of new knowledge in various production processes. More often, this has necessitated the establishment of links or partnerships among the science, engineering, and technology institutions (SETIs), and the private sector; thus substantiating it as a *highly networked* and *collaborative* field.

Biotechnology can make an important contribution to Africa's national priorities, particularly in the area of human health (including HIV/AIDS, malaria, and TB), food security, mining, and environmental sustainability. In its pursuit to address these priorities, Africa is fortunate in that she can beneficially be guided by the historical experiences of the developed countries. For instance, we know that to achieve success, a country requires a government agency to champion Biotechnology, to build human resources proactively, and to develop scientific and technological capabilities. In addition, successful commercialization of public sector-supported research and development (R&D) requires strong linkages between institutions within the National System of Innovation. Furthermore, there is also need for a vibrant culture of innovation and entrepreneurship, assisted by incubators, supply-side measures, and other supporting programs and institutions. It is, therefore, imperative for Africa to create deliberate strategies that target those areas that are of national economic priority and organize its R&D activities in such a way as to gainfully exploit the scientific expertise and technical infrastructures across the institutional landscape.

Overall, the potential of Biotechnology to improve the quality of our lives and the quality of our environment is very substantial. It could bring huge advances in health, nutrition, and remediation of the environment, to name but a few. However, in the realization of these benefits, Africa will have to be judicious and selective; avoiding those technologies that challenge our ethical value systems (such as human cloning) and focusing instead on those which can provide significant advances with minimal risks or ethical/moral conflicts.

In the subsequent sections, a number of Biotechnology applications that have the potential to enhance economic development in Africa are discussed. These include applications in mining industry, water treatment, biofuel production, agriculture, and the environment.

2.1 Biotechnology and the Mining Industry: Metal Recovery

The African continent is richly endowed with abundant reserves of strategic and economically important minerals, but not well harnessed and in most cases, exported as raw materials to the developed world; and later re-exported back to Africa as finished products with prohibitive price tags on them. The economic development of these minerals is very vital because of, among other benefits, their significant role in employment creation and earning of foreign exchange. Most of the currently available conventional mineral processing technologies (e.g., flotation, smelting, and high pressure acid leaching) in most African countries are costly and not technologically up-to-date to competitively meet the modern market needs in the mining industry. Above all, most of these conventional techniques are not environmentally friendly. Of significant note, however, is that most of Africa's mineral industries are export oriented, and thus exposed to the world market fluctuations. Low commodity or metal prices, can make the mine uneconomic to operate leading to its closure thus resulting in loss of employment. It is thus important to develop and utilize appropriate technologies that are simple to apply, provide low capital and operational costs, comply with environmental regulations, and yet are highly productive. Biotechnology is regarded as one of the most promising and certainly, one of the most revolutionary technological solutions to some of these problems, compared to say conventional pyrometallurgy or chemical metallurgy [6].

The branch of Biotechnology that deals with metal extraction is termed Biohydrometallurgy. Biohydrometallurgy is a natural process that uses microorganisms to enhance the dissolution of metals from mineral ores, especially sulfide ores, by making them more amenable to dissolution in aqueous solutions [7]. It is a multidisciplinary field involving hydrometallurgy, mineral processing, chemistry, the environment, and microbiology [8]. It has been globally applied to the recovery of base and precious metals such as copper and gold, is now an established industrial technology for the pretreatment of refractory ores [9], and is piloted and commercialized in many countries all over the world. It is characterized by low operational costs, lower environmental impact, simplicity of plant operation and maintenance and, nowadays, it is the most successful alternative in many mining projects. Overall, biohydrometallurgy is an acceptable practice considered as a successful and expanding area of Biotechnology [10]. The various bioprocessing technologies include bioremediation, biosorption, bioaccumulation, biooxidation, and bioleaching. However, biooxidation and bioleaching are the most studied for metal recovery. Both are oxidation processes, but where the metal to be recovered is extracted into solution the process is known as bioleaching, whereas when the metal remains in the mineral the process should strictly be referred to as biooxidation [11]. Current worldwide biooxidation and bioleaching research and commercial operations remain focused essentially on gold and copper production, respectively.

In Africa, however, there are only a few commercial operations where biooxidation technology is utilized. Notable ones include the following gold producing plants:

- Fairview in South Africa,
- Bogosa/Prestea in Ghana [12], and
- the Ashanti's Sansu plant near Obuasi in Ghana (most successful gold plant in Africa).

The cobalt plant at Kasese in Uganda uses bioleaching [13]. On the copper front, Zambia is the major African player in the production of copper. Although there have been some demonstration plants set up to try and implement the bioleaching technology for copper production in Zambia, there are, however, currently no large-scale commercial copper bioprocessing plants in operation. In addition, South Africa and Zimbabwe also produce a notable amount of copper, however, there are also no copper processing plants currently using bioleaching [12].

The bioleaching processes have been shown to be suitable not only for copper, but also for zinc (BioZINC^{TM}) and nickel ores (BioNIC^{TM}) [14]. These ores can be found on the African continent and application of the bioleaching technology will lead to greater economic and environmental rewards. In addition, past research has shown that this technology can also be used for coal desulfurization. South Africa and Zimbabwe have significant coal deposits that can benefit from the ready application of this technology [12].

The microorganisms which are important in the biohydrometallurgical processes concerned with metal extraction can be divided into two groups on the basis

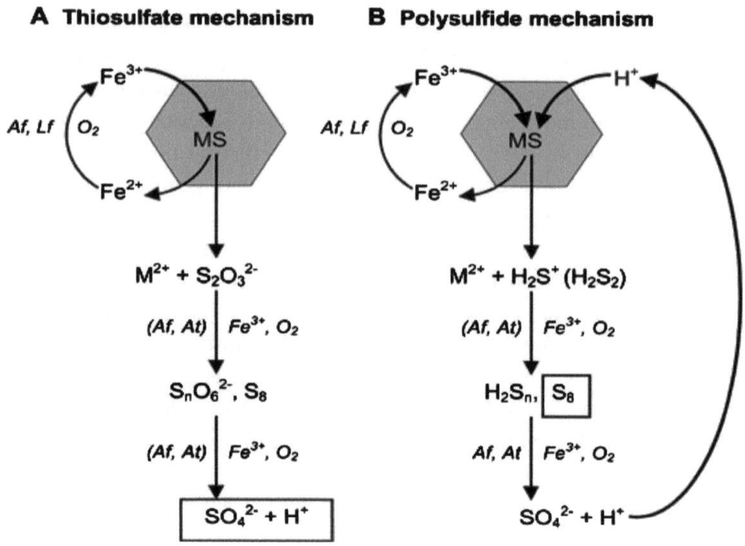

Fig. 1 Hypothetical scheme for the oxidation of pyrite or metal sulfide by iron- and sulfur-oxidizing acidophilic bacteria producing sulfuric acid [22, 23]

of their nutritional requirements, i.e., autotrophic and heterotrophic microorganisms. The most widely studied of these microorganisms in the bacterial assisted leaching of sulfides or minerals containing reduced iron (such as gold and copper ores) are chemolithotrophs. For non-sulfidic ores such as nickel laterites, the heterotrophs are used. While the commercialization of chemolithotrophic bioleaching has been largely successful as noted from the above mentioned operational plants in Africa, commercial mineral bioprocessing using heterotrophic microorganisms have so far not been viable because of the vast amount of carbon source (e.g., molasses) required to support the growth of these particular microorganisms and the production of greater amounts of leaching agents [15].

However, recently, Simate et al. [16–21] have shown that an alternative approach would be to use chemolithotrophs microorganisms to recover these non-sulfidic metal values. This can be achieved by allowing the primary oxidation of pyrite, or similar iron/sulfur minerals to provide sulfuric acid solutions, which solubilize the metal content. The production of sulfuric acid is schematically illustrated in Fig. 1.

This area of research has opened up a new era to the potential application of these chemolithotrophic microorganisms for the commercial processing of the difficult-to-process low grade nickel laterite ores and other non-sulfide metal containing resources such as silicate ores, oxidic converter furnace slags, and refractory oxides. It has highlighted the diverse and unconventional, but yet effective process routes in which biohydrometallurgy can be effectively harnessed to treat diverse ores for the sustainable development of the African minerals industry. This is important because in developing countries, there is a need to develop and utilize technologies that will likely maintain a competitive edge over other developed nations. Zambia, for example, has large piles of oxidic slag in its Copperbelt region which can, cost-effectively and safely, be processed biotechnologically. The current technology of using direct arc (DC) furnace for processing slag [24] is very expensive because of its energy intensity and is also prone to several runaway accidents; thus safety is also an issue with the conventional DC technique.

As mentioned earlier in the introductory section, the technological transfer of Biotechnology can help in the processing of low and lean ore grades cheaply. In general, the capital cost of a bioleaching operation is about half that of a conventional smelting/refining operation, which is very competitive compared with the unit costs of conventional smelting/refining [25]. With further refinements and use of high temperature tolerant microorganisms to accelerate the reaction rates and shorten the leaching time, costs should further be optimized with the bioleaching processes, especially if genetic modification of the bioleaching microorganisms is successfully engaged [25]. Furthermore, compared to the traditional high pressure leaching and smelting processes, Biotechnology is simple to operate with conditions of low pressure and low temperature.

Although the initial development of biohydrometallurgy was for metal extraction, the increased environmental awareness has led to serious exploitation of this technology in the environmental control processes and applications such as bioremediation and water treatment. These aspects are discussed in the subsequent three sections that follows.

2.2 *Biotechnology and the Mining Industry: Bioremediation and Phytoremediation*

Past and recent exploitation practices, mining activities, industrial production, and practices of solid waste disposals are some of the primary sources of metals, metalloids, and radionuclides contamination of soils and natural reservoirs of surface and groundwater aquifers [26]. This contamination of soils, sediments, surface, and groundwaters with different inorganic and organic substances is a great environmental problem, particularly in countries where mining and processing of ores is intense. Such contaminants can result in radioactive, in some cases, and chemical exposures that directly and/or indirectly impact human and/or animal health.

The remediation of contaminated waters and sediments in the deposits of these heavy metals is a common practice. Various methods based on both active and passive treatment systems have been applied. However, current remediation techniques which are based on chemical or physicochemical extraction techniques are costly, time consuming and environmentally destructive [27]. In some cases, the treatment is connected not only with the remediation of the contaminated

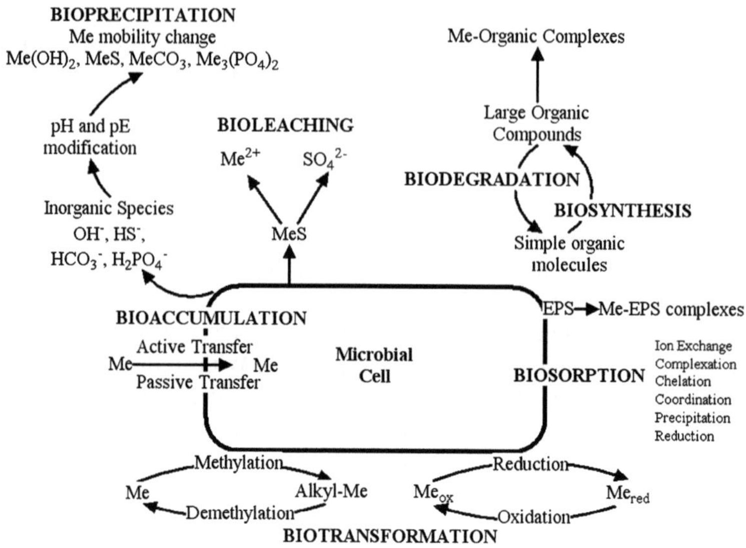

Fig. 2 Interaction of microbial cell metals [26]

waters and sediments but also with the recovery of some valuable components from them, e.g., of uranium and some non-ferrous metals. In other words, traditional efforts to manage contaminated soils or water often focus on their removal by extraction followed by off-site treatment/disposal [28]. These ex-situ management techniques are more expensive and more risky than in-situ management techniques [29], and thus there is a preference for in-situ techniques.

Some in-situ remediation methods are based on the activity of some soil or water microorganisms. This is termed bioremediation. Bioremediation processes include mainly the use of microorganisms, which may interact with targeted species in order to reduce their mobility and toxicity [30]. In fact, bioremediation processes may result both in the mobilization and the immobilization of the targeted species and present important advantages such as the following [26]:-

- They can be pollutant specific.
- They can be in many cases most efficient compared to other common physicochemical processes.
- They can be applied in-situ and ex-situ conditions.
- They are less energy and materials consuming.
- They often do not produce large volumes of secondary wastes such as sludge.

Figure 2 shows an illustration of various mechanisms in which both living-metabolising and non-metabolising microbial cells may interact with soluble species to immobilize them.

Another technique called phytoremediation is based on the ability of some plants to treat contaminated soil, ground water, or wastewater via their root systems.

Phytoremediation is an emerging technology for cleaning up contaminated sites, which is cost-effective, and has esthetic advantages and long-term applicability [27]. Plants can be used in the following ways to assist in remediation [31]:-

- Plants and microorganisms usually have symbiotic relationships, making the root zone, or rhizosphere, a very active area of microbial activity. Plants can help moderate the environment in the root zone, providing ideal conditions for bacteria and fungi to degrade organic pollutants.
- Certain plants hyperaccumulate metals, radionuclides, or organic compounds. These plants can be placed in areas of contamination where they remove the contaminant. Afterwards, the plants can be harvested and the contaminant can be disposed, destroyed, or recovered.
- Many plants require a lot of water in their daily lives. These plants can be used to inexpensively 'pump' groundwater in certain shallow aquifer conditions.
- Some plants contain enzymes and other certain proteins that can destroy contaminants.

Overall, phytoremediation is a promising new technology which is attracting much attention. In some forms (phytoextraction, phytostabilization, rhizofiltration, phytovolatilization) it is already available and proven [27]. New applications are being tested and developed. Phytoremediation, like bioremediation, is relatively slow. However, its main advantage is cost saving compared to other processes.

2.3 Biotechnology and the Mining Industry: Acid Mine Drainage/Acid Rock Drainage

One of the most pressing effects of mining activities, especially in water-scarce African countries, is the issue of contaminated water emanating from underground workings of deep mines. Generally, contamination occurs when mining activities bring sulfidic rock into contact with water. For example, under oxidizing conditions, pyrite-containing rock produces sulfuric acid and dissolved iron. These acidic waters may then dissolve other metals contained in the rock, resulting in low-pH metal-bearing water known as acid mine drainage (AMD) or acid rock drainage (ARD). This is generally of minor importance when a mine is in active production and water tables are kept artificially low through pumping [32]. However, when the mines are closed and/or abandoned with no pumping activities, the rebound of the water table can lead to contaminated groundwater being discharged to the surface, sometimes in catastrophic events, such as happened at Wheal Jane mine in Cornwall, UK in 1992 [32].

In fact, mines can cause environmental devastation decades and even centuries after they are shut down. South Africa, for example, is famous for such related mining activities. Environmentalist have indicated that AMD (which is associated with these mining activities) has become the single most significant threat to South Africa's environment and the sustainable development of the country's mineral wealth.

Some of the cases of the effects of this phenomena (AMD) have actually been reported in the South African media, including the following two controversial ones:

(1) the decant from abandoned gold and coal mines in the Vaal River catchment affecting the water resources for many downstream water users, and
(2) the decant of mine water from the Western Basin in the Witwatersrand gold fields into the Tweelopiesspruit which is threatening a world heritage site.

The greatest challenge is, therefore, to find a cheaper and sustainable treatment technology that will be able to improve the quality of the mine water and that which will remain operational even after the mine has closed down. There are a number of methods for dealing with AMD. Some are unconventional such as raising the pH through liming, removing water, binding iron with organic wastes, etc. More conventional ones include the following: application of bactericides, biocontrol with other bacteria/archaea, off-site wetland creation, use of metal-immobilizing bacteria, etc.

Given the potential for serious environmental damage and the associated reclamation costs, it is practical to seek long-term, cost-effective treatments for AMD. The long-term nature of the AMD problem, has resulted in focused interest in biological treatment approaches, which offer low costs and sustainability. The application of sulfate-reducing bacteria (SRB) has been demonstrated to be effective for the treatment of such wastewaters [33]. The general purpose of using SRB in AMD treatment is to produce sulfides for metal sulfide precipitation, while generating alkalinity. In other words, biological sulfate reduction based on the applications of SRB has been identified as a potentially valuable process for the removal of contaminant metals from acidic wastewaters, given their role in the generation of insoluble metal sulfides and the neutralizing effect of the sulfate reducing process [33, 34].

At least two technologies using off-line sulfidogenic bioreactors have been described: the *Biosulfide* and the *Thiopaq* processes. The *Biosulfide* system has two components, one biological and one chemical, which operate independently [35]. A schematic illustration of this process is shown in Fig. 3.

According to Johnson [32], raw AMD enters the chemical circuit, where it comes into contact with hydrogen sulfide generated in the biological circuit. By careful manipulation of the conditions (i.e., pH and sulfide concentration), selective separation of a particular metal sulfide is possible. This may then be removed from the partially processed water ahead of further treatment. Some of the treated AMD enters the biological circuit to provide the sulfate source in the bioreactor, which contains a mixed culture of SRB. For the process to run optimally, additional alkali may be required beyond that produced by the SRB, in which case it is added in chemical form. The *Thiopaq* system differs from the *Biosulfide* process in that it utilizes two distinct microbiological populations and processes, namely [32]:

(i) conversion of sulfate to sulfide by SRB, and precipitation of metal sulfides and,
(ii) conversion of any excess hydrogen sulfide produced to elemental sulfur, using sulfide-oxidizing bacteria (SOB).

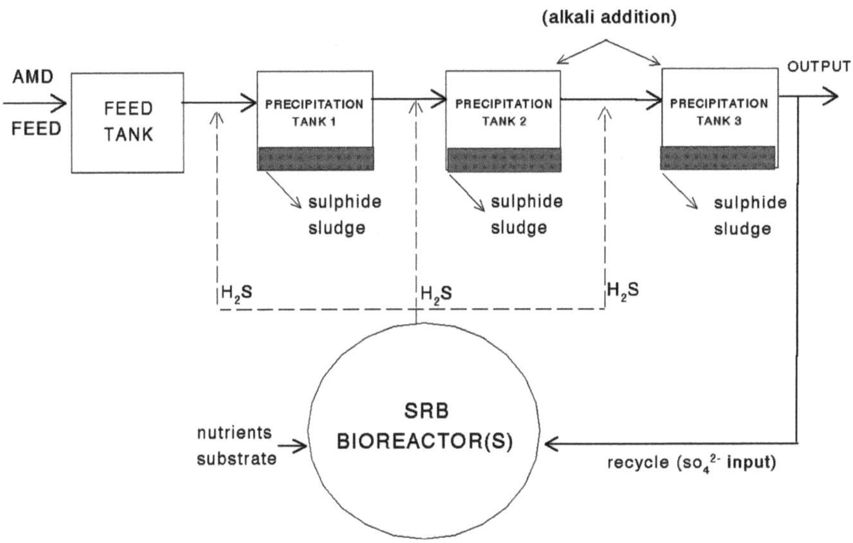

Fig. 3 Flow path of the "biosulfide" process for remedying acid mine drainage and recovery of heavy metals [32, 35]

In summary, the basis of bioremediation of AMD derives from the abilities of some microorganisms to generate alkalinity and immobilize metals, thereby essentially reversing the reactions responsible for the genesis of AMD [32].

2.4 Biotechnology and the Water Industry: Water Treatment

Mineral processing waste water usually contain metal elements such as copper, iron, lead, zinc, cadmium, molybdenum, arsenic, mercury, uranium, radium, gold, and silver. These elements must be removed and thus, this requires treatment of the waste water prior to its rejoining with the natural hydrologic system. Restrictive environmental legislations, ecological problems due to dispersion of heavy metals in natural environment, and the high cost of technologies for treat ment of effluents containing heavy metals at too low concentrations, but nevertheless exceeding the limits imposed by environmental legislation have stimulated the development of technologies to compete with or complement conventional techniques. Some of the conventional methods for removing the metals include precipitation, ion exchange, and electrolytic techniques. A satisfactory process would be one that achieves the goal of water purification by metal removal at a relatively low cost. Among the technologies under development, much is being researched about techniques involving the use of microorganisms and, above all, plant biomass as metal binding compounds.

In particular, the use of agrowastes in a pure or chemically modified form for the remediation of contaminants in aqueous solution and industrial effluents has been studied significantly in the recent past [36–42]. The metal adsorption capacities of different kinds of agrowastes have been identified as potential alternatives to the existing metal removal technologies. This is because agrowastes are:

- readily available;
- cheap;
- biodegradable;
- sludge free; and
- competitively accessible/proccessible, with a moderate initial cost and land investment.

Most of the agrowaste-related Biotechnology research has focussed on the use of algae, seaweed, alfalfa, sago waste, banana pith, sunflower, and cassava waste. For instance, cassava is a perennial woody shrub, and is a major source of low-cost carbohydrates for populations in the humid tropics in many parts of western and central Africa. It is actually a major staple food in Nigeria and Ghana. The preparation of cassava for consumption purposes produces a large volume of wastes such as peelings of the husks, which creates a lot of environmental problems. However, no effort has been made to control or manage the enormous wastes arising from processing cassava tubers into its various products, which are abundant and available in all seasons. The economic utilization of cassava tuber bark waste may not only provide a solution to its environmental nuisance, but will also create wealth and improve local economies, if properly harnessed. In addition, the world market for cassava starch and meal is limited due to the abundance of other carbohydrate substitutes, particularly in the developed world. Its alternative economic use would thus only benefit most of the African industries.

In summary, developing countries do not have the financial ability to invest in conventional wastewater treatment techniques that are expensive. The employment of natural resources (biomaterials) such as microorganisms and plant biomass can provide a cheaper and simpler alternative to serve a similar purpose. Since large amounts of excess plant biomass are produced by the agro-industry in Africa, it is desirable, therefore, to use this as a resource for sustainable bioremediation and biodegradation processes. If not used to generate a value-added product, the biomass would remain in the waste stream thus requiring expensive disposal or treatments.

2.5 *Biotechnology and the Fuel Industry: Biofuels*

Biofuel production is part of 'white' Biotechnology [43]. White Biotechnology (as known mainly in Europe) or industrial Biotechnology is the application of Biotechnology for industrial purposes, including manufacturing, alternative energy (or "bioenergy"), and biomaterials. It includes the practice of using cells or components of cells like enzymes to generate industrially useful products. Nowadays, the Organization for Economic Cooperation and Development (OECD) and a handful of big corporations argue that the fossil fuel era will come to an end [43]. White Biotechnology enjoys a positive social acceptance because it aims at being environmental friendly and contributes to sustainable development. By providing new materials and fuels that are not derived from petrochemical processes, and, by trying to use less fossil fuel energy, white Biotechnology may become acceptable to environmentalists.

Currently, ethanol fuel is the most common biofuel worldwide, particularly in Brazil. Alcohol fuels are produced by fermentation of sugars derived from wheat, corn, sugar beets, sugar cane, molasses, and any sugar or starch that alcoholic beverages can be made from (like potato and fruit waste, etc.). Another idea is to use the whole plant as a chemical feedstock. In this regard, Africa is blessed with a vast amount of virgin land to make this a reality.

Ethanol can be used in petrol engines as a replacement for gasoline; it can be mixed with gasoline to any percentage. Most existing car petrol engines can run on blends of up to 15 % bioethanol with petroleum/gasoline. For example, in the USA, nearly a tenth of all motor fuel sold is a blend of 90 % petrol and 10 % ethanol. Ethanol has a smaller energy density than gasoline, which means it takes more fuel (volume and mass) to produce the same amount of work. An advantage of ethanol is that it has a higher octane rating than ethanol-free gasoline available at roadside gas stations which allows an increase of an engine's compression ratio for increased thermal efficiency.

While the production, transport, and consumption of gasoline generate 11.8 kg of carbon dioxide per gallon (3.8 l), in the case of ethanol 7–10 kg of carbon dioxide is generated if conventional production processes are used, and only 0.06 kg if one relies on bioprocesses [44]. It is true that biofuels cost more than fuels derived from fossil energy, but the real cost of the latter does not integrate the heavy costs of shore and sea contamination by oil and oil-tanker wreckage, nor those of conflicts generated by oil exploitation. Biofuels have positive effects on employment, tax recovery, and supply reliability. The impact on local employment has been evaluated at 6–10 jobs created for every thousand tonnes produced [43].

2.6 *Biotechnology and the Agricultural Industry: Genetically Modified Crops*

Biotechnology in agriculture is not a new phenomenon. Although references to Biotechnology have been increasingly prominent in the past years, the use of biological processes to improve food production actually dates back to the time when humans started domesticating animals and growing crops, over 10,000 years ago.

Since the early origins of agriculture, farmers have worked to modify plants to increase yields and tolerate stresses. This began with basic hybridization and mutation, and then grew steadily with advances in technology. Today,

Biotechnology has become even more precise, allowing the transfer of beneficial genetic materials from one species to another. Biotech crops produced this way are sometimes called genetically modified (GM) crops or transgenic crops. GMs generally have increased crop yields while allowing a reduction of environmental impacts from agricultural activities.

In contrast to the 'green revolution' that only focused on three main food crops (rice, wheat, and maize), Biotechnology can be used to improve the characteristics of all target plants, which means that the genuine subsistence plants like cassava or potatoes could also be affected. While in the 1970s significant increases in agricultural output only became feasible when the specific agricultural environments were adapted to the needs of the newly developed, and standardized high yielding varieties (necessitating the installation of expensive irrigation systems as well as high inputs of fertilizer and pesticides), biotechnologies make it possible to improve the plants' adaptation to their specific geoclimatic surroundings. In this way, higher outputs, improved nutritional values, longer shelf-life capabilities, etc., can be achieved. This also means that salty areas, often the result of inadequate irrigation schemes, could be reused for agricultural purposes. This new approach to increasing agricultural productivity could be especially valuable for those African regions and social groups, which were never reached by the "Green Revolution"; whether for geoclimatic (no possibility to install irrigation schemes) or social (no access to credits in order to buy machinery and pesticides) reasons. Furthermore, Biotechnology could also contribute significantly to a pattern of agriculture which is more sustainable and ecologically sound as well as to the reforestation of desert and/or erosion-prone areas. For example, no-till agriculture (in limited use prior to 1996), is being widely adopted due to the superior weed control from biotech crops that are able to tolerate herbicides with low environmental impacts [45]. This has led to improved soil fertility, improved water retention capacity, reduced runoff, and reduced greenhouse gas emissions from agriculture.

In summary, Biotechnology can offer important economic and developmental opportunities to both the African farmers and the consumers. However, despite these benefits, the role of modern Biotechnology in spurring agriculture-led economic transformation and sustainable development in Africa is, more often, subject to furious scientific debate and intense public controversy. Furthermore, African governments face enormous uncertainty and pressure as they deliberate on national and regional policies, programs, and regulations that attempt to maximize the benefits and minimize the risks of Biotechnology products.

2.7 Biotechnology and the Environment

There are also issues of environmental standards that continue to stiffen, particularly regarding toxic wastes; so costs of ensuring environmental protection will continue to rise. This is a huge burden on the already financially strapped African countries. Biotechnology holds the potential of reducing environmental pollution because the biotech processes are carried out under mild conditions, usually without addition of toxic chemicals, and also, the products end up in an aqueous solution, which is more amenable to containment and treatment than gaseous wastes as in most conventional treatment techniques [6].

3 Nanotechnology

Nanotechnology is acknowledged world wide to be at the forefront of miniaturisation and one of the cutting edge state-of-the-art twenty-first century technology. This technology has many applications ranging from mining, computers, information technology, Biotechnology, electronics, aerospace, defense, manufacturing, environment, medicine, etc. In the sub-Saharan Africa, with the exception of South Africa, the response to this twenty-first century technology is still very low [46].

Iyuke et al. [46] in their paper outlined a comprehensive strategy that the South African government has followed in order to optimally use Nanotechnology to enhance its Global competitiveness and sustainable economic growth. One of these strategies involves a broad collaboration among the Government, the industry, and the academia to work together toward realizing the potential of Nanotechnology. Undeniably, it must be emphasized here that collaboration is one of the very strategic channel of effective knowledge dissemination among the interested players and participating partners; particularly where technology transfer is concerned.

In a nutshell, Nanotechnology involves miniature, stronger, cheaper, lighter, durable, and faster devices with greater functionality and efficiency, apparently using fewer raw materials input and consuming less energy, but with very high productivity output. These are all aspects which are very critical in most processing industries. In particular, the usage of fewer raw materials as input in Nanotechnology translates into greater quantitative output per unit volume or tonnage of the raw materials compared to other conventional technologies. With these characteristics, there is evidently no doubt that the future use and appropriate application of Nanotechnology in Africa, will inevitably contribute to the continent's success in its quest for sustainable development and economic vibrancy.

The subsequent sections below focuses on a limited number of Nanotechnology applications that may have potential to enhance development in Africa. These include water treatment, medical applications (e.g., malaria treatment), Global warming, energy, and agriculture.

3.1 Nanotechnology and the Water Industry: Water Treatment

Water is a very crucial and necessary element of human life, including all other living things. Water has been the foundation and sometimes the undoing of many great civilizations. Today, water is essential for agricultural, economic, and industrial activities that help society to develop. Less than a century ago, it was widely assumed that there were enough freshwater supplies in the world for everyone. Yet today, increased use of freshwater for industrial, agricultural, and domestic purposes has created acute water shortages in some areas of the world. These shortages are stimulating or worsening international conflicts over water, which has joined oil as a major commodity triggering wars. As water shortages and conflicts increase, water is increasingly being transformed into a privately owned commodity that can be sold and traded for profit. Furthermore, public pressure and stringent water quality requirements have been increasing in the past few decades to develop alternative treatment systems due to potential public health and environmental risks.

Fortunately, the advent of Nanotechnology has brought a lot of hope, and that Nanotechnology for water purification has been identified as a high priority area because water treatment devices that incorporate nanoscale materials are already available and human development needs for clean water are pressing. A range of water treatment devices that incorporate Nanotechnology are already on the market and others are in advanced stages of development. Some of these Nanotechnology applications include the following:

- nanofiltration membranes, including desalination technologies—nanofiltration (NF) is a cross-flow filtration technology which lies somewhere between ultrafiltration (UF) and reverse osmosis (RO). These membranes are able to remove particles below 100 nm in size. In addition, the transmembrane pressure (pressure drop across the membrane) required is considerably lower than the one used for RO, thus reducing the operating cost significantly;
- nanocatalysts—nanocatalysts have the advantage of very high reaction rates due to high specific surface areas and low mass-transfer restrictions, and can selectively target impurities;
- magnetic nanoparticles—these materials can easily be recovered from adsorbed heavy metals by utilizing magnetic separation; and
- nanosensors for the detection of contaminants—current methods in use for detecting and identifying contaminants in both air and water are relatively slow, and often require laboratory analysis. Also, they do not allow detection of many contaminants using a single sensor. Figure 4 is an illustration of a nanosensor.

In summary, Nanotechnology shows much potential to solve water quality challenges within the water sector and research efforts in this field could serve to ameliorate many of water problem. In other words, the water sector can apply Nanotechnology to develop more cost-effective and high performance water treatment systems, as well as instant and continuous ways to monitor water quality.

3.2 Nanotechnology and the Medical Industry: Malaria Treatment

Malaria, the most prevalent parasitic disease in the world, is caused by the apicomplex protozoan of the *Plasmodium* genus. Malaria is present all over the tropics, where four species, *Plasmodium falciparum* (most widespread and

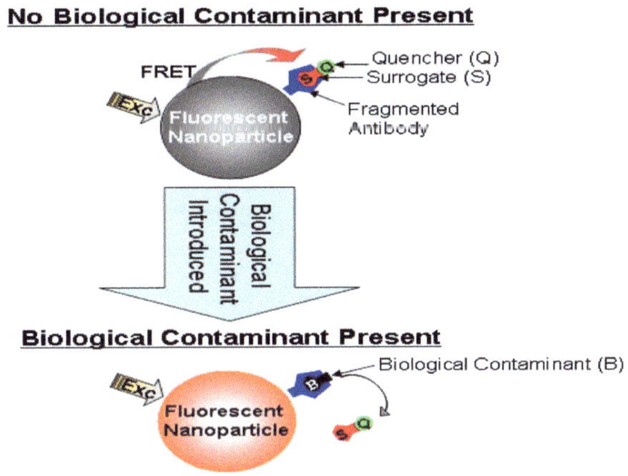

Fig. 4 FRET-based Nanosensor for biological contaminants [47]

Fig. 5 The mosquito

dangerous), *Plasmodium vivax*, *Plasmodium malariae*, and *Plasmodium ovale* are transmitted to humans by the bites of the female mosquito vector of the *Anopheles* genus [48] shown in Fig. 5. It kills over one million people each year, most of whom are children under 5 years, and almost 90 % of whom live in Africa, south of the Sahara [49]. Each year, there are over 300 million clinical cases of malaria that is five times as many as combined cases of TB, AIDS, measles, and leprosy [50]. The cheapest and safest malaria drug, chloroquine, is rapidly losing its effectiveness. In some parts of the world, malaria is resistant to the four leading frontline drugs (chloroquine, sulphadoxine-pyrimethamine or fansidar, meflo-quine, and quinine). In general, the main drawbacks of conventional malaria chemotherapy are the development of multiple drug resistance and the nonspecific targeting to intracellular parasites, resulting in high dose requirements and sub-sequent, intolerable toxicity [48].

In recent years, the production and applications of nanomaterials has become a widespread growing reality, both in the academia and industry. In the field of nanomedicine, one area that is receiving a lot of attention is in the drug delivery systems. Nanoparticulate drug delivery systems represent a promising approach for obtaining desirable drug-like properties by altering the biopharmaceutics and pharmacokinetics property of the drug molecule [51]. Recently, nanosized carriers

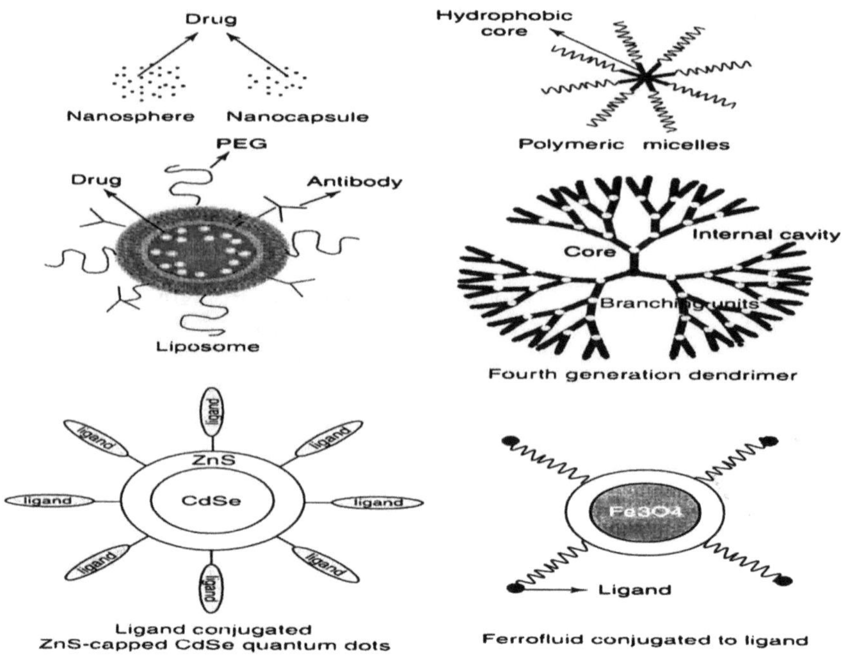

Fig. 6 Schematics of different Nanotechnology-based drug delivery systems [55]

have been receiving special attention with the aim of minimizing the side effects of drug therapy, such as poor bioavailability and the selectivity. The most important property of a nanocarrier in the context of malaria is the ability to remain in the blood stream for a long period of time in order to improve the interaction with infected red blood cells (RBCs) and parasite membranes [52]. Additional interesting properties are protection of instable drugs, cell-adhesion properties, and the ability to be surface-modified by conjugation of specific ligands of drugs [53, 54]. Several of these nanosized delivery systems have already proved their effectiveness in animal models for the treatment and prophylaxis of malaria [48].

Figure 6 is an illustration of different Nanotechnology-based drug delivery systems showing the following [55]: Nanoparticles are small polymeric colloidal particles with a therapeutic agent either dispersed in polymer matrix or encapsulated in polymer.

Polymeric micelles are self assembled block copolymers, which in aqueous solution arrange to form an outer hydrophilic layer and an inner hydrophobic core. The miceller core can be loaded with a water insoluble therapeutic agent. Liposomes are lipid structures that can be made 'stealth' by PEGylation and further conjugated to antibodies for targeting. Dendrimers are monodispered symmetric macromolecules built around a small molecule with an internal cavity surrounded by a large number of reactive end groups. Quantum dots are fluorescent

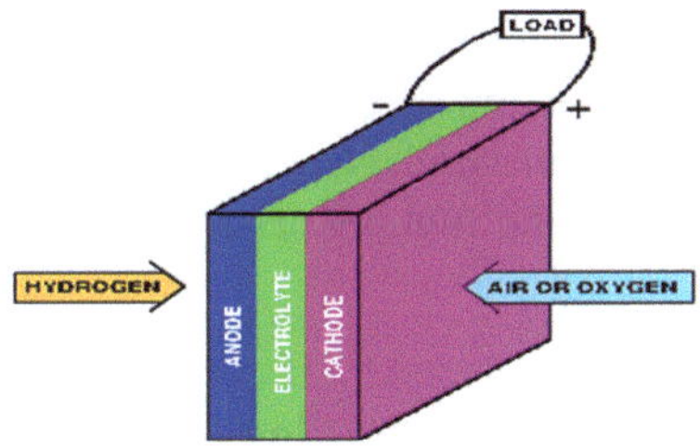

Fig. 7 An example of a fuel cell. Reaction taking place is $2H_2 + O_2 \rightarrow 2H_2O +$ Energy

nanocrystals that can be conjugated to a ligand and thus can be used for imaging purposes. Ferro fluids are colloidal solutions of iron oxide magnetic nanoparticles surrounded by a polymeric layer, which can be further coated with affinity molecules such as antibodies.

Overall, the advantages of using nanoparticles for drug delivery result from their two main basic properties [55]. Firstly, nanoparticles, because of their small size, can penetrate through smaller capillaries and are taken up by cells, which allows efficient drug accumulation at the target sites. Secondly, the use of biodegradable materials for nanoparticle preparation allows sustained drug release within the target site over a period of days or even weeks.

3.3 Nanotechnology and the Energy Industry: Fuel Cells

Energy is considered as the engine for industrial development in any country. In the coming years, few sectors of the Global economy will have changed as much as energy is changed. The current complete reliance on the combustion of fossil fuels as a source of energy for power generation in the industries and for running of vehicles is clearly disturbing the natural systems [56, 57]. In addition to the health and environmental concerns resulting from complete dependence on fossil fuels as a source of energy, a steady depletion of the world's limited fossil fuel reservoir also call for new energy technologies for energy conversion and generation. Such energy conversion technologies should be more efficient than the conventional heat energy with minimal or no pollution emissions and also compatible with renewal energy sources for sustainable development [58–60]. Fuel cells (Fig. 7) have been identified as one of the promising and potential clean energy technologies that meet all the requirements for energy security, economic growth, and environmental sustainability, and have attracted considerable attention as a possible replacement for power generation systems [61, 62].

However, fuel cells also face many obstacles, which researchers and industries must overcome before they can be widely introduced into the market [63–65]. The grafting of fuel cell membranes with carbon nanoparticles was found to improve their thermal stability, water uptake, porosity, methanol crossover, and more than 50 % increase in proton conductivity of the membrane [66]. This improvement in quality is anticipated to contribute to the reduction in cost of production of fuel cells and improve their efficiency, thus leading to more applications for fuel cells.

Summarized, Nanotechnology constitutes a vibrant technology for boosting and optimizing fuel cell based energy, and should be explored further.

3.4 Nanotechnology and the Agricultural Industry: Agrifood

Food, along with water and air, are among the most essential elements of life. On average, people eat three meals, mostly, a day without much thinking of it. There is currently a Global dilemma faced by humanity due to rising population pressures, constrained resources, and related food security issues. Agriculture is, undoubtedly, the backbone of most developing countries, with more than 60 % of the population reliant on it for their sustenance and livelihood.

For the developing countries where food shortage is a common occurrence, the drive is to develop drought and pest resistant crops, which also maximize yield. The potential of Nanotechnology to revolutionise the health care, textile, materials, information and communication technology, and energy sectors has been well publicized. However, the application of Nanotechnology to the agricultural and food industries was only first addressed by the United States Department of Agriculture roadmap published in September 2003 [67]. Nanotechnology has the potential to revolutionize the agricultural and food industry with new tools for the molecular treatment of food diseases, rapid and early disease detection, disease treatment delivery methods, protection of the environment, enhancing the ability of plants to absorb nutrients, etc. [68, 69]. Smart sensors and smart delivery systems will help the agricultural industry combat viruses and other crop pathogens, and in the near future nanostructured catalysts will be available which will increase the efficiency of pesticides and herbicides, allowing lower doses to be cost-effectively used [68]. This, in turn, will reduce pollution and make agriculture more environmentally friendly. Figure 8 depicts some of the potential applications of Nanotechnology in the food industry.

In summary, nanofood which encompasses Nanotechnology techniques or tools used during cultivation, production, processing, or packaging of food, is an important technology that can help Africa in food sustainability. Furthermore, Bio–Nanotechnology will, in the near future, take agriculture from the era of GM crops to the new world of atomically modified organisms.

3.5 Nanotechnology and the Environment: Global Warming

"Global warming" refers to the global-average temperature increase that has been observed over the last 100 years or more. Global warming is caused by several things, which include man-made or anthropogenic causes. Global warming is also caused by

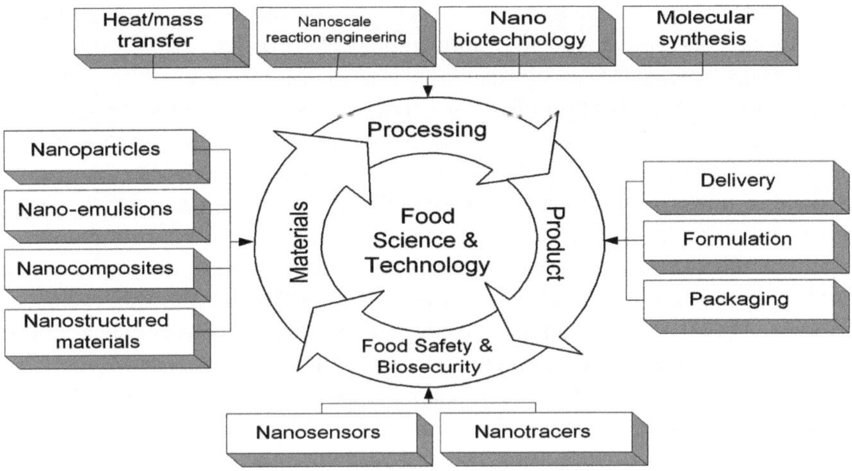

Fig. 8 Application matrix of Nanotechnology in food science [69]

natural causes. Earth's atmosphere contains natural greenhouse gases (mostly water vapor, carbon dioxide, and methane), which act to keep the lower layers of the atmosphere warmer than they otherwise would be without these gases. These greenhouse gases trap infrared radiation—the radiant heat energy that the Earth naturally emits to outer space in response to solar heating. Mankind's burning of fossil fuels (mostly coal, petroleum, and natural gas) releases carbon dioxide into the atmosphere, and this is believed to be enhancing the Earth's natural greenhouse effect. As of 2008, the concentration of carbon dioxide in the atmosphere was about 40–45 % higher than it was before the start of the industrial revolution in the 1800s [70].

The increase in global temperature will cause sea levels to rise and will change the amount and pattern of precipitation, probably including expansion of subtropical desert [71]. Warming is expected to be strongest in the Arctic and would be associated with continuing retreat of glaciers, permafrost, and sea ice. Other likely effects include changes in the frequency and intensity of extreme weather events, species extinctions, and changes in agricultural yields. Warming and related changes will vary from region to region around the globe, though the nature of these regional variations is uncertain. Political and public debate continues regarding global warming, and what actions (if any) to take in response. The available options are mitigation to reduce further emissions; adaptation to reduce the damage caused by warming; and, more speculatively, geoengineering to reverse global warming. Most national governments have signed and ratified the Kyoto Protocol aimed at reducing greenhouse gas emissions [72].

There are two main routes to slowing down and eventually stopping Global warming [73]. The first is to get more out of our current fuels—decreasing our consumption while increasing the efficiency of our technology. The second is to stop using fossil fuels altogether and make the transition to clean and renewable energy sources.

Nanotechnology is identified as an important contributor of meaningful solutions toward addressing Global warming problems. To make the most of our current fuels, researchers are designing nanoscale sieves that can filter out environmentally toxic molecules in fuels and fuel by-products. A promising area is in the custom designing of zeolites, porous nanoparticles that can extract more and cleaner gas from every oil barrel. Scientists are also designing nanosieves that can be fitted to power plants to capture carbon dioxide before it can enter the atmosphere. Furthermore, carbon dioxide acquired from natural reservoirs or recovered as a by-product of industrial chemical processes can be used in the production of carbon nanotubes [74]. Another approach is fuel additives—the U.K. company Oxonica (OXN.L), for example, already has its product, Envirox Fuel Borne Catalyst, incorporated into premium commercial diesel which reduces fuel consumption by up to 10 % and reduces carbon dioxide emissions by up to 15 %.

In summary, Nanotechnology is playing a big role in increasing the efficiency of current technology. Decreasing energy consumption and increasing efficiency are good solutions for now. Furthermore, the application of Nanotechnology through the production of nanomaterials using carbon sources such as CO_2 and methane is bound to beneficially reduce the atmospheric greenhouse gas pollution, ultimately decreasing the effects of Global warming. Without doubt, use of Nanotechnology as an environmentally friendly technology, will also indirectly address the 1997 Kyoto Protocol.

4 Challenges of Technology Transfer in Africa

This section of the chapter discusses some of the major challenges of technology transfer in Africa. Probable remedial measures to address some of these challenges are also discussed. Equally discussed, are the disadvantages of technology transfer in Africa.

4.1 Education: Skills Shortage

Technology transfer in the African perspective is both difficult and quite challenging. Most African countries do not have people with sufficient or rather the right scientific backgrounds, technically literate, innovative, nor entrepreneurial enough to readily adapt to the new technologies. Above all, established national scientific and engineering infrastructure institutes such as accredited laboratories and Global connectivity are non-existent [75]. To utilize and apply the technology effectively means having sufficient numbers of people possessing a range of appropriate competencies and significant levels of expertise in effectively applying that particular technology. It also means having the right people with a higher level of proficiency in redesigning or reengineering the technologies for local suitability and application. Lamentably, such characteristic skills are rare in most African countries. One way to address these problems is for the African countries to start by seriously investing in engineering education and research including setting up of the appropriate scientific and technological infrastructure.

Many African countries for instance, lack up-to-date engineering and scientific infrastructure in their universities; with Schools of Engineering, Science, and Technology graduating very few graduates per year. With such low graduation rate versus millions of people, it is definitely close to impracticality to address the prevalent engineering/scientific skills shortage; that is much needed to enhance technological advancement in these countries! In fact, South Africa is the only African country which has five universities that are ranked in the top 1,000 world universities [76]. Clearly, one possible solution would be to strategically reinvest and modernize Schools of Engineering, Science, and Technology so that these can positively contribute to meeting the engineering and scientific challenges of technology transfer.

The migration of skilled personnel to the developed countries as a result of economic and political factors prevalent in Africa is also another factor. This is a very significant problem faced by most African countries and as a result, Africa is a big exporter of skills especially engineering and science. African governments should devise methods of attracting skilled personnel such as equal pay for equal work.

4.2 Education: Inherent Resistance to Change

Most often, public officials prefer familiar solutions that limit unexpected consequences, i.e., the perceived human nature of resistance to change. If they are unfamiliar with a new technology or uncertain about its potential benefits, people would intuitively be reluctant to use it. A classical example is the refusal by the Zambian Government to use GM maize (corn) when the country had severe droughts in 2002. However, this is considered as an acceptable instinct of human nature, primarily due to insufficient knowledge about the technology and uncertainties of its collateral effects. Regrettably though, this kind of uninformed backlash against a new technology can actually stall development and economic prosperity.

Intensified and well structured educational awareness campaigns about the potential benefits of a new technology could be one of the probable solutions to this problem. Such campaigns should equally include any planned mitigation measures against the possible harmful or negative effects of the new technology. With this kind of awareness, people will definitely be more willing to accept and adapt to new technological changes.

4.3 Education: Technology Complexity and Literacy

The complexity of the technology itself is also one of the reasons for the non-adaptation or reluctance in technology transfer. Nanotechnology, for example, is a relatively new and not well-understood technology; with potential negative impacts on human health and the ecosystem [77–79]. To this end, education comes in hand as one of the effective means for technological advancement, particularly in African countries such as Burkina Faso where the illiterate rate is about 23.6 % [80] of the approximately 18.3 million people. At 68 % illiterate [81] rate of the approximately 11.7 million people, the situation is even much worse in Zambia. As a general indicator of the illiteracy level in most African countries, these numbers highlight the need and the importance of education as a critical stepping stone toward technological advancement.

4.4 Finance: Financial Constraints

Several economic and financial problems experienced by many African countries are some of the major factors contributing to the continent's technological backwardness. The cost of these technologies is often prohibitive and not readily affordable. To think of buying or investing in new technologies when most African countries' annual national budgets are at least 50 % financed by donor countries is thus a pipe dream.

Furthermore, most African countries have accumulated debts over the years and settling these debts is often one of the top priorities the moment financial resources are available; rather than investing in new technologies.

The perpetual food crisis in Africa is also an indirect and a major contributing factor to the slow pace of technological advancement in most African countries. Time and again, Governments are forced to divert all the available financial resources to averting hunger; which is often done at the expense of other investment sectors such as technology. Ironically, however, the agricultural industry of most Africa countries is equally not well developed and continues to indirectly impede on technological advancement.

In consideration of these financial hiccups, successful technological transfer to Africa can thus be attained only if the technology itself is affordable or the purchase terms are favorable. To this end, developed countries should equally play their role of promoting and facilitating smooth technology transfer by way of making their new technologies readily accessible and affordable.

4.5 Governmental Influence: Politics and Civil Instability

Perpetual political instability and cultural stigma in most African countries also contribute to adversely constrain the smooth technological inflow from developed countries. Zimbabwe is at worst, one typical example with an environment that has had political instability for the past decade. Once a thriving, self sustaining, and technological established country, it is now riddled with debt, poor infrastructure a collapsed service delivery, and technological investment is virtually at standstill.

Without doubt, political instability is in itself an obstacle toward technological advancement; as most of the "decision makers" would probably be engaged in political bickering instead of progressively thinking about economic and technological investments. Additionally, this is further compounded by the fact that even the technology investors themselves would not risk their presence in a politically unstable environment. So, it is apparent that political stability and peace must exist for technology transfer to successfully succeed.

Continuous civil fighting based on different political affiliations, religious associations, and/or tribal lines, such as in the Congo DR, Somalia, and Sudan, is also a big impedance to technological advancement. Instead of technologically empowering themselves, Governments and people in these civil war-torn areas are deeply engaged in gainless fights at the expense of economic development. Clearly, such unstable environments do not promote smooth technological advancement and needs to be avoided.

4.6 Governmental Influence: Priorities and Poor Planning

Lack of sustainable Governmental visions, poor long-term developmental plans, and miss-prioritizing of programs are also viewed as some of the common problems impeding technological transfer and economic growth in most African countries. Compared to the developed nations, most African countries lack long-term plans that are viable and rarely incorporate technology advancement. In this modern world, it is just imperative that African countries consider incorporating technological advancement in their long-term Governmental plans.

4.7 Governmental Influence: Poor Infrastructure

The slow application of telecommunication and information technologies in Africa, which are the main pipeline of technology transfer in many fields, also continues to contribute to the widening technological gap between technology intensive countries and African countries. Thus, it is important that African countries consider reprioritizing their long-term developmental plans if technological advancement is to be attained.

In particular, communication and transportation are some of the key infrastructures needing improvement so as to efficiently promote technology transfer in Africa. Without these infrastructures, it is highly doubtful for technological modernization to be successfully attained in Africa. One way to address this issue is for the local Governments to encourage the private sector to invest in infrastructures or invite investors to set up infrastructures in the African cities. This can be achieved through giving incentives such as tax concessions to investors and/or private business people who invest in long-term infrastructures such as roads, communication, buildings, etc.

4.8 Global Economic Effects: Regulations and Market Balances

Stringent regulations related to intellectual property rights and the fear that the technology could be inappropriately and dangerously used is another major obstacle encountered in technology transfer and adaptation. Furthermore, the marketing strategy of maintaining equilibrium and balance with respect to the principle law of supply and demand limits the equitable and smooth transfer of technology. It is a very simple economic and supremacy logic; if all were competitively at the same technological level; there will be no buyer nor seller nor boss nor junior. In these circumstances, technological stagnation of certain countries may in fact be viewed as necessary for economic and superiority balance. This is further compounded by the so called "holding back" tendency; that is technology is transferred only when and if it is within the Originator's best interests such as economic benefits, similar policy orientations, etc. Meanwhile, others would often hold on to their technologies just to maintain a competitive advantage on the edge and remain supreme.

Some countries may even attach unfavorable strings along with the technology transfer. Inevitably, Africa will continue experiencing these problems, which it has no control over, unless otherwise the continent starts researching and developing its own technologies; tailored specifically to meet the local challenges and needs.

It is also observed that the central protagonists of the new and highly effective technologies are no longer the semipublic international research centers of the international consultative groups, which were institutionally embedded in the UN system and made sure that access to their research results was open to everybody who was interested. For example, the central players in the new biotechnological innovations are the big chemical, pharmaceutical, mining, and food transnationals. These already dominate the international research arena, have the most qualified scientific personnel at their disposal, and play a central role in negotiations on how to shape the international framework for the application of the new Biotechnology, e.g., intellectual property rights.

Critics stress the fact that the present direction of development of most technologies is dominated by the research agenda of the industrialized countries. Thus, the main concern is not the realization of the potential offered by the new techniques to combat, e.g., hunger and malnutrition in a highly specific and target-oriented way, but almost exclusively the profit interests of the northern companies.

4.9 Global Economic Effects: Dumping Effect

The so called "dumping" tendency due to either greed or mere disposal may have negative consequences on both the human life and the environment. Outdated technology may also end up being transferred to Africa. Thus, while thirst for new technology, African countries should also be wary of buying obsolete technologies that may be dysfunctional (or even harmful) being dumped on their door steps. On the same basis, the technology exporters should also be humane enough not to rampantly dump obsolete technologies to poor countries.

4.10 Technology Application: Local Adaptability

Technology transfer despite its merits of economic development has setbacks in that imported technologies at times tend to hold performance and functionality that are applicable to the problems and needs of countries of origin [75]. This is because most developing countries have had bitter experiences of deploying scarce financial resources on inappropriate technologies that have failed to yield the expected results. In addition, new technologies and materials with the potential for improved performance may not meet existing conventional design specifications and standards. Thus, the ability to introduce innovation and reduce the costs can be stifled. A better example is bioleaching or bioxidation of minerals, which is relatively a slower process than the conventional methods of pyrometallurgy and chemical metallurgy.

As a remedy to this problem, receiving countries should carefully be selective as to which technologies they import, focusing on the ones that are better applicable to their needs. In a nutshell, technology importation should not be random (or unplanned for or all for every one), but rather be based on the needy, adaptability, applicability, and the perceived economic benefits. Some technologies may not work well in certain countries or may not really benefit the people. At any given time, each country's needs are often different from others, and so, are the technological needs.

4.11 Cultural Stigma: Ethical and Moral Issues

Developments in technology usually do not in themselves raise ethical or moral issues. Instead, it is the use to which these developments are put which raises basic questions, and from the perspective of history almost any use of a product of technological innovation can be made into a moral issue. So what kind of ethical issues should we be concerned with? Our moral concern is with the victims of technological change. Some uses of technological innovation have the possibility of having many more innocent victims than others. A good example has been the use of nuclear technology. It has great potential for helpful uses and also for tremendous harm. Another two-edged sword is the euphemistically called Biotechnology. A number of moral issues are raised by this about which humanists probably see no moral issue. A good example of this is in vitro fertilization. Although some religious groups have opposed this as contrary to the teachings of their religion, for humanists the issue does not pose a dilemma. Women unable to have children and willing to undergo in vitro fertilization through their own choice in order to do so, should be able to do so. Similar to the nuclear issue, however, is the problem of gene splicing. The danger here is that bacteria, in the process of being altered to meet some expected or anticipated problem, might develop more harmful effects. Since the potential of great profits is involved, it is essential that humanists be alert to the moral implications of various decisions.

4.12 Cultural Stigma: Perceived Job Loss Syndrome

Lastly, but not the least is the perceived job loss syndrome. In Africa, the majority of the population is illiterate or just semiliterate and thrive on the inherent philosophical concept of the so called traditional socialism and dependency culture. In a typical African family setup, there are often a lot of people (on the order of at least eight) that depend on one working literate individual. By nature, one of the inevitable and collateral ramifications of technological advancement is job losses. For instance, it would not be unprecedented to assume that if some mines in most of the African countries were to go biotech and nanotech, at least one or two miners would have their positions declared redundant. As a result of this, there is resentment toward technological advancement and most African countries are thus reluctant to invest or acquire new technologies.

Another snapshot example of this job loss syndrome is in the banking sector, where ATMs are becoming the norms of today's cash transactions. Without doubt, ATMs (although they desirably maximize efficiency) reduce the need for human tellers. In African countries such as Zimbabwe where unemployment rate remains perpetually high at about 80 % [82], this definitely becomes an issue and does indeed incite the resistance to new technological acquisitions. With its closest neighbor (Zambia) at 50 % [83], unemployment rate, these numbers do indeed signify the magnitude of the unemployment problem in Africa and how it may collaterally impede on technology transfer.

5 Summary

The strong link between sustainable development and economic prosperity, and technological innovation is not surprising. The challenge though is to develop mechanisms to improve the capacity to innovate and unleash the full capacity of science, engineering, and technology for the betterment of all. To this end, technology development, transfer, and adaptation play a pivotal role in setting the conditions for innovation and higher productivity; that ultimately leads to sustainable development and economic growth. In addition, in this emerging Global environment, technology transfer and adaptation also provides more valuable competitive skills.

As discussed in this chapter, technologies such as Biotech and Nanotech could be very instrumental in playing these roles particularly in the mining, food, energy, and medical industries, if properly harnessed. These technologies have a reputation for optimizing operational processes, increasing efficiency, maximizing productivity, and are more environmentally friendly compared to most conventional technologies.

However, technology transfer and its adaptation also come with its own challenges and limitations. For most African countries, some of these challenges include the following: lack of skilled experts with sufficient technical knowledge, illiteracy, financial limitations, poor infrastructure, political instability, poor economic policies, poor developmental plans and strategies, traditional and cultural stigma/inertia, dependence syndrome, etc. The perpetual food crisis is also a major indirect deterrence toward the rapid technological modernization of Africa. So, for Africa to be technologically up-to-date, she must first address some of these intrinsic challenges.

Unfortunately, developed nations may at times also not be so keen to facilitate a smooth-free technological transfer to Africa; for various reasons including the maintenance of a competitive advantage, supremacy, and the fear that the technology may fall into wrong hands and be recklessly used. Also the dumping tendency may often result in dysfunctional and/or obsolete technologies being dumped in Africa. Consequently, Africa should be wary of these constraints in its quest for technological advancement.

5.1 Disclaimer

The contents of this chapter reflect the views of the authors who are solely responsible for the facts and accuracy of the material presented herein and do not necessarily reflect the official views/policies of any agency/institute. This chapter does not constitute a standard nor is it intended for policy formulation purposes. Trade names were used solely for information and not for product endorsement.

References

1. Moriarity RT (1990) High–tech marketing: concepts, continuity and change. Eng Manage Rev 18(1):25–35
2. Palaniswami S, Bisho RC (1992) Operational technology transfer: can we calculate the behavioral cost. Seventh Int Work Semin Prod Econ 2:427–442
3. Guerin TF (1999) An Australian perspective on the constraints to the transfer and adoption of innovations in land management. Environ Conserv 26(4):289–304
4. Mashiri M, Kistan K, Marian B, Elrahman OA (2007) Uplifting developing communities through sustainable technology transfer. Proceedings: 3rd Africa technology transfer conference, Mangochi, Malawi, 22–25 May 2007

5. Falciola L (2009) Searching biotechnology information: a case study. World Pat Inf 31:36–47
6. Devasia P, Natarajan KA (2004) Bacterial leaching-biotechnology in the mining industry. http://www.ias.ac.in/resonance/Aug2004/pdf/Aug2004p27-34.pdf. Accessed Dec 2008
7. Deveci H, Akcil A, Alp I (2003) Parameters for control and optimisation of bioleaching of sulphide minerals. In: Kongoli F, Thomas B, Sawamiphakdi K (eds) Process control and optimization in ferrous and non ferrous industry. Materials Science and Technology Symposium, Chicago
8. Akcil A (2004) Potential bioleaching developments towards commercial reality: Turkish metal mining's future. Miner Eng 17:477–480
9. Hansford GS, Vargas T (2001) Chemical and electrochemical basis of bioleaching processes. Hydrometallurgy 59:135–145
10. Rawlings DE, Johnson DB (2007) The microbiology of mining: development and optimisation of mineral-oxidizing microbial consortia. Microbiology 153:315–324
11. Rawlings DE (2005) Characteristics and adaptability of iron- and sulphur- oxidizing microorganisms used for the recovery of metals from minerals and their concentrates. Microb Cell Fact 4:13
12. Ndlovu S (2008) Biohydrometallurgy for sustainable development in the African mineral industry. Hydrometallurgy 91:20–27
13. Brochot S, Durance MV, Villeneuve J, d'Hugues P, Mugabi M (2004) Modelling of the bioleaching of sulphide ores: application for the simulation of the bioleaching/gravity section of the Kasese Cobalt Company Ltd process plant. Miner Eng 17:253–260
14. Dew DW, Milleand DM (2001) Copper, nickel and cobalt recovery. US (2001) Patent 6,245,125, June 12
15. Burgstaller W, Schinner F (1993) Leaching of metals with fungi. J Biotechnol 27:91–116
16. Simate GS, Ndlovu S (2007) Characterisation of factors in the bacterial leaching of nickel laterites using statistical design of experiments. Adv Mater Res 20–21:66–69
17. Simate GS, Ndlovu S (2008) Bacterial leaching of nickel laterites using chemolithotrophic microorganisms: Identifying influential factors using statistical design of experiments. Int J Miner Process 88:31–36
18. Simate GS (2009) The bacterial leaching of nickel laterites using chemolithotrophic microorganisms. MSc (Eng) Dissertation, University of the Witwatersrand, South Africa. http://wiredspace.wits.ac.za/handle/10539/7098
19. Simate GS, Ndlovu S, Gericke M (2009) The effect of elemental sulphur and pyrite on the leaching of nickel laterites using chemolithotrophic bacteria. Hydrometallurgy conference 2009. In The Southern African Institute of Mining and Metallurgy
20. Simate GS, Ndlovu S, Gericke M (2009) Bacterial leaching of nickel laterites using chemolithotrophic microorganisms: process optimisation using response surface methodology and central composite rotatable design. Hydrometallurgy 98:241–246
21. Ndlovu S, Simate GS, Gericke M (2009) The microbial assisted leaching of nickel laterites using a mixed culture of chemolithotrophic microorganisms. Adv Mater Res 71–73:493–496
22. Schippers A, Jozsa PG, Sand W (1996) Sulphur chemistry in bacterial leaching of pyrite. Appl Environ Microbiol 62(9):3424–3431
23. Schippers A, Sand W (1999) Bacterial leaching of metal sulphides proceeds by two indirect mechanism via thiosulphate or via polysulphides and sulphur. Appl Environ Microbiol 65:319–321
24. Jones RT, Denton GM, Reynolds QG, Parker JAL, van Tonder GJJ (2002) Recovery of cobalt from slag in a DC arc furnace at Chambishi, Zambia. J South Afr Inst Min Metall January/February(Issue): 5–10
25. Mining (2008) http://www.richardcorfield.com/ pages/other_writing/research_cuts/Bugs_-banquet.pdf. Accessed December 2008
26. Tsezos M (2009) Metal-microbes interactions: beyond environmental protection. Adv Mater Res 71–73:527–532
27. Jadia CD, Fuleka MH (2009) Phytoremediation of heavy metals: recent techniques. Afr J Biotechnol 8(6):921–928

28. Groudev S, Spasova I, Nicolova M, Georgiev P (2009) In situ bioremediation of contaminated soils in uranium deposits. Adv Mater Res 71–73:533–540
29. Knox AS, Paller MH, Reible DD, Ma X, Petrisor IG (2008) Sequestering agents for active caps- immobilization of metals and organics. Soil Sediment Contam 17(5):615–632
30. Ehrlich H, Brierley CL (1990) Microbial mineral recovery. Mc Graw Hill, New York
31. Cornish JE, Goldberg WC, Levine RS, Benemann JR (1995) Phytoremediation of soils contaminated with toxic elements and radionuclides. In: Hinchee RE, Means JL, Burris DR (eds) Bioremediation of inorganics. Battelle Press, Columbus
32. Johnson DB (2003) Importance of microbial in the development of sustainable technologies for mineral processing and wastewater treatment. http://wiki.biomine.skelleftea.se/wiki/images/9/97/Microbiology_development_sustainable_technologies_mineral_processing_and_wastewater_treatment.pdf. Accessed Mar 2010
33. Molwantwa JB, Molipane NP, Rose PD (2000) Biological sulphate reduction utilizing algal extracellular products as a carbon source. WISA 2000 Biennial Conference, Sun City, South Africa, 28 May–1 June 2000
34. Postgate JR (1984) The sulphate reducing bacteria. Cambridge University Press, London
35. Rowley M, Warkentin DD, Sicotte V (1997) Site demonstration of the biosulfide process at the former Britannia mine. Proceedings of the 4th international conference on acid rock drainage. Vancouver, British Columbia. IV, pp 1531–1548, 31 May–6 June 1997
36. Abia AA, Horsfall M Jr, Didi O (2002) Studies on the use of agricultural by-products for the removal of trace metals from aqueous solution. J Appl Sci Environ Manage 6(2):89–95
37. Abia AA, Horsfall M Jr, Didi O (2003) The use of chemically modified and unmodified cassava waste for the removal of Cd, Cu and Zn ions from aqueous solution. Bioresour Technol 90(3):345–348
38. Gardea-Torresdey JL, Gonzalez JH, Tiemann KJ, Rodriguez O, Gamez G (1998) Phytofiltration of hazardous cadmium, chromium, lead and zinc ions by biomass of Medicago sativa (alfalfa). J Hazard Mater 57:29–39
39. Ho YS, Wase DAJ, Forster CF (1995) Batch nickel removal from aqueous solution by sphagnum moss peat. Water Res 29(5):1327–1332
40. Low KS, Lee CK, Leo AC (1995) Removal of metals from electroplating wastes using banana pith. Bioresour Technol 51:227–231
41. Quek SY, Wase DAJ, Forster CF (1998) The use of sago waste for the sorption of lead and copper. Water SA 24(3):251–256
42. Randall JM, Reuter FC, Waiss AC (1974) Removal of cupric ions from solution by contact with peanut skins. J Appl Polym Sci 19:156–171
43. Sasson A (2004) Biotechnologies: current achievements and prospects social acceptance of biotechnology-derived products. http://citeseerx.ist.psu.edu/viewdoc/download?doi=10.1.1.127.7413&rep=rep1&type=pdf. Accessed Feb 2010
44. Reverchon A (2002) Le vivant, nouveau carburant de l'industrie. Les bio-industries sontelles viables économiquement? Chez Genencor, les chercheurs forgent les outils des usines du vivant. Le Monde (Paris), 10 Sept 2002, p II
45. Agricultural Biotechnology Update (2009) Agricultural biotechnology's environmental success story. http://www.bio.org/foodag/Earth_Day_onepager.pdf. Accesses April 2010
46. Iyuke PO, Cross M, Iyuke SE, Potgieter HJ (2007) The role of a University in education and training in the field of nanotechnology-the case of the University of the Witwatersrand. Trans IChemE Part D Volume 2:56–67
47. Kumar A, Stephenson LD, Norton E, Hale J, Nelson A (2010) Nanosensors for detection of biological threat contaminants in critical buildings. http://www.stormingmedia.us/34/3475/A347505.html. Accessed March 2010
48. Santos-Magalhães NS, Mosqueira VCF (2010) Nanotechnology applied to the treatment of malaria. Adv Drug Deliv Rev 62:560–575
49. Gardella F, Assi S, Simon F, Bogreau H, Eggelte T, Ba F, Foumane V, Henry MC, Kientega PT, Basco L, Trape JF, Lalou R, Martelloni M, Desbordes M, Baragatti M, Briolant S,

Almeras L, Pradines B, Fusai T, Rogier C (2008) Antimalarial drug use in general populations of tropical Africa. Malar J 7:124
50. Malaria in Southern Africa (2008) http://www.malaria.org.za. Accessed April 2010
51. Devalapally H, Chakilam A, Amiji MM (2007) Role of nanotechnology in pharmaceutical product development. J Pharm Sci 96:2547–2565
52. Mosqueira VCF, Loiseau PM, Bories C, Legrand P, Devissaguet JP, Barratt G (2004) Efficacy and pharmacokinetics of intravenous nanocapsule formulations of halofantrine in Plasmodium berghei-infected mice, Antimicrob. Antimicrob Agents Chemother 48(4):1222–1228
53. Kayser O, Kiderlen AF (2003) Delivery strategies for antiparasitics. Expert Opin Investig Drugs 12:197–207
54. Date AA, Joshi MD, Patravale VB (2007) Parasitic diseases: liposomes and polymeric nanoparticles versus lipid nanoparticles. Adv Drug Deliv Rev 59:505–521
55. Sahoo SK, Labhasetwar V (2003) Nanotech approaches to drug delivery and imaging. Drug Deliv Today 8(24):
56. Odigure JO, Abdulkareem AS, Adeniyi OD, Amao FT (2005) Water quality management: a case study of water pollution in Minna and its environs. Botswana J Technol 14(1):31–35
57. Hussain MM, Baschuk JJ, Li X, Dincer I (2005) Thermodynamic analysis of a PEM fuel cell power system. Int J Therm Sci 44:903–911
58. Marr C, Li Á (1998) An engineering model of proton exchange membrane fuel cell performance. ARI Int J Phys Eng Sci 50:190–200
59. Inoue G, Matsukuma Y, Minemoto M (2006) Evaluation of the optimal separator shape with reaction and flow analysis of polymer electrolyte fuel cell. J Power Sour 154:18–34
60. Xianguo L (2006) Principles of fuel cells. Taylor and Francis, New York
61. Appleby AJ (1992) Fuel cell technology and innovation. J Power Sour 37:223–239
62. Klaiber T (1996) Fuel cells for transport: can the promise be fulfilled? Technical requirements and demands from customers. J Power Sour 61:61–69
63. Woo Y, Oh SE, Kanga YS, Jung B (2003) Synthesis and characterisation of sulphonated polyamide membranes for direct methanol fuel cell. J Membr Sci 220:31–45
64. Jung B, Kim B, Yang JM (2004) Transport of methanol and protons through partially sulfonated polymer blend membranes for direct methanol fuel cell. J Membr Sci 245:61–69
65. Choi W, Howze JW, Enjeti P (2006) Fuel cell powered uninterrupted power supply systems: design considerations. J Power Sour 157:311–317
66. Saka AA (2009) Design and development of proton exchange membrane (PEM) form synthetic rubber and carbon nanoparticles for PEM fuel cell. PhD Thesis, University of the Witwatersrand, Johannesburg, South Africa
67. CSREES/USDA (2003) Nanoscale science and engineering for agriculture and food systems. http://www.nseafs.cornell.edu/web.roadmap.pdf. Accessed Feb 2010
68. Joseph T, Morrison M (2006) Nanotechnology in agriculture and food. https://www.nanoforum.org. Accessed Feb 2010
69. Weiss J, Takhistov P, McClements J (2006) Functional materials in food nanotechnology. J Food Sci 71:107–116
70. Spencer R (2010) Global warming. http://www.drroyspencer.com/global-warming-natural-or-manmade. Accessed April 2010
71. Lu J, Vecchi GA, Reichler T (2007) Expansion of the Hadley cell under global warming. Geophys Res Lett 34:L06805. doi:10.1029/2006GL028443
72. Kyoto Protocol (1992). http://unfccc.int/resource/docs/convkp/kpeng.html. Accessed April 2010
73. Forbes/Wolf (2006) Nanotech report, Volume 5, Number 7. http://www.qsinano.com/pdf/ForbesWolfe_NanotechReport_July2006.pdf. Accessed April 2010
74. Simate GS, Iyuke SE, Ndlovu S, Yah CS, Walubita LF (2010) The Production of carbon nanotubes from carbon dioxide—challenges and opportunities. J Nat Gas Chem 19(5):453–460
75. Mangena M (2006) Keynote address by the Minister of Science and Technology. First all Africa conference on technology transfer and diffusion, Boksburg, Johannesburg

76. Ranking web of world universities (2009) http://www.webometrics. info/top4000.asp. Accessed Jan 2009
77. Moore MN (2006) Do nanoparticles present ecotoxicological risks for the health of the aquatic environment. Environ Int 32:967–976
78. Smart SK, Cassady AI, Lu GQ, Martin DJ (2006) The biocompatibility of carbon nanotubes. Carbon 44:1034–1047
79. Wiesner MR, Lowry GV, Alvarez P, Dionysiou D, Biswas P (2006) Assessing the risks of manufactured nanomaterials. Environ Sci Technol 40(14):4336–4337
80. UNDP (2008a) Human development reports—2007/2008 reports. http://hdrstats. undp.org/countries/data_sheets/cty_ds_BFA.html. Accessed Jan 2009
81. UNDP (2008b) Human development reports—2007/2008 reports. http://hdrstats. undp.org/indicators/3.html. Accessed Dec 2008
82. CIA World Factbook (2008a) Zimbabwe unemployment rate. https://www.cia.gov/library/publications/the-world-factbook/print/zi.html. Accessed Jan 2009
83. CIA World Factbook (2008b). Zambia unemployment rate. https://www.cia.gov/library/publications/the-world-factbook/geos/za.html#Econ. Accessed Dec 2008

Part III
International Collaboration: Relevance for Development in Africa

10. Mammo: The role of IPICS in enhancing research on the synthesis and the characterisation of conducting polymers at Addis Ababa University (25 pages)
11. Sundin: The International Programme in the chemical sciences (IPICS): 40 years of support to chemistry in Africa (17 MS pages)
12. Fawzi: International collaboration with a view to containing outbreak of emerging infectious diseases through bioprospection (25 pages)

Total (around 200 pages)

The Role of IPICS in Enhancing Research on the Synthesis and Characterization of Conducting Polymers at Addis Ababa University

Wendimagegn Mammo

Abstract Research in the area of conjugated polymers at Addis Ababa University (AAU) began in the early 1990s through grants obtained from the International Program in the Physical Sciences (IPPS), the physics wing of the International Science Programs (ISP). Since then, the continued support from ISP allowed for building research capacity and the training of scores of MSc and PhD candidates at the Departments of Chemistry and Physics of AAU. The program on the synthesis of conjugated polymers was launched in 1995. Since 2003, the activities of the synthesis group were supported by a research grant from the Internal Program in the Chemical Sciences (IPICS), the chemistry wing of ISP. This paper highlights the impact of sustained support by IPICS on research and postgraduate training in the synthesis and characterization of conducting polymers at AAU, Ethiopia.

1 Background

The International Science Programme (ISP) initiated, supported, and nurtured research in the Chemistry and Physics of conducting polymers at Addis Ababa University for nearly two decades. Two branches of ISP, i.e., the International Programme in Physical Sciences (IPPS) and the International Programme in the Chemical Sciences (IPICS) played vital roles in the identification of talented and motivated researchers to spearhead the organization of strong research teams at the Departments of Chemistry and Physics of the Addis Ababa University (AAU).

W. Mammo (✉)
Department of Chemistry, Addis Ababa University, Miazia 27 Square,
Arat Kilo, P.O. Box 1176, Addis Ababa, Ethiopia
e-mail: wmammo@chem.aau.edu.et

A. Gurib-Fakim and J. N. Eloff (eds.), *Chemistry for Sustainable Development in Africa*, DOI: 10.1007/978-3-642-29642-0_10,

In the period between 1990 and 2009, manpower and material capacity was built at both departments to conduct advanced research in the Chemistry and Physics of conjugated polymers and also the training of candidates at the postgraduate MSc and PhD levels. The support allowed for sandwich-type PhD-level training of Ethiopian candidates to be pursued between AAU and universities in Sweden. This also led to strong research collaborations between the departments of Chemistry and Physics and their Swedish counterparts.

2 Conjugated Polymer Research: The Initial Years

Research in the area of conjugated polymers started at the Department of Physics of AAU in the early 1990s. The International Program in the Physical Sciences and its leader at the time, Dr Lennart Hasselgren, played a pivotal role in initiating the research program. Dr Bantikassegn Workalemahu, the founder of the conducting polymers research at the Department of Physics, AAU, received his PhD training at the University of Linkoping, Sweden, under the auspices of the IPPS plan to organize a strong research program in conducting polymer research at AAU. In the ensuing years, IPPS availed the necessary funds to support and nurture manpower training, the acquisition of research-grade facility and the necessary supplies to study the electrical and optical properties of devices made of conducting polymers. Provisions were made for sandwich-type PhD-level training to be pursued between AAU and universities in Sweden. Efforts were also made to establish links between the Department of Physics, AAU and other departments of physics in the East African region. A notable achievement in this regard was the south–south collaboration that was established between the Department of Physics AAU and the University of Khartoum, Sudan, which allowed for Sudanese students to receive short-term training at the Department of Physics, AAU. IPPS also supported the exchange of scientists in the East African region.

Recognizing the fact that the synthesis of conducting polymers is an integral part of the research in organic semiconductors, IPPS helped to expand the scope of the research at AAU by initiating a program on the synthesis of conducting polymers. In 1995, an organic chemist was invited to join the research group and arrangements were made by IPPS for him to receive training in the art of polymer synthesis at the Chalmers University of Technology, Gothenburg, Sweden. A substantial proportion of the research funding was also allocated to purchase equipment, chemicals and supplies and to organize a modern synthetic organic chemistry laboratory in the premises of the Department of Chemistry of the AAU. The laboratory became functional in 1997 and it quickly elevated itself into the major supplier of polymeric materials to the conducting polymers research at the Departments of Physics and Chemistry, AAU. Studies conducted on the polymeric materials formed the basis for the MSc theses of several candidates in Physics and Chemistry. One Sudanese PhD candidate at the University of Khartoum also studied the electrical properties of some polythiophenes prepared in this laboratory. In addition, the properties of some of the materials were investigated at the Linköping University, Sweden.

After the initial activities of the synthetic organic chemistry group started showing fruitful results, a decision was reached by IPICS to further strengthen the group with adequate research funding and support. Thus, the group was invited to apply for research funding in 2002. This heralded the emergence of a full-fledged project on the synthesis of conducting polymers and IPICS effectively took over the funding and sustenance of this project starting from 2003.

3 The Aim of the Project

The main aim of the conducting polymer synthesis project was to design and prepare stable, soluble, and processable conjugated polymers (mainly polythiophenes and polyfluorenes), and to study the electrical and optical properties of these materials. The materials that are prepared are destined to find applications in organic solar cells, stable photodiodes, and other kinds of high-technology devices. In addition to building capacity in synthetic organic chemistry, the project aimed at bringing together chemists, physicists, and material scientists to solve research problems of common interest. The project also aimed at making meaningful impact on postgraduate MSc and PhD-level training in Chemistry and Physics at AAU.

4 Major Scientific Achievements

Over the years, we have made concerted efforts to synthesize and characterize a variety of different kinds of conducting polymers. We were mainly interested in polythiophenes and polyfluorene copolymers for solar cells and light emitting diodes. Herein, some of the works we have done in the synthesis and characterization of conducting polymers are described.

4.1 Polythiophenes

Over the years, we have prepared and studied the properties of a large variety of polythiophenes [1–17] Among the first series of polythiophenes we prepared were alkoxy-substituted poly(3-phenylthiophene)s and poly(3-phenyl-2,2′-bithiophene)s (Fig. 1). We studied the optical properties of some of these materials in great detail. We have also measured the photoluminescence quantum yields and photoluminescence lifetimes for a series of polythiophenes, in solution, and the quantum yields for the same polymers in thin films. We observed that increasing

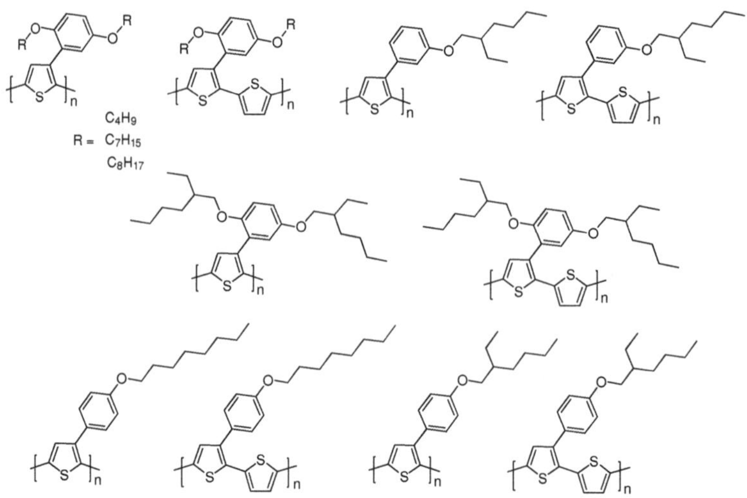

Fig. 1 The structures of some alkoxy-substituted poly(3-phenylthiophenes) and poly(3-phenyl-2,2′-bithiophenes)

the bulkiness of the substituent increases the quantum yield in solution. For spin-coated films, an increased ordering can either increase or decrease the quantum yield, depending on the separation of the conjugated backbones. Polythiophenes in which the electronic band gap is increased by steric hindrance showed very low quantum yields, both in films and in solution [4].

Recent years have witnessed a lot of development in the synthesis of a variety of conducting and electroactive polymers with a broad range of properties. Control on the absorption characteristics and the color of a polymer could be achieved by altering the extent of conjugation through introducing steric interactions. The electronic properties of polymers could be varied by the introduction of electron withdrawing and electron donating substituents.

The structural variations we created in polythiophenes have allowed us to study the structural effects that are responsible for high stability in the doped state and for tuning the color of the emission from polythiophenes for use in polymer light-emitting diodes [9]. We were therefore able to predict the criteria for the design and synthesis of polythiophenes with high luminescence efficiency for use in light-emitting diodes and lasers.

We have also studied the solar cell applications of poly[3-(2′,5′-dioctyloxyphenyl)thiophene], poly[3-(2′,5′-diheptyloxyphenyl)thiophene] and poly[3-(2′,5′-dibutyloxyphenyl)thiophene] prepared electrochemically from their monomers on nanocrystalline titanium dioxide (nc-TiO_2)-coated ITO-glass [15, 16]. We found out that the poly[3-(2′,5′-dialkoxyphenyl)thiophenes] sensitize nc-TiO_2 in liquid-state photoelectrochemical cells.

Figure 2 shows the structures of oligo(ethylene oxide)-substituted polythiophenes prepared in our laboratories [2, 14, 18]. Oligo (ethylene oxide)-substituted polythiophenes mixed with a salt act as light-emitting layer in light-emitting electrochemical cells (LECs). Under an applied bias, p-doping of the

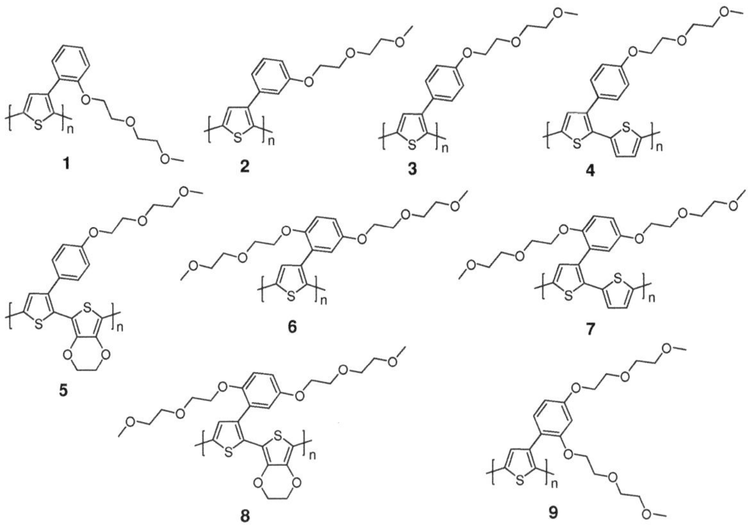

Fig. 2 Structures of some oligo(ethylene oxide)-substituted polythiophenes

electroluminescent polymer takes place at the anode while reduction takes place at the cathode. We have studied the doping processes of polymers **6** and **7** by in situ spectroelectrochemistry in both sandwich and planar electrochemical cells [7].

Most of the oligo (ethylene oxide)-substituted polythiophenes shown in Fig. 2 were investigated for their application in the roll-to-roll production of polymer-based electrochromic displays on flexible substrates [14]. These thiophene-based polymers and copolymers, which were thought to increase the contrast of displays based on poly (3,4-ethylenedioxythiophene)/poly(styrenesulfonic acid), were evaluated with respect to their contrast, switching speed, and reversibility in a water-based electrolyte. The results of the evaluation provided a basis for understanding what an aqueous electrolyte electrochromic display requires in terms of oxidation potential and material stability and the effect of chemical structure on the reversibility and switching speed.

Recently, we studied the solvatochromic and thermochromic behaviors of phenyl-substituted polythiophenes [13]. The pristine polymers, upon dissolution in chloroform, exhibited blue-shifted absorption. The solid films of the polymers showed significant blue-shifted as well as red-shifted absorptions when heated. The addition of methanol to the chloroform solutions of the polymers caused dramatic chromic changes and development of red-shifted spectra for many of the polymers investigated.

4.2 Polyfluorene Copolymers for Polymer Solar Cells

Polymer solar cells have attracted considerable attention due to their unique advantages, such as low cost, light weight, and potential use in flexible devices. Their cost of production is low and it may not be necessary to reach the performance of

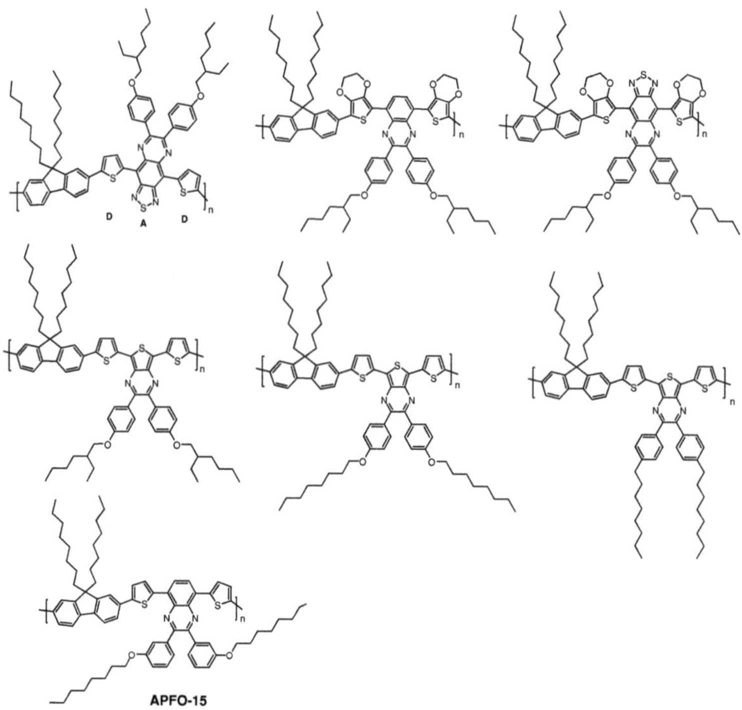

Fig. 3 The structures of some alternating polyfluorene copolymers containing D-A-D segments

traditional silicon-based solar cells to put such materials into the market. The main limitations to their applications are, however, low power conversion efficiency, smaller photocurrent and instability compared to silicon-based solar cells.

Polymer solar cells are fabricated by inserting an active layer between two electrodes with one electrode transparent to incident light. The active layer is usually composed of two materials with different electron affinities. The composition of the active layer can be polymer/polymer or polymer/molecule, where a material with lower electron affinity acts as electron donor and another material, with high electron affinity, acts as electron acceptor. In most polymer solar cells, the active layer is a combination of a polymer and a standard molecular electron acceptor, [6,6]-phenyl-C_{61}-butyric acid methyl ester (PCBM).

The main processes in a polymer solar cell are exciton generation, exciton dissociation, transport of electrons and holes, and free charge-carrier collection. First, excitons are generated in the active layer by absorbing incident light, then excitons diffuse in the active layer and dissociate at the interface between the two materials with different electron affinities to form free charge carriers, and, finally charge carriers are transported to the anode and the cathode, driven by a difference in chemical potential.

In our search for efficient polymer solar cells, we have prepared and tested a variety of low band gap alternating polyfluorene copolymers (Fig. 3), [19–31].

The polymers were designed for use in conjunction with C_{60} and C_{70} derivatives in bulk heterojunction solar cells. To absorb photons at long wavelengths, where more photon flux from emission of the sun is found, low-band gap polymers are necessary. However, at least two challenges could be identified for low-band gap polymers to harvest long-wavelength photons. One is the unavoidable reduction of the open-circuit voltage (V_{oc}) compared with high-band gap polymers, because the V_{oc} is related to the energy difference between the lowest unoccupied molecular orbital (LUMO) of the electron acceptor and the highest occupied molecular orbital (HOMO) of the electron donor. The other challenge is to have a large enough driving force for electron transfer from the polymer to the electron acceptor. This means that the LUMO of the electron donor must be closer in energy to vacuum than the LUMO of the electron acceptor in order to have enough driving force for exciton dissociation at the interface of the electron donor and acceptor. When decreasing the band gap of a polymer, both the LUMO level and the HOMO level are affected. The LUMO position of the polymers might be shifted away from the vacuum level so much that the use of new electron acceptors with lower LUMO levels than the commonly used PCBM is needed for preparing an efficient solar cell.

The alternating polyfluorene copolymers we synthesized (Fig. 3 and Schemes 1 and 2) were designed to have extended absorptions to cover the important regions of the solar emission. The band gaps could be lowered by the incorporation of planar conjugated segments with electron donor–acceptor–donor (D–A–D) functions in between the fluorene units.

Our studies have shown that, by careful design, it is possible to prepare polymers with similar HOMO positions but with significantly different absorptions. We have also shown that a minimal change in the nature and position of the substituents affects the energy levels of the polymers and hence the overall performance of the photovoltaic devices. The results we obtained clearly demonstrated that there is a correlation between the chemical structure and the HOMO levels determined from electrochemical studies as well as the V_{OC} values determined from photovoltaic devices. The polymers perform quite well in combination with PCBM in photovoltaic devices. We have demonstrated devices with fairly high quantum efficiency by preparing the active layer from blends of two different polymers with similar HOMO energy levels. Moreover, solar cells with efficiencies in excess of 3.5% have been realized from blends of APFO-15 (Fig. 3) and an electron acceptor molecule, PCBM [27].

Recently we synthesized two new black polymers (Fig. 4) with UV-Vis absorption spectra extending to approximately 850 nm [32]. These polymers were used as donors, and [6,6]-phenyl-C61-butyric acid methyl ester (PCBM[60] or PCBM[70]) as acceptors, in solar cell devices in various mixing ratios. The best combinations yielded an overall power conversion efficiency of 1.2% for **APFO-Black 1** and 1.5% for **APFO-Black 2**.

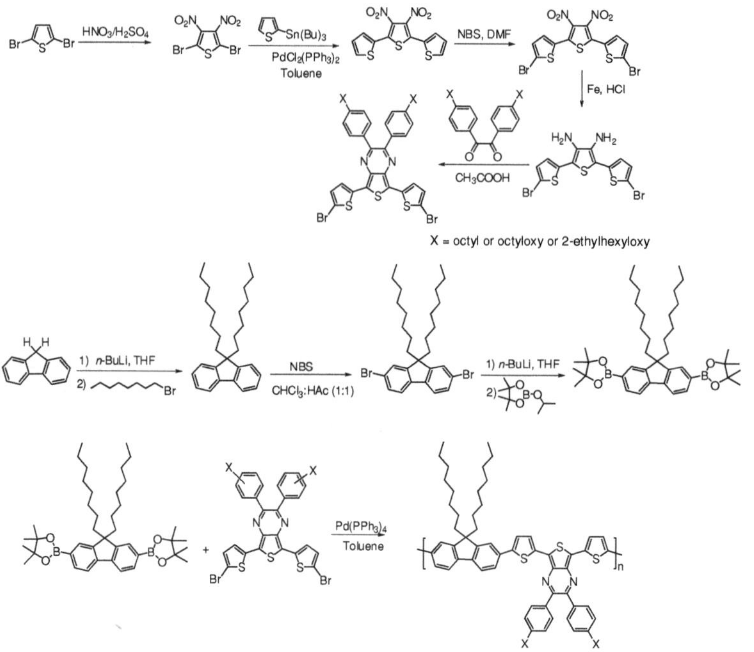

Scheme 1 Synthesis of alternating polyfluorene copolymers containing thienopyrazine units in the D–A–D segment

4.3 Miscellaneous Polyfluorene Copolymers

Figures 5, 6, 7 and 8 show the structures of a variety of different kinds of polyfluorene copolymers synthesized in our laboratories. The optical, electrochemical and light emitting characteristics of the anthracene- and benzothiadiazole-containing polyfluorenes **10**, **11** and **12** (Fig. 4) were investigated [25]. All three copolymers exhibited good thermal stabilities up to 296–350 °C as evaluated by thermogravimetric analyses (TGA) under nitrogen atmosphere. The polymers were characterized using cyclic voltammetry and the band gaps were determined. Polymer light emitting diodes (PLEDs) were fabricated from these polymers and the current density–voltage (J-V) characteristics of the PLED devices (ITO/PEDOT:PSS/polymer/LiF (0.5 nm)/Al(60 nm)) were studied. Blue light emissions were observed at about 3.5 V in devices made from polymers **10** and **12** with a more pronounced intensity for polymer **12**.

We have also prepared several polyfluorene copolymers containing oligo(ethylene oxide) side chains on the fluorene units as depicted in Fig. 6. The electrochromic (EC) behaviors of polymers **12, 13, 14** and **15** were investigated with the aim of increasing the contrast of EC displays based on PEDOT:PSS and water-

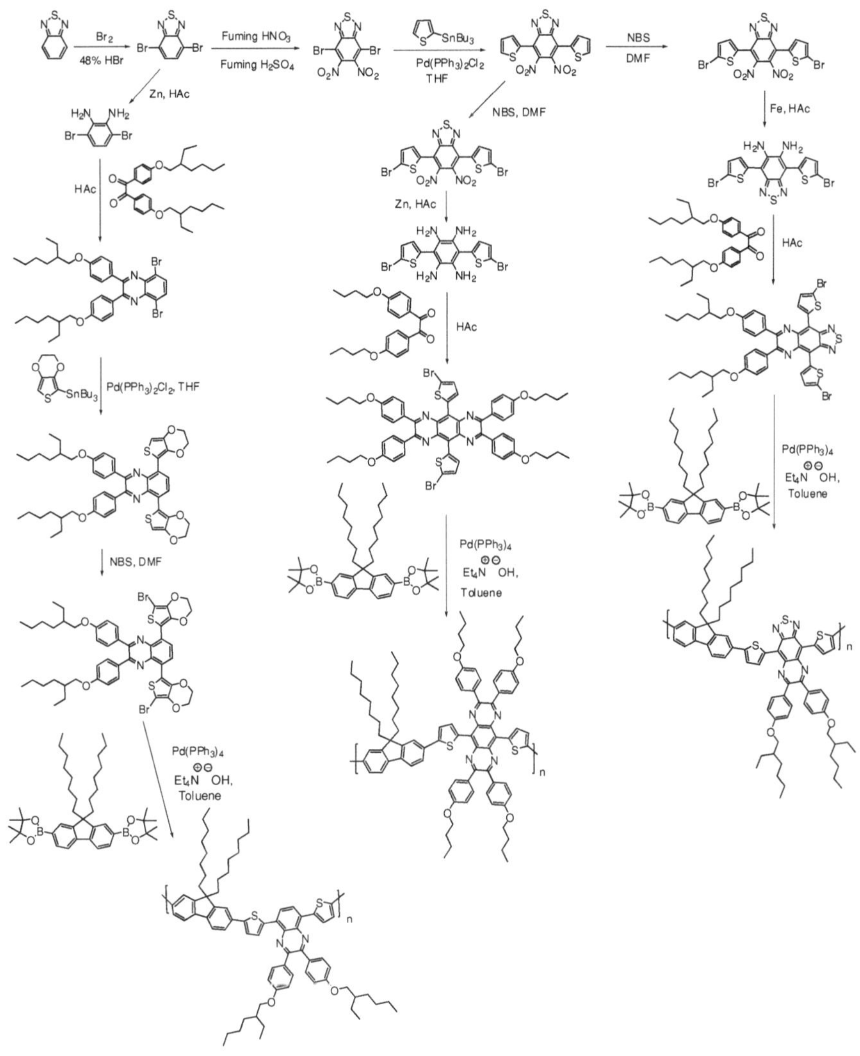

Scheme 2 Synthesis of alternating polyfluorene copolymers containing quinoxaline units in the D–A–D segment

based electrolytes [14]. Electrochemical measurements showed that only polymer **13** had a reversible switching in the measured potential range.

Recently, we have synthesized several polyfluorene copolymers containing bithiazole [33] and thiazolothiazole units (Fig. 7). The polymers were designed for possible applications in solar cells and polymer light emitting diodes. Preliminary studies have shown that some of these polymers have high photoluminescence efficiencies. Studies are currently underway to determine the applications of these materials in solar cells and light emitting diodes.

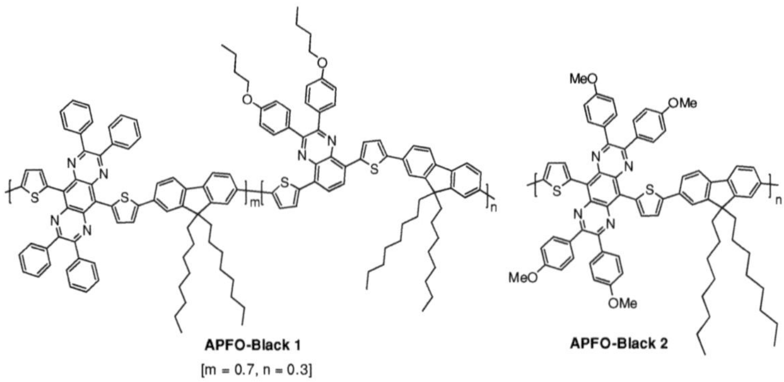

Fig. 4 The structures of **APFO-Black 1** and **APFO-Black 2**

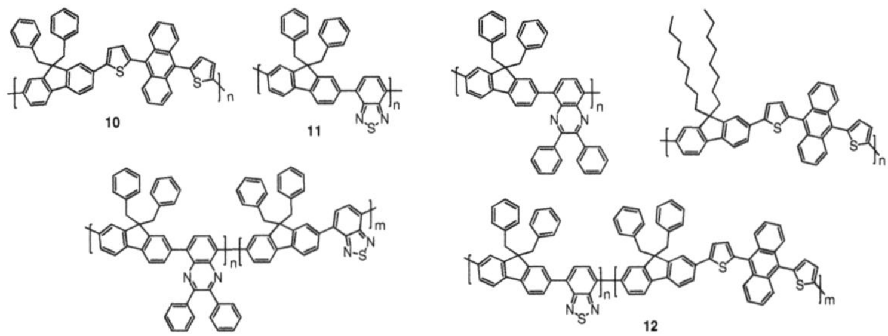

Fig. 5 Anthracene-, benzothiadiazole-, and quinoxaline-containing polyfluorene copolymers [36]

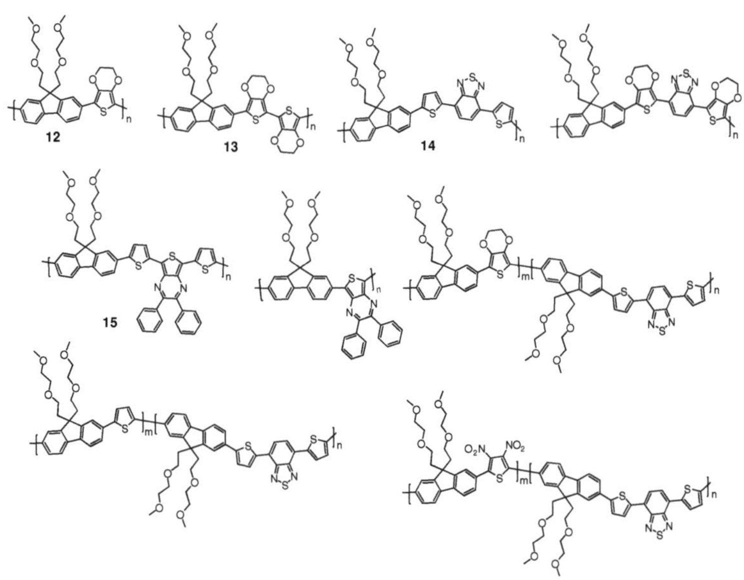

Fig. 6 Polyfluorene copolymers containing oligo (ethylene oxide) side chains on the fluorene units

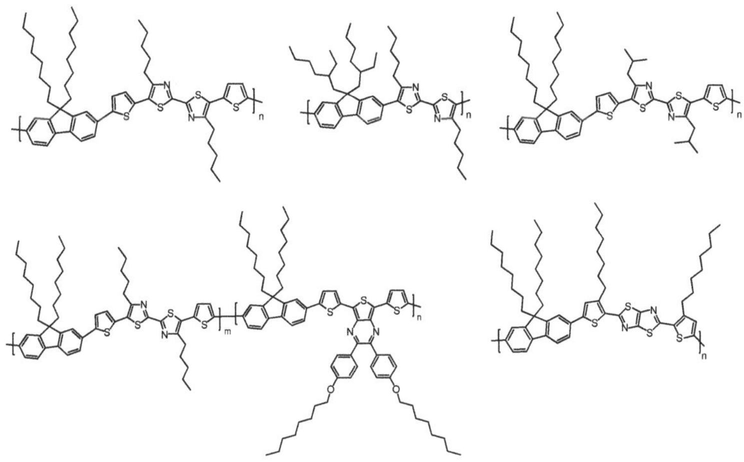

Fig. 7 Structures of polyfluorene copolymers containing bithiazole and thiazolothiazole units

Together with inorganic chemists at the Department of Chemistry, AAU, we have recently embarked on the synthesis of polymers capable of metal chelation [31].[1] Our aim is to study the effect of metal complexation on the optical, electrochemical and magnetic properties of the polymers. Such metal-complexed polymers may find applications in sensors, solar cells and light-emitting devices. Figure 8 shows the structures of the polymers we have synthesized for this purpose.

4.4 Oxadiazoles

Polythiophenes and polyfluorenes have a π-excessive nature, i.e., they are typical p-dopable polymers with much greater tendency for transporting holes than for transporting electrons. This charge imbalance is one of the key limitations to increasing the electroluminescence (EL) quantum efficiency of PLEDs. The EL quantum efficiency of PLEDs may be increased greatly by adding charge injection/transporting layers between a light-emitting polymer film and the electrodes. Thus, synthesis of light-emitting polymers with strong electron affinity (n-dopable) is necessary for fabricating highly efficient PLEDs.

As part of our interest to prepare electron injection/transport materials, we have prepared several oxadiazole-containing oligomers[2] (Fig. 9). The oxadiazole ring is believed to have a high electron affinity and these materials were designed to serve as electron-acceptors in polymer/polymer blend solar cells.

---

[1] W.Mammo and D. Antenehe, unpublished work.

[2] W. Mammo and M. Andersson, unpublished work.

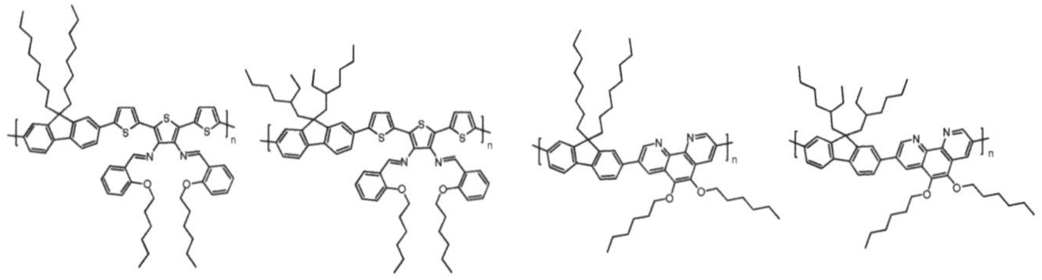

Fig. 8 Polyfluorene copolymers designed for metal chelation

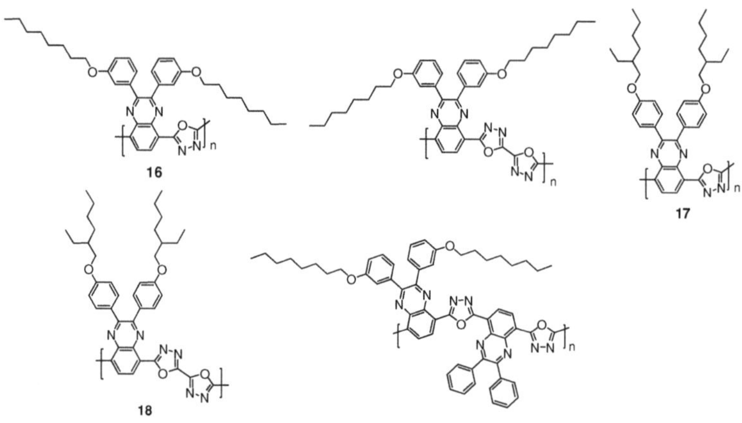

Fig. 9 1,3,4-Oxadiazole-containing oligomers

The absorption and emission properties of oligomers **16**, **17** and **18** were studied [34]. All three oligomers showed absorption (385–415 nm) and emission (500–522 nm), both from chloroform solutions and from solid films. The absorption and emission spectra taken from solution and film do not show significant shifts in the maxima indicating the non-flexibility of the backbone of the oligomer chain upon aggregation in the solid film. The cyclic voltammograms of the solvent-casted oligomers **16**–**18** on platinum disk electrodes showed an irreversible oxidation in the p-doping region and very good reversibility in the n-doping region. Cyclic voltammetry was also used to estimate the values of HOMOs, LUMOs and band gaps of the oligomers. Thus the band gaps of all three oligomers were estimated to be around 2.5 eV. The LUMO values were estimated to be around 3.6 eV indicating that these oligomers are n-dopable and may find possible applications in polymer/polymer blend solar cells and in PLEDs.

4.5 D–A–D Low-Band Gap Polymers

We have synthesized a series of low band gap conjugated polythiophene derivatives with quinoxaline and thienopyrazine units (Fig. 10) for possible use as active

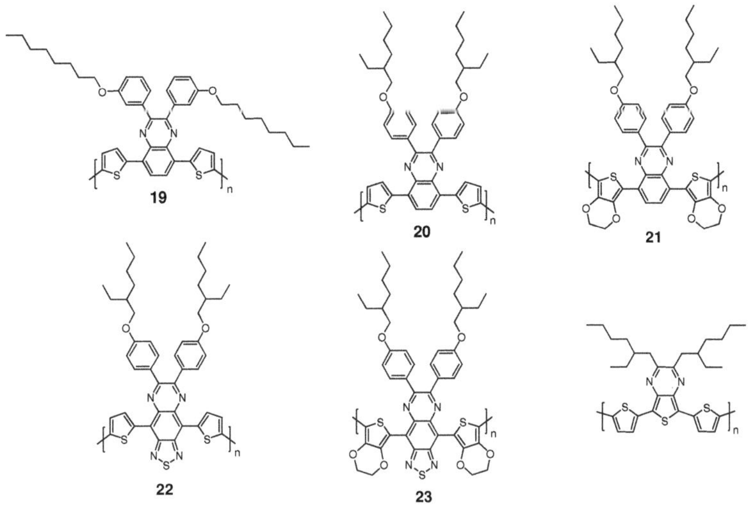

Fig. 10 Structure of low band gap D-A-D polymers

layers in polymer solar cells.[3] Such low band gap polymers are also very attractive due to their high intrinsic electrical conductivity, stability in the doped state, transparency in the neutral state and infrared absorption when p-doped.

The polymers we prepared were characterized optically and electrochemically [34]. The maximum absorption wavelength of the polymers ranged from 605 to 943 nm and the optical band gaps, determined from the onset of absorption, varied between 0.8 and 1.6 eV. Such a range of low band gap polymers was achieved by bringing together alternating donor–acceptor–donor (D–A–D) units. Polymer solar cell devices (ITO/PEDOT-PSS/Polymer:PCBM/LiF/Al) were fabricated and the photovoltaic characteristics of polymers **19–23** (Fig. 9) were studied.[4]

In general, the low band gap polymers were synthesized from the monomers by the oxidative polymerization methodology using anhydrous $FeCl_3$ in $CHCl_3$ or $Cu(ClO_4)_2 \cdot 6H_2O$ in $CHCl_3$ and CH_3CN. In a typical procedure, a slurry of the oxidant in $CHCl_3$ (for $FeCl_3$) or CH_3CN (for $Cu(ClO_4) \cdot 6H_2O$) is added in small portions, over an extended period of time, into a solution of the monomer in $CHCl_3$. After the addition is complete the mixture is stirred for some time and the polymer is precipitated by adding into MeOH.

Recently, we developed two new low band gap polymers (**P1TPQ** and **P3TPQ**, Fig. 11) and investigated their photophysical, electrochemical, and photovoltaic properties [35]. Bulk heterojunction solar cells were fabricated from **P1TPQ** and **P3TPQ** with a device architecture of glass/ITO/PEDOT:PSS/active layer/LiF/Al.

[3] W. Mammo and M. Andersson, unpublished work.

[4] W. Mammo, S. Admassie and F. Zhang, manuscript under preparation.

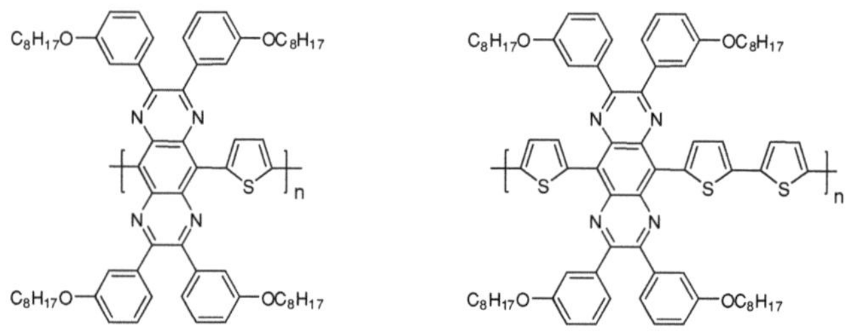

P1TPQ P3TPQ

Fig. 11 The structures of **P1TPQ** and **P2TPQ**

The solar cells based on **P3TPQ** and [6,6]-phenyl-C71-butyric acid methyl ester exhibited a power conversion efficiency of 2.1% and photoresponse up to 1.1 μm.

5 The Impact of IPICS Support

Postgraduate training in chemistry at Addis Ababa University started in 1978. Most research programs at the Department of Chemistry, AAU, were launched at about the same time as the emergence of the postgraduate training program. Even though the Department of Chemistry made significant contribution in manpower training in the last 32 years, the level of chemical research, however, has not grown very much because of financial constraints and poor administrative support. The University was unable to focus on and dedicate resources to long-term research endeavors that could have impact on nation building and advancing the basic science. The few research projects that also had significant impact on the graduate training program were those that enjoyed research funding from foreign sources such as SIDA-SAREC, IFS, TWAS, etc.

Chemical research in Ethiopia has been marred by serious predicaments such as:

- lack of research funding
- the prohibitively high cost of chemicals and scientific equipment
- poor research infrastructure
- poor administrative support for research
- poor procurement mechanism for chemicals and supplies
- lack of up-to-date scientific literature
- complete isolation of the researcher from the rest of the scientific world
- unavailability of chemical industries
- the low level of awareness of the importance of chemical science research in nation building
- poor information and communication technology infrastructure.

As a result, it was extremely difficult to launch viable research programs that could make an impact in a world-stage.

Most chemical research in organic chemistry at the Department of Chemistry, AAU, concentrated mainly on such limited areas as the isolation and characterization of secondary metabolites from plants. Although there are outstanding and world-famous research groups in the area of natural products chemistry at the Department of Chemistry, there was practically no research group dealing with advanced chemical synthesis. Thus, the challenge was immense for us to launch a research program in the synthesis and characterization of conducting polymers. If it were not for the sustained and long-term support by IPICS, we would not have been in a position to run such a high-level research program.

The support that was obtained from IPICS was multifaceted and helped to solve huge problems that have always hampered scientific research at AAU. Adequate research funding was availed and organizational support was provided by IPICS to facilitate:

- the acquisition of chemicals, equipment, and supplies;
- sandwich-type training programs;
- the exchange of researchers;
- participation in international conferences, workshops, and symposia;
- the acquisition of relevant scientific literature through journal subscription and purchase of books;
- travel and accommodation arrangements; etc.

We have managed to use the IPICS grant to organize an excellent synthetic organic chemistry laboratory and equip the laboratory with modern facilities. We have also managed to purchase research-grade equipment which supports the advancement of chemical research, not only in our specialized field of study but also in other areas of Chemistry. IPICS has also helped us to obtain, free of charge, a used gas chromatograph-mass spectrometer and fumehoods which were generously donated by the Swedish University of Agricultural Sciences, Umeå, and the University of Umeå, Sweden, respectively.

Recognizing the importance of electrochemistry in our conducting polymers research, we invested a significant amount of our research grant to acquire equipment and supplies to strengthen the electrochemistry laboratory in the premises of the Department of Chemistry. This investment enabled the group to conduct classical electrochemical research and to characterize conducting polymers.

Since its inception, our synthesis laboratory has been the main supplier of polymers to the conducting polymers research at the Departments of Chemistry and Physics. The studies of these materials by postgraduate students at the Departments of Chemistry and Physics allowed for 3 PhD and more than 40 MSc candidates to complete their studies. Two more PhD candidates are pursuing their studies in the area of materials synthesis while three other PhD candidates rely on the supply of materials by our laboratory for their theses works.

Other researchers at the Department of Chemistry have also benefitted from the very good research facility, such as solvent distillation system, water purification system, vacuum line, etc., we have organized in our laboratory. Our laboratory was also able to host one PhD candidate from the University of Gaborone, Botswana, for three months to pursue key aspects of the synthesis of natural products which he could not do at his home university.

IPICS has helped us to launch a sandwich-type PhD training program in the area of the synthesis of conducting polymers in collaboration with the Chalmers University of Technology, Gothenburg, Sweden. All arrangements (including visa, transport, and accommodation) for the candidate's stays at the Swedish university were made by IPICS. The cost of the "sandwich training" was borne by the research grant we obtained from IPICS.

The exchange of researchers scheme has helped scientists to break the isolation and to keep abreast with current developments in their respective areas of expertise. In addition, strong research collaborations could be established with researchers in Sweden. Members of our research group participated in identifying and prioritizing research areas together with their Swedish counterparts. They also spent time in Swedish laboratories and conducted scientific research. The joint efforts have led to the publication of scores of scientific articles in internationally recognized journals.

Part of the grant money that was obtained from IPICS could be used for journal subscription and for the acquisition of up-to-date scientific literature. In addition, IPICS assisted in the acquisition and transportation of a large collection scientific journals and books in Chemistry kindly donated by the Chalmers University of Technology, Gothenburg, Sweden. We have made the journals and books available for wider readership through the chemical information center of the Department of Chemistry and have helped advance graduate education at this department.

The success of our research endeavor largely hinged on the excellent working relationship we developed with the staff of IPICS. Communication with the staff of IPICS was often smooth and efficient. We forwarded our orders of chemicals and equipment to IPICS using electronic communication. The same electronic infrastructure was used to acquire proforma invoices and to handle negotiations. As the whole purchasing process did not have any bureaucratic hurdles, the timely acquisition of essential supplies could be guaranteed. Particularly important was the quick and timely acquisition of fine chemicals which allowed us to plan and execute specific synthetic tasks at specific times. As a result, the longstanding belief among organic chemists in Ethiopia that advanced organic synthesis is simply untenable and too complex to handle could be proved wrong. For the first time in the history of AAU, young and bright Ethiopians could receive adequate hands-on training in the art of modern organic synthesis.

6 Future Prospects

The ongoing effort in the synthesis of conjugated polymers for possible applications in solar cells, photovoltaic diodes, sensors, and polymer electronics will further be intensified in the coming years. Efforts will be directed at preparing polymers with improved properties such as low bandgap, high mobility, IR-luminescence, high stability, and processability, to make them suitable for of the destined applications. Conducting polymers that combine several attractive features in a single material will also be designed and synthesized aiming at improved properties. In the short term, we aim to realize materials with enhanced optical and electrical properties with an overall solar energy conversion efficiency of >8%.

We will strive to maintain state of the art research capacity and train high-level manpower at the MSc and PhD level in the area of material synthesis and characterization within the premises of the College of Natural Sciences, AAU. We will also continue to support research activities and postgraduate training at the Departments of Physics and Chemistry, AAU, by supplying polymeric materials for characteristic studies. In addition, our linkages with several research groups in Sweden in collaborative research and manpower training will be reinvigorated. This would allow for the establishment of a center of excellence in materials research in this part of Africa.

It is undeniable that research in the area of conjugated polymers is a very expensive venture. The initial investment to organize a good working environment is high, and once the research activity is underway, there must be a reliable supply of chemicals and consumables in order to ensure the sustainability of the research undertaking. We have shown above that our research achievements thus far are mainly credited to the strong and sustained backing that we received from IPICS. Unfortunately, the IPCS support was discontinued in 2009 as a result of a decision reached by the Swedish International Development Cooperation Agency (SIDA), the main funding agency for the IPICS activities. The fate of our research is therefore under threat from a combination of factors including inadequate research funding, poor research infrastructure, and poor administrative support at AAU. These factors could seriously hamper our capacity to conduct high-level research in the future.

Financial sustainability has become a major issue of concern for us since 2009. We strongly believe that the research funding that is going to be provided by AAU will not be adequate for the kind of multidisciplinary research we are planning to pursue. It is therefore imperative that we have to plan and provide for the changing needs of our research in a tight funding climate. We will therefore work hard to raise sufficient funds from sources outside the University in order to maintain an adequate level of investment to put our research onto a sustainable long-term footing.

AAU's commitment to the sustainability of our research is another area of concern. A number of initiatives have to be taken on the part of the University to address the issue of sustainability including adequate research funding, reforms of the purchasing and procurement mechanism, strong administrative support, attracting competent technical staff for employment, retention of its own qualified staff, and better management of university research assets. We are hopeful that AAU will improve upon its purchasing and procurement mechanisms in order for our efforts to become fruitful.

The long-term sustainability of our research will also depend on building excellent research capacity to support our missions. We need to reexamine the state of the research asset base in AAU and the way it is managed and come up with better and more efficient management practices. We are fully convinced that resource sharing is the only way to guarantee the sustainability of our research. The initiation and implementation of income-generating schemes will also be given serious considerations.

References

1. Roman LS, Mammo W, Pettersson LAA, Andersson MR, Inganäs O (1998) High quantum efficiency polythiophenes/C_{60} photodiodes. Adv Mater 10:774
2. Mammo W, Andersson MR (1998) New polythiophenes with oligo(oxyethylene) side chains. Bull Chem Soc Ethiop 12:141
3. Andersson MR, Thomas O, Mammo W, Svensson M, Theander M, Inganäs O (1999) Substituted polythiophenes designed for optoelectronic devices and conductors. J Mater Chem 9:1993
4. Theander M, Inganäs O, Mammo W, Olinga T, Svensson M, Andersson MR (1999) Photophysics of substituted polythiophenes. J Phys Chem B 103:7771
5. Andersson MR, Mammo W, Olinga T, Svensson M, Theander M, Inganäs O (1999) Synthesis of regioregular phenyl substituted polythiophenes with $FeCl_3$. Synth Met 101:11
6. Roman LS, Chen LC, Petersson LAA, Mammo W, Andersson MR, Johansson M, Inganäs O (1999) Multifunctional polythiophenes in photodidodes. Synth Met 102:977
7. Johansson T, Mammo W, Andersson MR, Inganäs O (1999) Light-emitting electrochemical cells from oligo (ethylene oxide)-substituted polythiophenes: evidence for in situ doping. Chem Mater 11:3133
8. Theander M, Zigmantas D, Sundstrom V, Mammo W, Andersson MR, Inganäs O (2000) Photoluminescence quenching at a polythiophene/C60 heterojunction. Phys. Rev. B 61:12957
9. Aasmundtveit KE, Samuelson EJ, Mammo W, Svensson M, Andersson MR, Petersson LAA, Inganäs O (2000) Structural ordering in phenyl-substituted polythiophenes. Macromolecules 33:5481
10. Johansson T, Mammo W, Svensson M, Andersson MR, Inganäs Olle (2003) Electrochemical band gaps of substituted polythiophenes. J Mater Chem 13:1316
11. Abdalla TA, Mammo W, Workalemahu B (2003) Electronic properties of poly[3-(2",5"-diheptyloxyphenyl)-2,2′-bithiophene]/Al junctions. SINET: Ethiop J Sci 26:11
12. Abdalla TA, Mammo W, Workalemahu B (2004) Electronic and photovoltaic properties of a single layer poly[3-(2″,5″-diheptyloxyphenyl)-2,2′-bithiophene] devices. Synth Met 144:213
13. Admassie S, Mammo W, Solomon T, Yohannes T (2005) Chromic transitions in phenyl-Substituted polythiophenes. Bull Chem Soc Ethiop 19:267
14. Tehrani P, Isaksson J, Mammo W, Andersson MR, Robinson ND, Berggren M (2006) Evaluation of active materials designed for use in printable electrochromic polymer displays. Thin Solid Films 515:2485
15. Sergawie A, Admassie S, Mammo W, Yohannes T, Solomon T (2007) Synthesis and characterization of poly[3-(2′,5′-diheptyloxyphenyl)thiophene] for use in photoelectrochemical cells. Bull Chem Soc Ethiop 21:405
16. Sergawie A, Admassie S, Mammo W, Yohannes T, Solomon T (2008) Effect of side chain length on the electrochemical and photo-response characteristics of poly[3-(2′,5′-dialkoxyphenyl)-thiophene]s. Synth Met 158:307
17. Antenehe D (2002) Synthesis of some polythiophenes. MSc thesis, June 2002, AAU
18. Getachew A (2007) Synthesis of thiophene-based conjugated polymers. MSc thesis, July 2007, AAU
19. Zhang F, Perzon E, Wang X, Mammo W, Andersson MR, Inganäs O (2005) Polymer solar cells based on a low band-gap fluorene copolymer and a fullerene derivative with photocurrent extended to 850 nm. Adv Funct Mater 15:745

20. Inganäs O, Zhang F, Wang X, Gadisa A, Persson NK, Svensson M, Perzon E, Mammo W, Andersson MR (2005) Alternating fluorene copolymer–fullerene blend solar cells. In: Sun S-S, Serdar N (eds) Organic photovoltaics: mechanisms, materials and devices, Sariciftci, Ch. 17, CRC Press, Boca Raton
21. Perzon E, Wang X, Zhang F, Mammo W, Delgado JL, de la Cruz P, Inganäs O, Langa F, Andersson MR (2005) Design, synthesis and properties of low band gap polyfluorenes for photovoltaic devices. Synth Met 154:53
22. Wang X, Perzon E, Mammo W, Oswald F, Admassie S, Persson NK, Langa F, Andersson MR, Inganäs O (2006) Polymer solar cells with low-band gap polymers blended with C_{70}-derivative give photocurrent at 1 μm. Thin Solid Films 511–512:576
23. Admassie S, Inganäs O, Mammo W, Perzon E, Andersson MR (2006) Electrochemical and optical studies of the band gaps of alternating polyfluorene copolymers. Synth Met 156:614
24. Zhang F, Mammo W, Admassie S, Andersson MR, Inganäs O (2006) Low band-gap alternating fluorene copolymer/methanofullerene heterojunctions in efficient near infrared polymer solar cells. Adv Mater 18:2169
25. Admassie S, Yacob Z, Zhang F, Mammo W, Yohannes T, Solomon T (2006) Synthesis, optical and electrochemical characterization of anthracene and benzothiadiazole-containing polyfluorene copolymers. Bull Chem Soc Ethiop 20:309
26. Mammo W, Admassie S, Gadisa A, Zhang F, Inganäs O, Andersson MR (1010) New low band gap alternating polyfluorene copolymer-based photovoltaic cells. Sol Energy Mater Sol Cells 2007:91
27. Gadisa A, Mammo W, Andersson LM, Admassie S, Zhang F, Andersson MR, Inganäs O (2007) A new donor-acceptor-donor polyfluorene copolymer with balanced electron and hole mobility. Adv Funct Mater 17:3836
28. Perzon E, Zhang F, Andersson M, Mammo W, Inganäs O, Andersson MR (2007) A conjugated polymer for near infrared optoelectronic applications. Adv Mater 19:3308
29. Lindgren LJ, Zhang F, Andersson M, Barrau S, Hellström S, Mammo W, Perzon E, Inganäs O, Andersson MR (2009) Synthesis, characterization, and devices of a series of alternating copolymers for solar cells. Chem Mater 21:3491
30. Gedefaw D, Zhou Y, Hellström S, Lindgren L, Andersson LM, Zhang F, Mammo W, Inganäs O, Andersson MR (2009) Alternating copolymers of fluorene and donor-acceptor-donor segments designed for miscibility in bulk heterojunction photovoltaics. J Mater Chem 19:5359
31. Melaku Y (2007) Synthesis of some fluorene-thiophene copolymers. MSc thesis July 2007, AAU
32. Zho Y, Gedefaw D, Hellström S, Krätschmer I, Zhang F, Mammo W, Inganäs O, Andersson MR (2010) Black polymers in bulk heterojunction solar cells. IEEE J Sel Top Quantum Electron 16:1565
33. Abdissa Z (2007) Synthesis of alternating copolymers of fluorene and bithiazole. MSc thesis, July 2007, AAU
34. Admassie S (2006) Electrochemical and optical characterization of conjugated polymers for use in electronic devices, PhD thesis, May 2006, AAU
35. Wang E, Hou L, Wang Z, Hellström S, Mammo W, Zhang F, Inganäs O, Andersson MR (2010) Small band gap polymers synthesized via a modified nitration of 4,7-dibromo-2,1,3-benzothiadiazole. Org Lett 12:4470
36. Yacob Z (2004) Synthesis of some polyfluorene copolymers. MSc thesis, June 2004, AAU

The International Programme in the Chemical Sciences (IPICS): 40 Years of Support to Chemistry in Africa

Peter Sundin

Abstract The International Science Programme at Uppsala University, Sweden, started in 1961 with the inception of the International Seminar in Physics, stimulating the participation of scientists from developing countries in training, and research in physics at the university. Based on the good experience of the seminar in physics, the International Seminar in Chemistry was started in September 1970. In 1988, major changes in the mode of operation of the programs were implemented, and they were collected under the common name the International Science Programmes. In 2002, the International Programme in the Mathematical Sciences was added. The operation of the International Science Programme (ISP) is today made possible by funding from the Swedish government authority Sida, but other organisations including IAEA and UNESCO have been important contributors. Uppsala University is the scientific and administrative home of ISP and has provided substantial funding since 1988. The International Seminar in Chemistry started similarly to the seminar in physics to announce for individuals interested in training in Uppsala. After 1988 the programs, now named the International Programme in the Physical Sciences and the International Programme in the Chemical Sciences, respectively, changed strategy to focus on long-term support to selected research groups rather than to individuals. Research activities supported were to be of high relevance to the country or region concerned, and long-term support required to assist in the process of building up sustainable research environments, generating useful scientific results to be disseminated, and implemented for the development of the country or region. Contacts with relevant host laboratories were also supported to facilitate development of activities. In 2008, as a result of a change in Swedish policy for development support,

P. Sundin (✉)
International Science Programme, Uppsala University,
P.O.Box 549, SE-751 21 Uppsala, Sweden
e-mail: peter.sundin@isp.uu.se
www.isp.uu.se

A. Gurib-Fakim and J. N. Eloff (eds.), *Chemistry for Sustainable Development in Africa*, DOI: 10.1007/978-3-642-29642-0_11,

a considerably reduced number of countries became available for ISP cooperation. ISP celebrates its 50-year anniversary in 2011 and IPICS its 40-year anniversary in 2010. The outcome of 40 years of IPICS support to African research groups in chemistry can be exemplified with 70 PhD and 164 MSc examinations, more than 900 scientific publications, and in the period 1996–2009, 49 scientific meetings. Besides this, considerable development of instrumental and intellectual resources has been accomplished by the supported research groups.

Abbreviations

IAEA	International Atomic Energy Agency
IPICS	International Programme in the Chemical Sciences
IPMS	International Programme in the Mathematical Sciences
IPPS	International Programme in the Physical Sciences
ISP	International Science Programme
IUPAC	International Union of Pure and Applied Chemistry
IYC	International Year of Chemistry
kSEK	Thousands of SEK (Swedish currency units)
SAREC	Sida Department for Research Cooperation
Sida	Swedish International Development Cooperation Agency
UFR	Unité de Formacion et de Recherché
UNESCO	United Nations Educational, Scientific, and Cultural Organization

1 The International Science Programme at Uppsala University, Sweden

Uppsala University is a modern, high-ranking university with its origin dating back to the Middle Ages. It was founded in 1477 and is the oldest university in the Nordic countries. Famous scientists, such as Carl Linnaeus, Anders Celsius, and Olof Rudbeck are some of Uppsala's renowned figures from the past. Eight Nobel prizes have been awarded to researchers at Uppsala University, two of them in physics and two in chemistry [5].

As a natural consequence of this, many young scientists from around the world have since long been attracted to spend some time at Uppsala University. However, in the past practically none of them came from developing countries. This was noted, and in the late 1950s an idea appeared at the Institute of Physics, that there would be a special organization stimulation the participation of scientists from developing countries, and facilitating and providing contacts, travels, fellowships, accommodation, and medical and social care in Sweden. As a result, the International Seminar in Physics was launched in 1960, inviting scientists with priority given to developing countries, and a first batch of trainees arrived at the start of the activities in 1961. A similar program in chemistry was started in 1970.

In 1988, major changes in the mode of operation of the programs were implemented, and they were collected under the common name the International Science Programmes. This development has been described in detail in Lindqvist [10]. In 2002, the International Programme in the Mathematical Sciences was added.

The operation of the International Science Programme (ISP) is today made possible by funding from the Swedish government authority Sida, which took this responsibility from its inception in 1965. In 1978, the agency SAREC took over the funding, first independently and from 1993 from its position at Sida, whereto it was transferred the same year. In October 2008, SAREC was resolved and ISP funding was again administered by Sida, through its Secretariat for Research Cooperation, which organizationally replaced SAREC. Sida and SAREC have been the most prominent collaborators, discussion partners, financing bodies, and drivers in developing ISP to its present position. Uppsala University is the scientific and administrative home of ISP and has also provided substantial funding since 1988. Among earlier financial contributors were IAEA and UNESCO.

1.1 The International Seminar in Chemistry 1970–1988

Based on the good experience of the first years of the seminar in physics, discussions started in the mid 1960s to launch a similar program in chemistry [8]. Professor Rune Liminga, Institute of Chemistry, University of Uppsala, was engaged in the planning and was then selected to lead the program. The first International Seminar in Chemistry was announced in the fall of 1969 to start in September 1970. It was agreed with Sida that highest priority should be given to universities in 10 countries of Africa, a few in Asia and a few in Latin America. The aim was described as *to initiate the creation of research groups or to provide assistance to already existing research groups at universities or national laboratories in developing countries. The assistance, which if proving successful may continue through several years, is given in order to improve the conditions and prospects of local research work.*

At this time, still, individual scientists applying were subject to training, a condition which continued for the next 10 years. There were no clear criteria for the definition of a "research group", aimed to be the subject of the program. In the early 1980s, however, discussions with SAREC and Sida led to development plans including concentration of support to a selected group of institutions in a limited number of countries through a more restricted announcement, and to the initiation of regional exchange of scientists (which was started by the Chemistry Seminar already 1981).

1.1.1 Introduction of a Goal-Oriented Approach

Following an evaluation in 1986, the designations of the programs were changed to those used currently, the International Programme in the Physical Sciences (IPPS), and the International Programme in the Chemical Sciences (IPICS), with the collective name being the International Science Programmes (ISP). Also, a change in operation of the programs was induced in order to better meet the objectives, following advice from in particular leading scientists in developing countries. In the case of IPICS the aim was to be to *select goal-oriented projects, of high relevance to the country or region concerned, for long-term support to assist in the process of building up sustainable research environments, generating useful scientific results to be disseminated and implemented for the development of the country or region.* This change was initiated in 1988.

The risk of "brain-drain" was a matter of concern in the 1970s, and the chemistry program lost a few participants in particular due to drastic political changes in the participants' home countries. When the program was later developed to address long-term support to goal-oriented research groups and "sandwich" postgraduate training, this problem was largely eliminated.

1.2 The International Programme in the Chemical Sciences 1988–1995

According to the ordinance given in 1988 by the Swedish government, the ISP has the task to initiate and support long-term collaboration in research of foremost Swedish institutions with institutions in developing countries. The purpose herewith shall be to increase the research capacity of universities and research institutes in the (at that time so called) Third World. ISP shall also encourage regional collaboration amongst countries in the Third World in their respective field of the program.

The previous mode of announcement of the chemistry program through Swedish embassies was abandoned and replaced by a grant application system under direct control of IPICS scientific staff. The implementation of the changes was carried out by Professor Liminga, continuing his engagement but now as the director of IPICS. In 1988, 38 different projects were selected among those getting support in chemistry at the time, 12 of these in Africa. However, with the new mode of operation it was necessary to invest more funding in each project and the number had to be reduced. Over the next 7 years the total number was decreased to 23, 10 of which in Africa [8].

There were several important advantages with the new mode of operation:

- transfer of more responsibilities to the supported research groups for planning of the activities and handling of the funds,
- better and more advanced planning of the activities in each project,
- more efficient use of funds, when each research group takes responsibility and has to make priorities,
- less administration in the application procedure,
- monitoring of progress carried out more easily than was possible earlier.

1.2.1 Host Laboratories to Provide Expertise

From the beginning the International seminars were programs concentrated to Uppsala. Almost all participants were located at institutions at Uppsala University, with a few at the Uppsala campus of the Swedish University of Agricultural Sciences. As activities expanded, more scientific expertise was needed than was available in Uppsala, resulting in a widening of the cooperation to a national level with institutions at other Swedish universities. The philosophy was to meet the demand of expertise required by the participants, and from 1988 in particular the need by the supported research groups, independent of where the most appropriate host laboratory was to be found. When called for, host laboratories also outside Sweden were used, including leading laboratories in the regions. If contact was not already established, host laboratories were usually selected by the research groups concerned after being supported to visit tentative cooperation partners in the relevant scientific fields.

1.2.2 Policy and Working Principles in 1995

The policy and working principles of both IPPS and IPICS in 1995 were to assist third-world countries to strengthen their domestic research and postgraduate education capacity by providing long-term project oriented support for developing active and sustainable research environments in selected countries in Africa, Asia, and Latin America. Research of strong relevance to the countries or regions concerned was selected for support and the work carried out by the selected research groups in close cooperation with one or more host laboratories.

Identification of new groups to support was typically made in connection with visits by staff to countries in cooperation with ISP, and through proposal by scientists in Sweden or in developing countries. Invitation to submit an application for ISP was preceded by thorough investigation and planning. An invited group would need to appoint a group leader to take responsibility for the research planning, the application and reporting procedures, the management of research and training, and to be accountable for the grants allocated.

Long-term support is important because it takes many years to develop scientific competence and to form sustainable environments for scientific research and postgraduate education. For this reason commitments typically lasted for a decade or more, with agreements signed for three- to five-year periods and regular monitoring to follow development progress by recording e.g.:

- publications,
- postgraduate training,
- arrangement of courses, workshops and conferences,
- expansion into interdisciplinary work, and
- attraction of funding from other organizations.

The gender distribution of participants was recorded in order to stimulate increased participation of scientists of the underrepresented sex.

Long-term commitment also implied that the major parts of the ISP funds were not available to new research groups, because the turn-over was slow. With this development the program ceased to be announced for open competition for funding. Another reason for active selection of research groups to be invited for funding, at the occasions funding was available, is that support was often initiated to environments that were not scientifically strong enough to compete in open processes. The termination of support because development was not satisfactory sometimes was the case, but the desirable outcome would be that the supported group became strong enough to attract other funding to sustain activities, meaning that ISP funding could be directed to new, promising groups.

Other important components of collaboration were exchange of scientists, "sandwich" type postgraduate education and regional collaboration by forming networks of scientists in the same fields, or sharing scarce instrumental resources.

Decisions on who to support and the size of funding were taken by the ISP Board consisting of scientists at Uppsala University, from 1988 representatives for participating Swedish Universities, a representative of IAEA, up to 1993 a representative for UNESCO and from 1994 for developing countries, a representative for the student union, from 1970 a representative for the university administration, and up to 2000 one member nominated by the Swedish donor. The Board was chaired by the Vice Chancellor of Uppsala University.

In essence, these principles are valid also today, with a few changes.

1.3 The International Programme in the Chemical Sciences 1996–2010

ISP today consists of three main activities:

- The International Programme in the Physical Sciences (IPPS), which started in 1961 as the International Seminar in Physics.
- The International Programme in the Chemical Sciences (IPICS), which started in 1970 as the International Seminar in Chemistry.
- The International Programme in the Mathematical Sciences (IPMS), which started its program activities in 2002.

The common features of IPPS, IPICS, and IPMS are still that they provide long-term support to research groups in selected countries, and South–South collaboration in the form of scientific networks, to assist in developing and promoting scientific research and postgraduate education in the scientific fields covered by the program, and relevant to the needs of the countries targeted to build competence to address developmental problems on their own terms.

The development of the chemistry program up to 2003 is described by Liminga [8, 9], Lindqvist [10], as well as in Project Catalogues for 1988, for each year 1990–1995, and for 1997 and 2003. In 1997, Professor Rune Liminga, left IPICS for another position and a new program director, Assoc. Prof. Malin Åkerblom, was recruited. She continued to develop IPICS until her retirement in 2006, when she was replaced by Assoc. Prof. Peter Sundin. In 2007, Dr. Sundin was appointed head of ISP.

1.3.1 Changes in the Composition of the ISP Board

In comparison to the previous period some changes of the ISP Board took place after 1995. From 2000, Swedish donors were no longer represented, and from 2005 the Board was chaired by the Vice Rector of the Disciplinary Domain of Science and Technology rather than by the Vice Chancellor.

In 2008, the IAEA representative at the ISP Board retired, but IAEA did not find reason to appoint a replacement. After considering the matter thoroughly, the ISP Board approached the African Union Commissioner of Human Resources, Science and Technology, with an invitation to serve as ISP Board member. The commissioner accepted, and was in 2009 appointed to the ISP Board by the Vice Chancellor of Uppsala University.

1.3.2 Introduction of Scientific Reference Groups

The most significant change to ISP during the period was the introduction of formal scientific reference groups, the task of which are primarily to yearly meet to evaluate applications from research groups and scientific networks for new or continued support, but also to be an advisory body to the program director. The IPICS reference group was established in 2001 and is since then nominally composed of three renowned chemistry professors from developing countries and three from the Nordic countries. The composition of the reference group is intended to cover the competence needed to evaluate applications in the different fields in the programs, keeping in mind that development potential may be a criterion as important as scientific achievement, depending on how long an activity has been supported.

1.3.3 Changes in the Selection of Countries to Support

Another prominent change regards the selection of countries to support. The initial scope of countries addressed was rather wide, but it was soon restricted to fewer countries with priority given to developing countries. From 1973 the beneficiaries were mainly in least developed countries, primarily Sida program countries [8].

In 2008, as a result of a new Swedish development support policy issued in August 2007, target countries for ISP support was decided by Sida to be shifted to a group of 12 "focus countries" (so-called category 1-countries) decided by the Swedish government [6].

Nine of the twelve focus countries are in Africa, being Burkina Faso, Ethiopia, Kenya, Mali, Mozambique, Rwanda, Tanzania, Uganda and Zambia, and of these Kenya, Mozambique, Rwanda, and Zambia were new to the chemistry program.

In the ISP agreement signed for the period 1 July 2008–30 June 2010 (now extended into 2011) it was stated that *the objective of the Program is to strengthen research and postgraduate education in the field of basic sciences in developing countries with focus on category 1-countries. The support is long-term and directed toward research groups, networks, and resource centres. In countries with Sida supported bilateral university programs, the research groups shall be incorporated into these programs gradually and country wise*. In practice, the latter condition implied that any occurring ISP support should be phased out when a new Sida bilateral program in science is initiated in a focus country, or when an existing bilateral program enters a new agreement phase. The ambition expressed by Sida was that in such cases, formerly ISP-supported groups would be incorporated in the bilateral support programs.

The agreement required that networks and resource centers supported by ISP had to be further defined by the time when the next ISP application for funding to Sida was due. ISP support to scientific networks and resource centers, most of them in the field of chemistry, has therefore recently been comprehensively accounted for [7].

1.3.4 Consequences of the Changes to IPICS Support in Africa

With regard to ISP support to research groups in Africa, there were several concrete consequences of the new Swedish policy. IPICS support to two research groups in Malawi, not being a Swedish focus country, had to be phased out prematurely by 2010. More seriously, IPICS support to one research group in Ethiopia (described by its group leader, Professor Wendimagegn Mammo, in another section of this chapter), one research group in Tanzania, and two research groups in Uganda had to be phased out from IPICS support already at the end of 2008—with less than 6 months notice—since new Sida bilateral program phases were expected to start in these focus countries in 2009. Because of delays in starting the new program phases were feared to cause obstructions in the research and postgraduate education activities by these groups, ISP proposed and managed to implement provisional support in 2009, financially covered by unused grant money from the concluded bilateral program phases. By 2010, the two research groups in Uganda were adopted by the new phase of the Sida bilateral program with Makerere University, and the one research group in Tanzania nominally by the new phase of the Sida bilateral program with University of Dar es Salaam. However, no support at all within the frame of the new phase of the Sida bilateral program was given to continue the activities in 2010 of the previously IPICS-supported research group at Addis Ababa University in Ethiopia. IPICS therefore intends to resume the cooperation in 2011.

The Sida decision thus prevented ISP from supporting research groups in focus countries having agreements with Sida on bilateral support to research development in science, being Ethiopia, Mozambique, Rwanda, Tanzania, and Uganda. In Burkina Faso, IPICS initiated pilot support to the Department of Chemistry at University of Ouagadougou in 2008, and since the Faculty of Science (UFR/Sciences Exact et Appliquées) at that time was not embraced by the Sida bilateral program with the country, IPICS support is expected to continue until the next phase of the Sida bilateral program starts, in which the faculty is hopefully included.

1.3.5 An ISP Strategy for the New Situation

In October 2008, ISP adopted a working strategy ([4], replacing the Strategy Plan 2003–2007, [2]), outlining the development of the program under the current agreement. With regard to research groups, it was established in the working strategy that *one consequence of the Sida decision is that ISP can continue operating with direct support to capacity development and enhancement to research and postgraduate education in the basic sciences physics, chemistry, and mathematics—but limited to five of the twelve focus countries: Kenya, Mali, Zambia, Bangladesh, and Cambodia. This can continue at least to the point in time when Sida establishes bilateral support programs in these countries. The working strategy for ISP activities with regard to research groups in these countries is proposed to be to develop capacity in physics, chemistry, and mathematics as far as possible. The line of this work should be in agreement with the relevant policies and strategies on national as well as on the university level, and with Sida policy, guidelines, and thematic focus. The purpose of ISP support is in this case to contribute to strengthening the capacity in basic sciences in order to provide as solid base as possible once Sida bilateral support programs are to be established.*

For IPICS support to research groups in Africa this implied that the existing support to a research group in Mali could continue for the time being, and support to research groups in Kenya and Zambia could be initiated. Indeed, after thorough discussions and repeated visits to Kenya and Zambia in 2009 and 2010, applications were received from two research groups at the Department of Chemistry of University of Nairobi, and one from the Department of Chemistry of the University of Zambia. The applications were favorably assessed by the IPICS scientific reference group in its meeting in September 2010, and support is proposed to start from 2011 provided that funding is made available to ISP.

Zimbabwe does not belong to the 12 countries which can benefit from long-term development cooperation with Sweden, according to the current Swedish policy. However, to help keep scientific capacity, IPICS has agreed to continue supporting a few groups at the institutional level.

2 The International Program in the Chemical Sciences 40 Years 2010

ISP as such celebrated its 50-year anniversary in 2011. IPICS has been in operation for 40 years in 2010, and this chapter can be regarded as an anniversary account of IPICS support to scientists in Africa, the publication of which just following the International Year of Chemistry 2011 (IYC, see www.chemistry2011.org).

After a resolution for IYC was approved by the UNESCO Executive Board (179 EX 47), the 63rd UN General Assembly adopted the resolution in December 2008 proclaiming 2011 as the International Year of Chemistry, placing UNESCO, and the International Union of Pure and Applied Chemistry (IUPAC) at the helm of the event. The draft resolutions were submitted by Ethiopia both to UNESCO and the UN General Assembly (Dr. Julia Hasler, UNESCO, personal communication 2010).

The present overview section briefly accounts for IPICS support to African scientists since 1970, and is followed by two scientific sections exemplifying the outcome and impact of ISP long-term support to chemistry in Africa. One is summing up the IPICS support to biochemistry and molecular biology in Zimbabwe since 1990, focusing on the latest years' development in the field of drug discovery and associated research. The section is kindly provided by Dr. Collen Masimirembwa, Director of the African Institute of Biomedical Science and Technology, with Head Office and Laboratory in Harare, Zimbabwe. The other describes the development of research on the synthesis of conjugated polymers in Ethiopia, supported by IPICS since 2002 (and before that by IPPS since 1995). The section is kindly provided by Professor Wendimagegn Mammo, Department of Chemistry, Addis Ababa University, Addis Ababa, Ethiopia. Indeed, it is suitable that a section of development of chemistry research in Ethiopia is included considering the fact that the draft resolution of the IYC 2011 was submitted by Ethiopia.

2.1 IPICS Support to Chemistry in Africa 1970–2010

In the period 1970–2010 IPICS provided support in the field of chemistry to scientists from 20 different African countries (Table 1). Up to 1988, the International Seminar in Chemistry provided training in Sweden to individual scientists applying for scholarships made available by the program. From 1988, the support was given to invited research groups applying for funding to develop activities at their own terms. Support has typically been provided to one or two research groups in each country. The longest term cooperation has been with scientists in Cameroon, Ethiopia, Nigeria, Tanzania, and Zimbabwe.

Table 1 African countries with IPICS support 1970–2010

Country	1970–1974	1975–1979	1980–1984	1985–1989	1990–1994	1995–1999	2000–2004	2005–2010
Botswana	–	–	–	X	–	–	–	–
Burkina Faso	–	–	–	–	–	–	–	X
Cameroon*[a]	–	–	X	X	X	X	X	X
Egypt	X	X	–	–	–	–	–	–
Eritrea	–	–	–	–	X	X	–	–
Ethiopia*[b]	–	X	–	X	X	X	X	X
Ghana	–	X	–	X	–	–	–	–
Kenya	X	X	X	X	–	–	–	–
Mali	–	–	–	–	–	–	X	X
Malawi*[c]	–	–	–	–	–	X	X	X
Mauritius	X	–	–	–	–	–	–	–
Nigeria*[d]	X	X	X	X	X	X	X	X
Somalia	–	–	X	X	–	–	–	–
Sudan	X	–	X	X	X	X	–	–
Tanzania*[b]	X	X	X	X	X	X	X	X
Tunisia	X	–	–	–	–	–	–	–
Uganda*[b]	X	–	–	–	–	X	X	X
Zambia	X	–	–	–	–	–	–	–
Zaire	–	–	–	X	X	–	–	–
Zimbabwe	–	–	–	X	X	X	X	X
Total	9	6	6	11	8	9	8	9

X 5-year intervals; Up to 1988 under the designation "The International Seminar in Chemistry", with focus on individual participants. From 1988, under the designation "The International Programme in the Chemical Sciences, and focusing on research groups." Countries supported exclusively through scientific networks and resource groups are not included
*Support phased out [a] 2009; [b] 2008; [c] 2010; [d] 2005

2.1.1 Outcome of IPICS Support 1970–1995

Statistics for the period 1970–1995 is accounted for in Liminga [8], and only briefly summarized here. Since support to postgraduate studies was not applied systematically until from the mid 1980s, the outcome with regard to graduations is concentrated to the latter part of the period. In 1995, eight research groups in Africa, having been supported on average for 8.4 years, had produced 9 PhD theses and 48 MSc theses. The number of international, peer reviewed publications was 111, and the number of international reports 56. The total funding to these groups during the period was 11,912 kSEK.

The average outcome per year and per million SEK invested is given in Table 2. It should be noted that a nominal investment of 1,000 kSEK provided the total outcome listed. Thus, 0.8 PhD theses, 4 MSc theses, 9.3 peer reviewed publications, and 4.7 international reports were obtained for each 1,000 kSEK invested. In addition, a multitude of other results not registered in the statistics

Table 2 Average outcome of IPICS support to eight African research groups up to 1995 in terms of PhD and MSc examinations, peer reviewed publications, and internationally published reports

Outcome	Number per year	Number per 1,000 kSEK
PhD examinations	1.0	0.8
MSc examinations	5.7	4.0
Peer reviewed publications	13.2	9.3
International reports	6.7	4.7

Table 3 Gender of IPICS supported African participants 1970–1995

	1970–1974	1975–1979	1980–1984	1985–1989	1990–1995	Total
Female	1	1	1	6	10	19
Male	11	22	23	20	35	111
Total	12	23	24	26	45	130
% Female	8	4	4	23	22	15

were achieved, including the acquisition if material resources and strengthening of intellectual capacity besides what can be accounted for by academic degrees.

The gender statistics point to a situation where the minority of participants is female (Table 3). The situation appears to improve from the period 1985–1989, most likely because of the changes in the IPICS mode of operation applied from 1988. With regard to the leadership of the eight groups supported at the end of the period, only one group leader was female [8].

2.1.2 Outcome of IPICS Support 1996–2009

Statistics for the period 1996–2009 has been acquired from the IPICS project catalogues 1997 and 2003 ([1] and [3], respectively), the ISP Annual Reports for the years 1996–2007, and from summary tables (unpublished) underlying the ISP Annual Reports 2008 and 2009 [11] and [12], respectively.

The data regard research groups supported by IPICS in African countries during the period. The effects of the changes required in ISP operation after 2008 is visible in the data for 2009. It should be noted that support to two research groups in Burkina Faso commenced in 2009, after pilot support in 2008. The outcome given for this country in 2009 is a result of previous activities and is not included in the summary statistics regarding the outcome of IPICS support to African research groups. IPICS support to two groups in Nigeria and two research groups in Cameroon was phased out in 2005 and 2008, respectively, as part of the normal processes with ISP. IPICS support to one research group in Ethiopia, one in Tanzania and two in Uganda was phased out in 2008, following the Sida agreement with ISP in 2008.

Tables 4, 5, 6, 7, 8, and 9 show the outcome in terms of PhD examinations of students in IPICS-supported African research groups, on "sandwich" (Table 4) or

Table 4 Awarded PhD degrees 1996–2009 (sandwich programs) by IPICS supported research groups in Africa

Country	96	97	98	99	00	01	02	03	04	05	06	07	08	09	T
Burkina Faso	–	–	–	–	–	–	–	–	–	–	–	–	–	0	0
Cameroon	0	3	1	1	1	0	2	2	2	4	1	1	0	–	18
Ethiopia	–	–	–	–	–	–	–	0	0	0	0	0	0	–	0
Malawi	–	–	–	–	–	–	0	0	0	0	1	0	1	1	3
Mali	–	0						0	0	0	0	0	0	0	0
Nigeria	0	0	1	0	0	1	1	1	0	0	–	–	–	–	4
Tanzania	0	1	0	0	0	0	0	1	1	2	0	1	1	–	7
Uganda	–	–	–	0	0	0	0	0	0	0	0	0	1	–	1
Zimbabwe	0	0	0	0	0	0	1	0	1	0	0	0	1	2	5
Total	0	4	2	1	1	1	4	4	4	6	2	2	4	3	38

T total

Table 5 Awarded PhD degrees 1996–2009 (local and other programs) by IPICS supported research groups in Africa

Country	96	97	98	99	00	01	02	03	04	05	06	07	08	09	T
Burkina Faso	–	–	–	–	–	–	–	–	–	–	–	–	–	2	2
Cameroon	1	0	0	2	0	0	0	0	4	0	0	2	3	–	12
Ethiopia	–	–	–	–	–	–	–	0	0	0	0	0	0	–	0
Malawi	–	–	–	–	–	–	0	0	0	0	0	0	0	0	0
Mali	–	–	–	–	–	–	0	0	0	0	0	0	0	1	1
Nigeria	1	0	0	1	1	0	0	1	0	0	–	–	–	–	4
Tanzania	0	0	0	0	0	0	0	0	0	0	0	0	0	–	0
Uganda	–	–	–	0	0	0	0	0	0	0	0	0	1	–	1
Zimbabwe	0	0	0	1	1	0	0	0	0	0	0	0	1	0	3
Total	2	0	0	4	2	0	0	1	4	0	0	2	5	3	23

T total

other (Table 5) programs, MSc examinations (Table 6), peer reviewed publications (Table 7), other publications (Table 8), and arranged meetings (Table 9).

The total expenditure by IPICS supported research groups in Africa was close to 43 million SEK in total for the period 1996–2009. The average outcome per year and per million SEK invested is given in Table 10. It should be noted that a nominal consumption of 1,000 kSEK provided the total outcome listed. Thus, 1.4 PhD theses, 2.7 MSc theses, 5.9 peer reviewed publications, 12 other publications, and 1.1 arranged meetings were obtained for each 1,000 kSEK used by the groups during the period.

The gender statistics for the later period (not available for 1996 and 1997) still shows a situation where the minority of participants is female. Among the staff in the IPICS-supported African research groups the proportion of female participants oscillates around 20% (Table 11). Among PhD students, the situation appears to have changed to an increasing participation of female students toward the end of

Table 6 Awarded MSc/MPhil degrees 1996–2009 (sandwich and other programs) by IPICS supported research groups in Africa

Country	96	97	98	99	00	01	02	03	04	05	06	07	08	09	T
Burkina Faso	–	–	–	–	–	–	–	–	–	–	–	–	–	0	0
Cameroon	3	5	4	11	8	7	6	8	2	3	4	6	1	–	68
Ethiopia	–	–	–	–	–	–	–	0	1	0	0	6	7	–	14
Malawi	–	–	–	–	–	–	0	0	0	1	0	1	1	0	3
Mali	–	–	–	–	–	–	0	0	1	0	2	4	0	0	7
Nigeria	3	1	0	1	2	2	0	1	1	1	–	–	–	–	12
Tanzania	0	0	0	0	0	0	0	0	0	1	0	0	0	–	1
Uganda	–	–	–	0	0	0	1	0	0	0	0	3	2	–	6
Zimbabwe	0	0	0	0	0	0	3	0	0	0	0	1	1	1	6
Total	6	6	4	12	10	9	10	9	5	6	6	21	12	1	117

T total

Table 7 International peer-reviewed publications by IPICS supported research groups in Africa 1996–2009 (*T* total)

Country	96	97	98	99	00	01	02	03	04	05	06	07	08	09	T
Burkina Faso	–	–	–	–	–	–	–	–	–	–	–	–	–	1	1
Cameroon	7	4	6	5	1	1	9	7	6	11	10	9	7	–	83
Ethiopia	–	–	–	–	–	–	–	1	1	2	9	3	2	–	18
Malawi	–	–	–	–	–	–	2	0	0	2	4	2	3	1	14
Mali	–	–	–	–	–	–	0	2	0	0	1	1	0	3	7
Nigeria	3	5	2	5	7	0	4	4	1	4	–	–	–		35
Tanzania	0	3	4	5	3	7	2	4	6	3	4	1	0	–	42
Uganda	–	–	–	0	0	0	0	0	0	0	1	0	0	–	1
Zimbabwe	3	4	2	2	2	4	5	6	2	0	3	3	11	5	52
Total	13	16	14	17	13	12	22	24	16	22	32	19	23	10	253

T total

Table 8 Other publications including conference reports IPICS supported research groups in Africa 1996–2009

Country	96	97	98	99	00	01	02	03	04	05	06	07	08	09	T
Burkina Faso	–	–	–	–	–	–	–	–	–	–	–	–	–	2	2
Cameroon	9	10	17	12	9	9	10	17	8	5	8	2	8	–	124
Ethiopia	–	–	–	–	–	–	–	1	0	1	1	1	3	–	7
Malawi	–	–	–	–	–	–	3	3	2	4	6	0	4	2	24
Mali	–	–	–	–	–	–	1	1	0	2	4	3	4	5	20
Nigeria	2	3	8	5	1	1	2	1	9	2	–	–	–	–	34
Tanzania	1	1	2	7	1	2	2	6	8	10	3	2	4	–	49
Uganda	–	–	–	0	0	0	0	1	2	6	15	6	6	–	36
Zimbabwe	17	25	7	16	12	12	14	17	8	4	20	15	16	23	206
Total	29	39	34	40	23	24	32	47	37	34	57	29	45	32	502

T total

Table 9 Meetings arranged by IPICS supported research groups in Africa 1996–2009

Country	96	97	98	99	00	01	02	03	04	05	06	07	08	09	T
Burkina Faso	–	–	–	–	–	–	–	–	–	–	–	–	–	0	0
Cameroon	0	2	2	1	2	1	1	4	1	3	1	1	1	–	20
Ethiopia	–	–	–	–	–	–	–	0	0	0	0	0	0	–	0
Malawi	–	–	–	–	–	–	0	0	0	0	0	0	0	0	0
Mali	–	–	–	–	–	–	0	0	0	0	0	1	0	1	2
Nigeria	0	1	1	0	1	0	1	0	0	0	–	–	–	–	4
Tanzania	0	0	0	0	0	0	1	1	0	2	1	0	2	–	7
Uganda	–	–	–	0	0	1	2	1	1	3	0	0	1	–	9
Zimbabwe	0	0	0	1	0	0	1	2	0	0	0	1	0	2	7
Total	0	3	3	2	3	2	6	8	2	8	2	3	4	3	49

T total

Table 10 Average outcome of IPICS support to African scientists 1970–1995 in terms of PhD and MSc examinations, peer reviewed publications, other publications, and arranged meetings

Outcome	Number per year	Number per 1,000 kSEK
PhD examinations	4.2	1.4
MSc examinations	8.4	2.7
Peer reviewed publications	18	5.9
Other publications	36	12
Arranged meetings	3.5	1.1

Table 11 Staff gender in IPICS supported research groups in Africa 1998–2009

	98	99	00	01	02	03	04	05	06	07	08	09
Female	8	9	7	9	15	11	11	11	15	17	19	21
Male	27	30	30	33	40	57	55	46	61	62	66	38
Total	35	39	37	42	55	68	66	57	76	79	85	59
% Female	23	23	19	21	27	16	17	19	20	22	22	36

Table 12 PhD student gender in IPICS supported research groups in Africa 1998–2009

	98	99	00	01	02	03	04	05	06	07	08	09
Female	1	4	2	5	9	7	10	11	10	11	10	9
Male	17	28	22	26	26	21	20	23	26	22	33	13
Total	18	32	24	31	35	28	30	34	36	33	43	22
% Female	6	12	8	16	26	25	33	32	28	33	23	41

Table 13 MSc student gender in IPICS supported research groups in Africa 1998–2009

	98	99	00	01	02	03	04	05	06	07	08	09
Female	4	6	7	7	9	6	4	6	12	14	11	4
Male	13	13	15	11	7	5	14	19	23	26	31	8
Total	17	19	22	18	16	11	18	25	35	40	42	12
% Female	24	32	32	39	56	55	22	24	33	35	26	33

the period (Table 12). The situation with regard to MSc students has been more stable, although female dominance can be noted 2002–2003 (Table 13). With regard to the leadership of the totally 16 groups supported in the period, one female group leader could be noted, but only in 2009.

References

1. Anonymous (1998) International Programme in the Chemical Sciences IPICS. Project Catalogue 1997. International Science Program, ISP. Universitetstryckeriet, Uppsala
2. Anonymous (2003) Research co-operation with developing countries. Strategy plan 2003–2007. International Science Program, ISP. http://www.isp.uu.se/Strategy.pdf. Accessed 27 November 2010
3. Anonymous (2004) International Programme in the Chemical Sciences IPICS. Project Catalogue 2003. International Science Program, ISP. Universitetstryckeriet, Uppsala
4. Anonymous (2008) Proposed ISP working strategy in 2009–2010 in perspective of the Sida decision in 2008 on two years of continued support, until June 2010, and required changes in working mode. http://www.isp.uu.se/WS.pdf. Accessed 27 Novemb 2010
5. Anonymous (2010) Uppsala University in brief. http://www.uu.se/filedownload.php?id=4243 . Accessed 27 November 2010
6. Government Offices of Sweden (2007) The new development policy. http://www.sweden.gov.se/sb/d/9382/a/86595. Accessed 27 November 2011
7. Kiselman C (ed) (2010) Proceedings of the international conference on regional and interregional cooperation to strengthen basic sciences in developing countries, United Nations Conference Centre, Addis Ababa, Ethiopia, 1–4 Sept 2009. Acta Universitatis Uppsaliensis, Skrifter rörande Uppsala universitet, C. Organisation och Historia, 90 (In press)
8. Liminga R (1996) Uppsala University, international program in the chemical sciences, 1970–1995: summing up—looking into the future. Repro HSC, Uppsala
9. Liminga R (1998) International program in the chemical sciences 1970–1997. In: Niemayer H (ed) IPICS 1970–1997: results, lessons learned, and prospects for development of sustainable research environments in developing countries. Proceedings of a meeting held at Termas El Corazón, Chile, 19–23 Oct 1997. Repro HRC, Uppsala
10. Lindqvist T (ed) (2001) International science programme, Uppsala University 1961–2001. Historical review and participants' experiences. Acta Universitatis Uppsaliensis, Skrifter rörande Uppsala universitet, C. Organisation och Historia, 71. Elanders Gotab, Stockholm
11. Sundin P, Abrahamsson L, van Groningen E, Sjöblom L, Kristófi Z (2009) International science programme. Annual report 2008. www.isp.uu.se/AR%202008.pdf. Accessed 27 November 2011
12. Sundin P, Abrahamsson L, van Groningen E, Sjöblom L, Kristófi Z (2010) International science programme. Annual report 2009. http://www.isp.uu.se/AR%202009.pdf. Accessed 2 May 2011

International Collaboration With a View to Containing Outbreak of Emerging Infectious Diseases Through Bioprospection

Mohamad Fawzi Mahomoodally

Abstract The past few decades have witnessed an evergrowing trend of international scientific collaboration and partnership as favored means of building capacity and transfer of technology with respect to bioprospecting in developing countries like Africa. It is frequently stated that Africa is a famous biological hotspot with unique biodiversity and traditional knowledge but inherently lack proper state-of-art technologies, technical expertise, and funding to extensively probe and add value to them. This chapter therefore, endeavors to survey the significance of international collaborations as an instrument for bioprospection in the African regions, geared toward containing eventual outbreaks of emerging infectious diseases (EIDs). The author also focuses on how outbreaks of EIDs in the Indian Ocean has resulted in international collaborations and transfer of technology among developing African member states and between some developed and developing countries. One case study overviewed herein is the recent epidemic outbreak of the chikungunya fever in the Indian Ocean which resulted in the setting of an international center for research and intelligence on EIDs in the Indian Ocean for collaborative effort, and which subsequently endowed the PHYTOCHIK consortium. Furthermore, mandates of important African organizations and partnerships such as AAMPS (Association for African Medicinal Plants Standards), CRVOI (International center for research and intelligence on emerging diseases), IMRA (Malagasy Institute of Applied Research), ICIPE (International Centre of Insect Physiology and Ecology), NAPRECA (Natural Product Research Network for Eastern and Central Africa), ICBGP (The International Cooperative Biodiversity Groups Programs), ISP (International Science Programme), NCI-US (The US National Cancer Institute), and the Malaria

---

M. F. Mahomoodally (✉)
Department of Health Sciences, Faculty of Science,
University of Mauritius, Reduit, Mauritius
e-mail: f.mahomoodally@uom.ac.mu

A. Gurib-Fakim and J. N. Eloff (eds.), *Chemistry for Sustainable Development in Africa*, DOI: 10.1007/978-3-642-29642-0_12,

Consortium, which have been on the forefronts of developing international collaborations on bioprospection in Africa, are also addressed. Taken together, multi-international partnerships have nurtured panoply of regional and European consortiums geared toward identification and development of standard extracts and potential pharmacophores from the local biodiversity in compliance with the Convention on Biological Diversity and World Trade Organization. In conclusion, international joint ventures in Africa have fostered important development programs, capacity building, led key transfer of technology, mutual sharing of bioresources with the idea of access and benefit sharing, encouraged exchange, and networking between European and African bioprospectors.

1 Emerging Infectious Diseases in African Regions

Infectious diseases, such as HIV, malaria, tuberculosis, diarrheal diseases, pneumonia, and measles, still remains one of the major causes of human mortality and morbidity. Moreover, recent bulletin of the World Health Organisation [1] reveals that the striking effect of antimicrobial resistance is a serious problem in controlling infectious diseases. The result of such a natural response by microorganisms has the potential to halt or even bring progress to a standstill. Well regulated health systems like those of European countries are already witnessing an increase in specific pathogenic resistance. Developing countries from Africa are definitely not spared amidst these tycoons. Accordingly, the gap created in the development of new antibiotics in the past decades is vital for surfeit of reasons including treatment of chronic infections. Moreover, each year the death toll from these infectious diseases is more than 11 million worldwide and is mostly among young parents and children and hence remains a major origin of illness and death in many parts of Africa [1–3]. Currently, infectious diseases are responsible for one in two deaths in developing countries like Africa, where poverty, limited access to health care, drug resistance and a changing environment make populations particularly vulnerable (http://www.scidev.net).

On the other hand, emerging infectious diseases (EIDs) represent another significant strain on human health as well as the capacity to generate sustainable livelihoods, food security, and promote environmental conservation. Examples of EIDs in Africa include Ebola, Monkeypox, Marburg, Avian Influenza, and more recently chikungunya. Growing human populations and associated pressures imply that livestock, wildlife, and people are increasingly forced into greater proximity, thereby enhancing diseases transfer. Economically, EIDs can be devastating on both the local and national level, especially for livestock-dependent populations. Ultimately, EIDs may create cycles of illness, malnutrition, and poverty, further impacting sustainable land use, natural resource management, and conservation initiatives.

To this effect, research scientists have been compelled to venture for substitute curatives to combat growing resistances to antibiotics and to cope with EIDs. It is indeed of paramount importance to unveil new therapies directed at novel targets as budding to alternatives to antibiotics as well as validation of traditional remedies. A plethora of studies has emerged to the natural products in addressing the dearth of current therapies via bioprospecting natural resources.

2 Bioprospection in African Regions

At the beginning of the 1960s the discovery of the anticancer drugs, vincristine and vinblastine from *Catharanthus roseus* (Apocynaceae) originally from Madagascar has highlighted the significance of plants as the raw materials in the quest for natural and alternative drugs from Africa. Many pharmaceutical companies have since then showed immense interest in bioprospection of natural resources as a source of novel biological compounds. At the fundamental level, bioprospecting or biodiversity prospecting is the exploration, development and commercialization of biodiversity and the exploration of new biological resources and their derivatives with potential social and economic value. It has been carried out by a wide variety of industries that include pharmaceuticals, botanical medicines, crop protection, cosmetics, horticulture, agricultural seeds, environmental monitoring, manufacturing, and construction among others [4].

In this respect, Africa is fortunate to be innately gifted with a rich biodiversity and traditional medical knowledge that could pioneer successful bioprospecting by companies in a broad range of sectors. It is now established that African countries are well endowed with both variety and abundance of living things, particularly in terms of plants—with a rough estimate of overall plant richness at species, genus, and family level in the African mainland between 40,000 and 60,000 plant species, of which approximately 35,000 are endemic. Therefore, it is tempting to argue that Africa roughly contains 25 % of the world's biodiversity, and at a crude estimate the combined total value of all products derived from the world's genetic resources lies between $500 and $800 billion annually [5, 6]. It is clear among the international community that Africa is a rich, profitable seam of raw material, and knowledge for the development of new medicines, foods, cosmetics, and other products from biodiversity. With a good policy, bioprospecting has numerous benefits that include scientific, environmental, educational, economic, and institutional. Taking advantage of opportunities provided by these benefits requires building both individual and institutional capacity at regional and international level [4].

One area that has attracted panoply of potential bioprospectors is traditional herbal medicines of Africa. Indeed, since time immemorial natural products have been a potential source for new drugs and as tools for pathway screening and target identification and only recently, pharmaceutical industries have set much emphasis and intensified their investments in the development of phytomedicines derived from the African pharmacopeia. The use of plant extracts and phytochemicals of known biological properties has been of great significance in the treatment of several diseases in countless communities of Africa. The intensifying significance in medicinal herbs and food plants is mainly due to the apprehension regarding the toxicity, safety of modern drugs and benefits of the complementary systems. Undeniably, the growing occurrence of resistance to antimicrobial agents and EIDs has fueled the necessity to explore for newer antibiotic sources which are inexpensive, safe, and effective [7]. Natural products from botanical sources utilized from African pharmacopeia may offer a new route against multidrug-resistant microbial infections through the elucidation of biological compounds with novel mode of actions [8, 9]. Phytocompounds from the African continent are considered prospective sources of novel antibiotics, anticancer and anti-microbial agents among other pharmaceutical agents and it is anticipated that African biodiversity may be hosting cures for existing EIDs not excluding existing impending pandemic such as HIV, cancer, Hepatitis B, and many other chronic diseases and potential EIDs. For instance, plant derived phytoalexin, isothiocyantes, allicins, anthocyanins, essential oils, tannins and polyphenols have exhibited antibacterial and/or antifungal activities. Examples of some botanical drugs employed in traditional medicine which gave rise to effective modern drugs are: *Adhatoda vasica*, and *Harpagophytum procumbens*. *Adhatoda vasica* was indigenously used as antispasmodic, antiseptic, insecticide, against fish poison, uses in biomedicine include antispasmodic, and cough suppressant [10]. Three examples of phytochemicals with known anti-infective activities isolated and studied from the African flora are;

- *Veronica colorata*—compounds isolated sesquiterpene lactones which were found to possess significant in vitro antischistosomal, plasmodicidal, and leishmanicidal activities. The lactones also showed anthelmintic and amebicidal properties. Leafs extracts of this plant also showed antimalarial activities and potent antibacterial potential.
- *Warburgia salutaris*—the drimane sesquiterpenoids, such as warburganal and polygodial have been reported to be highly active in low concentrations against candida. Polygodial has been found to potentially useful in clinical medicine as an adjunct to treatment with antibiotics and antifungals that have poor membrane permeability.
- *Voacanga africana*—several alkaloids (voacamine, voacangine among others) of biological activities have been isolated. Root and bark extract containing these alkaloids were found to have antibacterial, antiamoebic, and antispasmodic activity in vitro [8]).

3 The Need for Bioprospecting Partnerships in Africa

It is estimated that the African continent has many mega-diverse biodiversity rich countries that have the potential to develop their economies through organized bioprospecting activities. To this effect, there has always been a pressing and immediate necessity for these member states to endeavor multilateral cooperation and joint venture collaborations with developed and other developing countries. It is of no denying that African member states are extremely rich in biodiversity and traditional medical knowledge but lack the initial state-of-art technologies, technical expertise and financial funding to evocatively explore and add value to existing local biodiversity. In contrast, the developed countries are well-off in state-of-art technologies and financial resources but lacks in the natural resources and the traditional medical knowledge [11]. Thus, it has always been wise to come mutually in well-established and with the idea of benefit-sharing partnerships in a challenge to bridge these gaps and differences, and hence come up with effective cures against EIDs. In fact, one of the best initial alternatives for African nations is to work in partnership with the developed member states and interested pharmaceutical companies alike and jointly explore them strategically, synergistically and wisely. Interestingly, bioprospecting consortiums are increasingly supported by international and national laws and self-regulation procedures, including codes of ethics, high-quality contracts, and transparent institutional policies that result in mutual benefit-sharing. In doing so, the models of international collaboration has been such that it builds the science infrastructure within, preserve and protects the local traditional medical knowledge reducing the brain drain, and equally share the outcome of the joint projects. While bioprospecting should gear-up, strengthening the African traditional medicine should hence be a simultaneous effort since it has multi-pronged benefits to prospective partners. Firstly, it had significant contribution to the health of the Africans; has a long standing tradition of use of medicinal plants and other raw materials benefiting from local biodiversity; third, it has been providing employment opportunities; and fourth, it has a rich source of medicinal materials that could be useful for bioprospecting [11].

Furthermore, bioprospecting partnerships have been mimicking the role of the central nervous system in promoting ethical and sustainable utilization of bioresources to benefit indigenous people, business community, human health, and the environment in African countries. Also, it has been a vital tool for the discovery of new pharmacophores that could save millions of lives. It is therefore imperative that there is continuing support for protecting, promoting, and propagating this traditional medicine and at simultaneously preserving the rich biodiversity [12, 13].

4 International Organizations for Collaborative Work on Bioprospection

Researchers from scientifically and technologically advanced countries collaborating with developing country counterparts in bioprospection report that international collaboration is the preferred method of building scientific capacity in developing countries. Indeed, these activities are building international-level scientific capacity in many countries, including Africa. Available reports tend to show that the amount of collaborative research between advanced and African scientists is rising.

A review of such existing international collaborations can be categorized in a number of forms, including sharing of research data, joint experimentation, conferences and other meetings, building of databases, standards-setting, and equipment sharing. In the case of infectious disease control, success can also be achieved via can international collaboration, followed by the location of specific resources, unique expertise, and the location of large-scale equipment within regional or international institution or organizations [12, 14] as elaborated in the following sections.

4.1 Malagasy Institute of Applied Research

IMRA, a Malagasy institution is another important key center actively involved in infectious diseases research, particularly malaria and chikungunya, and which has been relying heavily on international and regional collaborations in the Eastern African regions. Because of its integration of Western science and Malagasy cultural traditions, IMRA has set a positive example for African and other organizations interested in bioprospecting. It has been actively searching for international partners to support bioprospecting activities, notably through European Union grants aimed at promoting bioprospecting in Madagascar.

Indeed, Professor P. Rasoanaivo, IMRA's CSO has always believed that it is difficult for researchers in developing countries to have in one single laboratory with all the required expertise, facilities, and human resources to maximize the conduct of drug discovery programs involving natural products. A network of partnerships is indispensable to achieve successful outcomes. Hence, IMRA has always ventured in developing relationships with European companies, for instance to build a new production facility where phyto-drugs are manufactured according to international standards. Along with supporting bioprospecting, strategic partnerships and collaborations have greatly uplifted training of Malagasy researchers during the last few decades. The foremost source of international collaborations on bioprospecting for IMRA are summarized below:

- In July 2004, the CRNS in France signed a memorandum of understanding with the University of Antananarivo, with the participation of IMRA. Under this bioprospecting project, more than 800 diverse plants were collected since 2004–2009. Plants were reportedly collected with and extracts exported for screening after an agreement was signed with the Direction de la Préservation de la Biodiversité.
- A second project called PHYTOCHIK, funded by CRVOI has attempted to search the area's biodiversity for compounds targeting the chikungunya virus on the Indian Ocean islands. It involves multilateral cooperation between Mauritius, Madagascar (IMRA), Réunion Island, France, and Belgium.
- Funding through a third project from the International Cooperative Biodiversity Groups (ICBG) reportedly involved agreements with local villagers, and has aided the development of infrastructure in Madagascar, including the construction of a new laboratory at the University of Antananarivo and equipment upgrades at a laboratory at the University of Fianarantsoa.

- At regional level, recently a frame contract of partnerships was signed in August 2010 by IMRA, CEPHYR (Centre for Phytotherapy Research–Mauritius), and CYROI (Reunion Island) with the objective of adding value to the marine and terrestrial biodiversity of the 3 islands [15, 16].

4.2 Association for African Medicinal Plants Standards

Another more recently established organization which has constantly fostered regional/international and successful collaboration is the AAMPS. Located in Mauritius, AAMPS was established following recommendation(s) from about 28 regional/international experts on African herbal medicines from 17 different countries. Since then, this non-profit NGO has been very vigorous regionally and via international sponsorship from various institutions and collaborators in overcoming the major barrier toward regional and international trade and addressing specific technical specifications and quality control standards for African medicinal plants and herbal medicines. In fact, AAMPS has flourished following technical and financial support from international collaborators such as PRO€INVEST, which is a program of the group of African, Caribbean and Pacific Group of States (ACP) and the European Commission (EuropeAid) for the promotion of investment and technology transfers in ACP countries. Additionally, AAMPS has received much support from SADC secretariat through the implementation and coordination of agricultural research and training projects with support from the European Union [17]. Ever since its existence, founding members of AAMPS have been dynamic (being the first to launch the first African herbal pharmacopoeia) in the region with several major and pioneering projects in the pipeline encompassing a plethora of activities; ranging from regional training of African fellows to plant databases and monographs focused on the African pharmacopeia as summarized hereunder:

- Preparation of AAMPS laboratory manual
- Database of African medicinal plants
- Development of the AAMPS/International Foundation for Science fellowship program
- Preparation of additional AAMPS African medicinal plants and essential oil standards
- Preparation of AAMPS African herbal product standards
- AAMPS audit and review of African testing laboratories
- Development of AAMPS accredited laboratory network and inspection and accreditation service
- Development of AAMPS regional training workshops in the application of AAMPS African medicinal plants standards (http://www.aamps.org).

4.3 International Centre of Insect Physiology and Ecology

ICIPE, is another key example of a non-governmental institute of advanced research and training in bioprospecting. The core ICIPE's mission has been to help alleviate poverty, ensure food security and improve the overall health status of peoples of the tropics by developing and extending management tools and strategies for harmful and useful insects, while preserving the natural resource base through research and capacity building. This has been achieved mainly from various regional and international collaborations. ICIPE has been focusing on building awareness and capacity about important biodiversity issues among East African institutions. It does this through workshops, and the coordination of two key projects focused on bioprospecting via major collaborations. The first, funded by WHO, the World Bank, and UNDP, involves research institutions from Tanzania, Uganda, Ethiopia and Kenya in a bioprospecting initiative focused on mosquito-repellent and insecticidal plants in East Africa. The second project, funded by the UN-affiliated International Centre for Scientific Culture, aims at providing, free of charge, mass spectral services to African scientists who are investigating natural products from plants and animals but who do not have access to such facilities. They also offer a fellowship program to African scientists. Biodiversity research, conservation and rural development are priorities in this regard. ICIPE has also been collaborating with Belgian NGO [the International Organisation for Chemical Sciences in Development (IOCD)] in preparing a bioprospecting program of workshops and training courses. The IOCD spearheads a "Biotic Exploration Fund", which raises money to "help developing countries build local scientific and entrepreneurial capabilities for bioprospecting". A program is being planned for Uganda, following work in Kenya and South Africa. This initiative would not have been possible without financial assistance by international collaborators such as Monsanto, Ciba-Geigy, Novartis, and several other agrochemical and pharmaceutical corporations [5, 6] (http://www.iocd.unam.mx/biodiver.htm).

4.4 The International Cooperative Biodiversity Groups Programs

The International Cooperative Biodiversity Groups Programs (ICBGP) has been addressing the interdependent issues of drug discovery, biodiversity conservation, and sustainable economic growth in many developing countries, and particularly Africa. The ICBGP mandates are based on the belief that discovery and development of pharmaceutical and other useful agents from natural products can, under appropriate circumstances, promote scientific capacity development and economic incentives to conserve the biological resources from which these products are derived. ICBG programs have also been addressing discovery of safe new agents for crop protection, veterinary medicines and EIDs. One such example is the

Axxon Biopharm Inc. project which was established by a drug development program. The company has commercialized nearly 100 leads in association with the Bioresources Development and Conservation Programme based in Nigeria. Another example ICBGP-supported work is the Phyto Nova project which was established to research, develop, and market safe and affordable medicines for African wasting diseases, opportunistic infections, and other public health needs; to promote African traditional medicine, to scientifically validate natural products in order to ensure safety, efficacy and quality; and to ensure the sustainability of the supply of raw materials through conservation and local rural development. ICBGP has also been supporting multinational collaborations such as (1) the Biodiversity Utilization in Madagascar and Suriname project (from 2003–2008) which has involved several partners (Virginia Polytechnic Institute and State University, Missouri Botanical Garden, Conservation International; Madagascar National Centers for Pharmaceutical, Environmental and Oceanographic Research, Pharmaceutical Distribution, Organization of Suriname; Bristol–Meyers Squibb, Eisai Research Institute and Dow Agrosciences) and the (2) Drug Development and Conservation of Biodiversity in West and Central Africa project (from 1994–2003) involving mainly the Nigeria and Cameroon tropical rainforest plants (http://www.icbg.org/; [18]).

4.5 International Science Program

Nordic and European collaborators have always been supporting bioprospecting projects in developing nations. One such program is the ISP at Uppsala University, Sweden, which has long been supporting institutional capacity building in research and higher education in developing countries like Africa, within the basic sciences chemistry, mathematics, and physics. ISP has been providing long-term support to research groups and scientific networks which have been carried out in close cooperation with research groups at more advanced host institutions. ISP has always believed that regional and inter-regional cooperation is one way of overcoming financial drawbacks in bioprospecting. To this effect, ISP has been active in organizing strategic workshops, one of which has been the international conference of regional and inter-regional cooperation to strengthen basic sciences in developing countries—where it was concluded that international collaboration and scientific networks are important starting step in countries where resources are limited [19]. Additionally, in 2010 ISP supported 28 research groups in 10 African and 4 Asian countries; 5 groups in Kenya, 4 in Sri Lanka, 3 each in Laos and Zimbabwe, 2 each in Bangladesh, Burkina Faso, Cambodia, and Malawi, and 1 each in Ghana, Mali, Nigeria, Senegal and Zambia. Moreover, 15 regional scientific networks were supported, 12 in Africa, 2 in Asia, and 1 in Latin America [19, 20].

4.6 Natural Product Research Network for Eastern and Central Africa

Natural Product Research Network for Eastern and Central Africa (NAPRECA), affiliated to UNESCO since 1989 has been very active to gather and mobilize scientists in natural product research in the East and Central African sub-region. NAPRECA currently branches several African countries such as Ethiopia, Sudan, Tanzania, Kenya, Rwanda, Zimbabwe, Uganda, Cameroon, Botswana, DR Congo, and Madagascar among others. In line with international collaboration, NAPRECA has been establishing scholarship schemes, funded by the International Education Exchange Program (DAAD) of Germany with the aim of enabling young researchers in NAPRECA member countries to embark on postgraduate training leading to masters and/or doctoral degrees in non-home country universities but within the NAPRECA region. Other organizations such as OPCW and IFS have also funded NAPRECA activities which has created strong interlinkages between researches in the region. Besides research capacity development at the participating universities, the development of human resources and the science of natural products, this scheme has contributed toward strengthening research linkages among universities in some NAPRECA member countries. In addition, a wealth of scientific knowledge has been created in the course of the postgraduate research training activities. Examples to this effect include the establishment of chemical structures of a great variety of natural products at the University of Dar es Salaam, such as bioactive heptenolides, imidodrimanoids, and terpenoids [21].

4.7 Malaria Research Consortiums

According to the world malaria report 2010, there were 225 million cases of malaria and an estimated 781000 deaths in 2009, a decrease from 233 million cases and 985000 deaths in 2000. Most deaths occur among children living in Africa where a child dies every 45 s of malaria and the disease accounts for approximately 20 % of all childhood deaths. To this effect, much emphasis has been laid during the past decades to explore the potential of the African biodiversity in the development of potential antimalarial drugs. One such unit that has been actively engaged in this war is the Malaria Research Unit (MRU). MRU has always fostered national, regional and continental (Africa) malaria related research as its priority. Nonetheless, a major obstacle to effective malaria control has been the emergence and spread of insecticide resistant vectors. Resistance to all commercially available insecticides has been detected. Hence, as part of the Novel Drug Development Platform, the MRU has been involved in the screening and testing of indigenous plants as potential mosquitocidal agents. Also as part of the Innovative Vector Control Consortium, a Gates funded project, the MRU has been working with the private sector insecticide companies to evaluate new formulations of insecticides for use in indoor residual spray programs. The MRU has also been conducting laboratory and field trials to determine the efficacy and durability of these new bioformulations. The Malaria Consortium is committed to sustainability and high impact work on the ground, thanks to varied partnerships that have been integral to the organization's achievements over the years. Indeed, the aims of Malaria Consortium have been most readily achieved through successful working in partnerships to maximize effect and reach. Recently, this organization has been focusing on 4 key types of partnerships: with communities, the commercial sector, local governments and academic institutions. All these invaluable relationships have strengthened crucial elements of Malaria Consortium's work and have provided long lasting impact through strengthened capacity (http://www.mrc.ac.za).

Action for natural medicine (Anamed) and research initiative on traditional antimalarial methods (RITAM) are other examples of international research initiatives that aim to develop new antimalarial lead from the African biodiversity. Interestingly, Anamed has promoted the cultivation and use of an artemisia hybrid in over 70 countries so that rural people affected by malaria have an affordable antimalarial option. However, [22] claimed that more research is needed to develop an evidence base for the use of *Artemisia annua* and other traditional antimalarial plants to support the work of Anamed and similar grassroots organizations. A network which seeks to do this has been the RITAM comprising of over 200 researchers from at least 30 countries working together to validate the use of medicinal plants for malaria. These are small initiatives with tiny budgets but they provide a glimmer of hope in the desperate struggle against malaria [22].

4.8 The US National Cancer Institute Bioprospecting Partnership in Africa

Since decades the NCI has been screening thousands of plant-derived extracts for antitumoral activity in the in vitro and in vivo assays based on three cell lines. This effort resulted, for example, in the discovery of taxol (paclitaxel and docetaxel) and camptothecin (topotecan and irinotecan). In the 1970s, the discovery of artemisinin, a potent drug to treat cerebral malaria, isolated from *Artemisia annua*, a Chinese medicinal plant, was another important event in the search for plant-derived drugs and confirmed the potential of secondary metabolites as leading new molecules. The global demand to preserve the biodiversity in developing countries has been widely cited as a mean to promote economic development through incentives for sustainable uses of the genetic resources. A five-year International Cooperative Biodiversity Group (ICBG) program was launched in 1993 by NIH-USA, directed at bioprospection in Latin America and Africa aiming to search for compounds to treat or prevent cancer, infectious diseases, including AIDS, cardiovascular diseases, malaria and several parasitic disturbances, among others. NCI has been collecting plant material through contracted collectors; in 1986 contract was awarded to the Missouri Botanical Garden for collections of plant material in Africa and Madagascar. During this international venture, NCI has been active in the development of the anti-HIV compound michellamine B which was derived from *Ancistrocladus korupensis*—a species with pharmaceutical potential from Cameroon.

4.9 International Center for Research and Intelligence on Emerging Diseases

Following the chikungunya outbreak, international centers have emerged for collaborative research, and data sharing on potential emerging diseases. One such setup is CRVOI in the Indian Ocean established in 2007 at Reunion Island. Indeed, prior to this epidemic outbreak, regional surveillance of emerging infections in many parts of the Eastern–African countries were still in its infancy. A number of infectious diseases were notorious by name, but no formal systematic laboratory-based and epidemiological surveillance system was established to confirm diagnoses or to subtype microorganisms for epidemiology purposes. To this effect, CRVOI has taken the pioneering lead and fostered a multidisciplinary research collaboration following the sheer magnitude of the 2005–2007 outbreaks on infectious diseases of regional and African interest with the following four mandates.

1. Investigation of infectious diseases with multidisciplinary interest in the Indian Ocean region; through calls for proposals and conducted by teams located in the region, in collaboration with partners and international metropolitan,
2. Coaching and training for master students, doctoral, and young researchers,
3. Ensure scientific and technical conduct of an activity of "data mining" routine on infectious diseases of regional interest,
4. Regional cooperation reinforced on EIDs with the other countries of the Indian Ocean.

CRVOI has also been on the forefronts of collaborative work and behaved as an important platform for training, networking and organization of workshops in the Indian Ocean and African countries. Important regional training in 2010 has involved the recent organization of the 2nd international conference on neuroinfection and worldwide impact in Reunion Island. This was preceded by a joint ARC-WERC IBRO UNESCO school on "Neurobiology of Infectious Diseases: A view for global neuroscience" where researches and young fellows from Africa, Asia, Europe and Indian Ocean region (Reunion, Mauritius, Mayotte, Madagascar, Comoros and Seychelles) were trained on the pathogenesis of EIDs and addressed important gaps in knowledge of some of the most prevalent infectious diseases in the region.

4.9.1 The PHYTOCHICK Consortium

The recent outbreak of chikungunya fever in the islands of the Indian Ocean has drawn much attention to chikungunya virus (CHIKV, genus Alphavirus, family Togaviridae), first identified in the 1950s in Africa. Intriguingly, it was initially classified as a neglected tropical disease and it was only the sheer magnitude of the 2005–2007 CHIKV outbreaks that brought this virus into the limelight of both the scientific community and the general public [23]. CHIKV has since then been associated with the urban *Aedes aegypti* mosquito (possibly supplemented by *Aedes albopictus*) in an epidemiologic cycle resembling that of dengue and characterized by the absence of an animal reservoir, direct human–human transmission by urban mosquitoes, and the potential for major epidemics [23, 24]. Furthermore, since Indian Ocean islands are popular tourist destinations, this further complicates the management and control of CHIKV. Recently, CHIKV-infected travelers returned home to countries where competent vectors are indigenous, which raises serious concern for potential disease spread. At present, neither vaccine nor a selective antiviral drug is available for the prevention or treatment of this debilitating viral infection. Chloroquine is active in cell culture and may alleviate the symptoms of arthritis by acting as an anti-inflammatory agent, although this latter is still under investigation.

To this effect, in 2008 the CRVOI has funded several international research programs aims toward bioprospecting of local/regional biodiversity against EIDs. One of particularly interest has been the PHYTOCHIK consortium geared toward harnessing biodiversity in an attempt to combat emerging viruses in the Indian Ocean with main aim as selection of natural drug candidates to fight the CHIKV. In addition, this consortium has also focused on capacity building/transfer of technology to lesser developed countries. PHYTOCHK consortium has indeed been a clear illustration of international collaboration that has brought together four areas of expertise: the use of plants as natural source for the treatment of disease (ethnopharmacology), the identification and classification of plant species biodiversity, the purification and identification of unique molecules/secondary metabolites of natural origin, and the identification and characterization of selective inhibitors of virus replication.

Three Eastern African teams (Table 1), located on Reunion Islands (Laboratoire de Chimie des Substances Naturelles Université de la Réunion (LCSN), Mauritius (Faculty of Science, University of Mauritius), and Madagascar (IMRA), together with the France (ICSN), have build a unique sample library and database consisting out of more than 1500 crude plant extracts, fractions and pre compounds that have been evaluated for selective antiviral activity against CHIKV in Leuven (Laboratory of Virology, Rega Institute for Medical Research, Belgium and Marseille (Unité des Virus Emergents, Faculté de Médecine, Marseille). So far, a number of promising leads have been discovered from Mauritius, Madagascar and Reunion and presently several bio-assay-guided purification/fractionation of pure substances are underway, yielding promising preliminary results. Concomitantly, enzymatic assays are being developed in Marseille to evaluate and possibly

Table 1 PHYTOCHIK main partnerships

Partner name	Country	Nature of collaboration
LCSN laboratory–Laboratoire de Chimie des Substances Naturelles, Faculté des Sciences et Technologies, Université de la Réunion	Réunion Island	Collection, identification, preparation of crude extracts/fractions and isolation of pure compounds from local biodiversity
IMRA–Malagasy Institute of Applied Research	Madagascar	Collection, identification, preparation of crude extracts/fractions and isolation of pure compounds from local biodiversity
Faculty of Science, University of Mauritius, Reduit	Republic of Mauritius	Collection, identification, preparation of crude extracts/fractions and isolation of pure compounds from local biodiversity
ICSN–CNRS–Centre de Recherche de Gif (FRC 3115), Institut de Chimie des Substances Naturelles UPR 2301 du Centre National de la Recherche Scientifique - Gif-sur-Yvette	France	Overall coordination, cytotoxicity evaluation on human cell line
Laboratory for Virology and Chemotherapy, Rega Institute for Medical Research (LVC-RIMR)–Katholieke Universiteit Leuven	Belgium	Antiviral evaluation
Unité des Virus Emergents, Faculté de Médecine, Marseille	France	Characterization of selective inhibition of CHIKV

characterize in detail the selective inhibitory effect of these phytocompounds. Additionally, researchers and students from Mauritius and Madagascar have benefited extensively from exchange programs and training attachment to more developed laboratories such as ICSN (Gif-sur-Yvette, France), Belgium and Réunion University.

On the overall, more than 1554 crude and filtered extracts, 22 pure compounds have been sent to France and Belgium partners for cytotoxicity and CHIV evaluation during the first 2 years of the PHYTOCHIK partnership. A total of 13 and 8 hit extracts were recorded for Madagascar and La Reunion partners respectively. Interestingly, 12 extracts have proved to be potent (super-hit against the CHIKV) from Mauritius belonging to the family of Celastraceae, Sapindaceae, Ebenaceae, Meliaceae, Sterculiaceae, Rubiaceae, and Apotaceae. Additionally, 5 plants from Mauritius were initially selected for further fractionation, phytochemical analysis and anti-CHIKV evaluation. Promising leads were found from 4 fractions which showed a maximum inhibition of 88.8 % at 20 μg/mL; 3.9 % at 4 μg/mL; 100 % at 20 μg/mL and 95.3 % at 20 μg/mL against the CHIKV virus respectively. Also, during the last annual meeting in Belgium, it was decided that this international consortium will continue via signing of material transfer agreements between the Mauritian and ICSN partners for the sending and evaluation of extracts on other targets than CHIKV and will explore opportunities to continue their collaboration through international networks.

5 Conclusion and Future Trend

The chapter has the aim to illustrate how international collaboration between African member states and developed counterparts have been an alternative route for bioprospecting at the grassroots level in an endeavor to enhance the local livelihoods of communities and promoting their empowerment and capacity building. It is of no denying that bioprospecting in the African regions had experienced some golden age as supported by the plethora of success stories. Nonetheless, recent report tend to show that the total funding for international collaborative research and development on neglected tropical diseases and EIDs has suffered major cuts from international collaborators in the wake of the global financial crisis. Appreciatively, this impact is curd down by a substantial increase in private sector funding.

Taken together, panoply of international organizations and funding agencies now support bioprospecting and related programs in the African continent. However, there is still a need for substantial support at the national level as well as regionally to promote medicinal plants, traditional medicines and ethnopharmacology, and to assure that bioresources of the Africa are being harnessed at a sustainable level in compliance with the Convention on Biological Diversity and World Trade Organization.

References

1. WHO (2010) Containing antimicrobial resistance: a renewed effort, bulletin of the World Health Organization [Online], 88(12):877–953. <http://www.who.int/bulletin/volumes/88/12/10-084236/en/>. Accessed 15 Dec 2011]
2. Hotez PJ, Molyneux DH, Fenwick A, Ottesen E, Ehrlich-Sachs S, Sachs J (2007) Incorporating a rapid-impact package for neglected tropical diseases with programs for HIV/AIDS, tuberculosis, and malaria. PLoS Med 3(5):e277. doi:10.1371/journal.pmed.0040277
3. Hotez PJ, Molyneux DH, Fenwick A, Kumaresan MB, Ehrlich-Sachs S, Sachs J, Savioli L (2007) Control of neglected tropical diseases. New Engl J Med 357:1018–1027
4. Kavaka WM (2010) How to make fortunes from bioprospecting. Prota Newsl
5. Wynberg R, (2000a) Benefit-sharing in South Africa: fact or fiction? Forthcoming in: biodiversity and traditional knowledge: equitable partnerships in practice. In: Laird SA (ed) A WF/UNESCO/Kew people and plants conservation manual, Earthscan Publications
6. Wynberg R (2000b) Privatizing the means for survival: the commercialization of Africa's biodiversity. Global Trade and Biodiversity in Conflict, Issue no. 5, Biowatch, South Africa with contributions from GAIA/GRAIN
7. Engelsen CD, Werf CVD, Matute AJ, Delgado E, Schurink CAM, Hoepelman AIM (2009) Infectious diseases and the use of antibiotics in outpatients at the emergency department of the University Hospital of Leon, Nicaragua. Int J Infect Dis 13:349–354
8. Brendler T, Eloff JN, Gurib-Fakim A, Philips LD (2010) African herbal pharmacopoeia. Graphic press limited, Mauritius
9. Quave CL, Pieroni A, Bennett BC (2008) Dermatological remedies in the traditional pharmacopoeia of Vulture-Alto Bradano, inland southern Italy. J Ethnobiol Ethnomed 1–10
10. Borchardt JR, Wyse DL, Sheaffer CC, Kauppi KL, Faulcher RG, Ehlke NJ, Biesboer DD, Bey RF (2008) Antimicrobial activity of native and naturalized plants of Minnesota and Wisconsin. J Med Plants Res 2(5):98–110
11. Wangchuk P, Wangchuk D, Aagaard-Hansen J (2007) Traditional Bhutanese Medicine (gSo-ba rig-pa): An integrated part of the formal health care services. Southeast Asian J Trop Med Public Health 38:161–167
12. Wangchuk P (2007) Herbal remedies and utilization of medicinal resources in Bhutan. Book of abstracts of the international workshop on herbal medicinal plants and traditional herb remedies, Hanoi, p 1
13. Wangchuk P, Dorji Y (2007) Historical roots, spiritual significance and the health benefits of mkhempa–ljong gnyes Tshachu in Lhuentse. J Bhutan Stud 16:112–127
14. Wagner C, Brahmakulam I, Jackson B, Wong A, Yoda T (2001) Science and technology collaboration: building capacity in developing countries? MR-1357.0-WB. World Bank. RAND Science and Technology
15. Puri M, Masum H, Heys J, Singer P (2010) Harnessing biodiversity: the Malagasy Institute of Applied Research (IMRA). BMC International Health and Human Rights, 2010, 10:S1–S9. http://www.biomedcentral.com/1472-698X/10/S1/S9
16. Rasoanaivo P (2011) Single-constituent drugs vs multi-component phytomedicines: from research to policy. Book of abstracts of the international workshop on bioprospecting, policy and practice. Conservation and use of medicinal plants of the small island developing states (SIDS) of the Indian Ocean and Madagascar, April 20–22 2011, Mauritius
17. Gurib-Fakim A, Brendler T, Philips D, Eloff JN (2010) Green gold. Successs stories using southern African plant species. AAMPS publishing. Caractere LTD, Mauritius
18. Beattie J, Barthlott W, Elisabetsky E, Farrel R, Kheng CT, Prance I, Rosenthal J, Simpson D, Leakey R, Wolfson M, Kate K (2005) New products and industries from biodiversity. In: Hassan R, Scholes R, Ash N (eds) Ecosystems and human well-being: current state and trends, Island Press, USA, 1–921
19. Sundin P (2011) ISP support to basic sciences in developing countries–importance of scientific networks. Book of abstracts of the international workshop on bioprospecting, policy

and practice. Conservation and use of medicinal plants of the small island developing states (SIDS) of the Indian Ocean and Madagascar. April 20–22, 2011, Mauritius
20. International Science Programme (ISP) (2010) Annual report to Sida
21. Midiwo OJ (2011) The impact of natural products research network in East and Central Africa on natural products research in the region. Book of abstracts of the international workshop on bioprospecting, policy and practice. Conservation and use of medicinal plants of the small island developing states (SIDS) of the Indian Ocean and Madagascar, April 20–22, 2011, Mauritius
22. Challand S, Willcox M (2008) Affordable antimalarials. Rapid Response-British Med J 337:a2495. http://www.bmj.com/cgi/eletters/337/nov12_1/a2495#205024
23. Powers AM, Logue CH (2007) Changing patterns of chikungunya virus: re-emergence of a zoonotic arbovirus. J Gen Virol 88:2363–2377
24. Leyssen P, Litaudon M, Guillemot J, Rasoanaivo P, Smadja J, Gurib-Fakim A, Canard B, Gueritte F (2011) PHYTOCHIK: biodiversity as a source of selective inhibitors of CHIKV replication. Antiviral Res 90:A1–A20

About the Editors

Ameenah Gurib-Fakim a Mauritian national, is currently the Managing Director of the Centre for Phytotherapy Research (CEPHYR) and Professor of Organic Chemistry with an endowed chair at the University of Mauritius. She has recently been elected Chairperson of the International Council for Scientific Union—Regional Office for Africa for the period (2011–2014). Since 2001, she has served successively as Dean of the Faculty of Science and Pro Vice Chancellor (2004–2010). She has also worked at the Mauritius Research Council as Manager for Research (1995–1997).

Ms Gurib-Fakim earned a BSc in Chemistry from the University of Surrey, UK (1983) and a PhD from the University of Exeter (1987) at which time she began working at the University of Mauritius. Between 1987–1992, she served as Project leader for the first regional research project on the Inventory and study of

A. Gurib-Fakim and J. N. Eloff (eds.), *Chemistry for Sustainable Development in Africa*, DOI: 10.1007/978-3-642-29642-0,

medicinal and aromatic plants of the Indian Ocean, funded by the European Development Fund under the aegis of the Indian Ocean Commission. During 2000–2002, she served as the national coordinator for the 'Indian Ocean University' funded by the European Union. She has participated in several consultation meetings on environmental issues organised by the World Bank and most recently as the lead coordinating author on the international Assessment of Agricultural Knowledge, Science and Technology for Development (IAASTD) spearheaded by the World Bank.

As a Founding member of the Pan African Association of African Medicinal Plants, she co-authored the first ever African Herbal Pharmacopoeia funded by the CDE and ProInvest (ACP), Brussels. She is member of the Reference Group of the International Science Programme (IPICS) of Uppsala University, Sweden and serves as scientific advisor to the International Foundation of Science (IFS), Stockholm. She has also served as member of the Expert Panel on 'Infectious diseases' of the Special Programme convened by the UNDP, UNICEF, WHO and the World Bank. She also serves as a member of the Scientific Advisory Council of NTembi and Nuclear Energy Commission of South Africa (NECSA); member of the International Advisory Committee for the Cyclotron project of the Indian Ocean based in La Reunion (France) and the international representative on the assessment panel of the Canadian AUCC-CIDA projects. Ms Gurib-Fakim has authored and/or co-edited 26 books and several book chapters. Within the field of biodiversity conservation and sustainable development, she has authored numerous scientific articles and has lectured extensively across the world. She is a member of the editorial boards of major journals, having served on technical committees in various capacities, including the chair of several national committees in Mauritius.

Ms Gurib-Fakim has been elected Fellow of the Linnaen Society of London in 2007, Fellow of the Islamic Academy of Science, Jordan in 2009 and Fellow of the African Science Institute in 2010. Ms Gurib-Fakim received the 2007 l'Oreal-UNESCO Prize for Women in Science and Laureate of the National Economic and Social Council. She is recipient of the special prize from the CTA/NEPAD/AGRA/RUFORUM and the African Union Commission Award for Women in Science both in 2009. She was elevated to the Order of the Commander of the Star and Key by the Government of Mauritius in 2008 and admitted to the Order of the Order of the Chevalier dans l'Ordre des Palmes Academiques by the Government of France in 2010.

www.cephyr-recherche.com
www.uom.ac.mu

Jacobus Nicolaas (Kobus) Eloff was professor at Universities Free State, Cape Town and Pretoria, Executive Director National Botanic Gardens (Head Office Kirstenbosch) and Research Director National Botanical Institute. Currently leader interdisciplinary Phytomedicine Programme, Veterinary Science, University of Pretoria (www.up.ac.za/phyto)

Promoting of 44 MSc and 27 PhD students completed. Currently 5 MSc and 15 PhD students.

Editor of several scientific journals and books. Reviewed manuscripts for 93 different scientific journals.

More than 170 peer evaluated scientific publications and more than 300 papers at scientific meetings including many plenary lectures on all continents.

Internationally acclaimed researcher category evaluation by the National Research Foundation.

Appointed by European Union's agent Centre for the Development of Enterprise to develop Pan-African quality control standards and monographs for African Medicinal plants, leading to production of African Herbal Pharmacopoeia.

Leadership role in many national and international professional societies.

Several national and international awards. Several patents registered and products licensed to Industrial companies.

Index

A. Gurib-Fakim and J. N. Eloff (eds.), *Chemistry for Sustainable Development in Africa*, DOI: 10.1007/978-3-642-29642-0,

MIX
Papier aus verantwortungsvollen Quellen
Paper from responsible sources
FSC® C105338

If you have any concerns about our products,
you can contact us on
ProductSafety@springernature.com

In case Publisher is established outside the EU,
the EU authorized representative is:
Springer Nature Customer Service Center GmbH
Europaplatz 3, 69115 Heidelberg, Germany

Printed by Libri Plureos GmbH
in Hamburg, Germany